WAVELETS
and OTHER
ORTHOGONAL
SYSTEMS

SECOND EDITION

Studies in Advanced Mathematics

GILBERT G. WALTER
Department of Mathematical Sciences
The University of Wisconsin–Milwaukee
Milwaukee, Wisconsin

XIAOPING SHEN
Department of Mathematics and Computer Sciences
Eastern Connecticut State University
Willimantic, Connecticut

WAVELETS and OTHER ORTHOGONAL SYSTEMS

SECOND EDITION

CRC Press
Taylor & Francis Group
Boca Raton London New York

CRC Press is an imprint of the
Taylor & Francis Group, an **informa** business

A CHAPMAN & HALL BOOK

CRC Press
Taylor & Francis Group
6000 Broken Sound Parkway NW, Suite 300
Boca Raton, FL 33487-2742

First issued in paperback 2019

© 2001 by Taylor & Francis Group, LLC
CRC Press is an imprint of Taylor & Francis Group, an Informa business

No claim to original U.S. Government works

ISBN-13: 978-1-58488-227-5 (hbk)
ISBN-13: 978-0-367-39781-4 (pbk)
Library of Congress Card Number 00-050874 1

Library of Congress Cataloging-in-Publication Data

Walter, Gilbert G.
     Wavelets and other orthogonal systems / Gilbert G.
Walter, Xiaoping Shen.— 2nd ed.
        p. cm.— (Studies in advanced mathematics)
    Includes bibliographical references and index.
    ISBN 1-58488-227-1 (alk. paper)
    1. Wavelets (Mathematics)   I. Shen, Xiaoping. II. Title. III. Series.

QA403 .3 .W34 2000
515'.2433—dc21

00-050874

Visit the Taylor & Francis Web site at
http://www.taylorandfrancis.com

and the CRC Press Web site at
http://www.crcpress.com

Preface to first edition

The subject of wavelets has evolved very rapidly in the last five or six years—so rapidly that many articles and books are already obsolete. However, there is one portion of wavelet theory that has reached a plateau, that is, the subject of orthogonal wavelets. The major concepts have become standard, and further development will probably be at the margins. In one sense they are no different than other orthogonal systems. They enable one to represent a function by a series of orthogonal functions. But there are notable differences: wavelet series converge pointwise when others don't, wavelet series are more localized and pick up edge effects better, wavelets use fewer coefficients to represent certain signals and images.

Unfortunately, not all is rosy. Wavelet expansions change excessively under arbitrary translations—much worse than Fourier series. The same is true for other operators such as convolution and differentiation.

In this book wavelets are presented in the same setting as other orthogonal systems, in particular Fourier series and orthogonal polynomials. Thus their advantages and disadvantages can be seen more directly.

The level of the book is such that it should be accessible to engineering and mathematics graduate students. It will for the most part assume a knowledge of analysis at the level of beginning graduate real and complex analysis courses. However, some of the later chapters are more technical and will require a stronger background. The Lebesgue integral will be used throughout. This has no practical effect on the calculation of integrals but does have a number of theoretical advantages.

Wavelets constitute the latest addition to the subject of orthogonal series, which are motivated by their usefulness in applications. In fact, orthogonal series have been associated with applications from their inception. Fourier invented trigonometric Fourier series in order to solve the partial differential equation associated with heat conduction and wave propagation. Other orthogonal series involving polynomials appeared in the 19th century. These too were closely related to problems in partial differential equations. The Legendre polynomials are used to find solutions to Laplace's equation in the sphere, the Hermite polynomials and the Laguerre polynomial for special cases of Schrödinger wave equations. These, together with Bessel functions, are special cases of Sturm-Liouville problems, which lead to orthogonal series, which are used to solve various partial differential equations.

The arrival of the Lebesgue integral in the early 20th century allowed the development of a general theory of orthogonal systems. While not oriented to applications, it allowed the introduction of new systems such as the Haar and Walsh systems, which have proven useful in signal processing. Also useful in this subject are the sinc functions and their translates, which form an orthogonal basis of a Paley-Wiener space. These are related to the prolate spheroidal functions, which are solutions both to an integral equation and a Sturm-Liouville problem.

The orthogonal sequences of wavelets, which are generalizations of the Haar system and the sinc system, have a number of unique properties. These make them useful in data compression, in image analysis, in signal processing, in numerical analysis, and in acoustics. They are particularly useful in digitizing data because of their decomposition and reconstruction algorithms. They also have better convergence properties than the classical orthogonal systems.

While the Lebesgue integral made a general theory of orthogonal systems possible, it is insufficiently general to handle many of the applications. In particular, the delta "function" or impulse function plays a central role in signal processing but is not a square integrable function. Fortunately a theory that incorporates such things appeared in the middle of the 20th century. This is the theory of "distributions," due mainly to L. Schwartz. It also is related to orthogonal systems in that it allows representation of distributions by orthogonal systems and also allows representations of functions by orthogonal distributions.

The body of the book is divided into 13 chapters of which the first 7 are expository and general while the remaining are more specialized and deal with applications to other areas. Each will be concerned with the use of or properties of orthogonal series.

In Chapter One we present two orthogonal systems that are prototypes for wavelets. These are the Haar system and the Shannon system, which have many, but not all, of the properties of orthogonal wavelets. They will be preceded by a section on general orthogonal systems. This is a standard theory that contains some results that will be useful in all of the particular examples.

Chapter Two will give a short introduction to tempered distributions. This is a relatively simple theory and is the only type of generalized function needed for much of orthogonal series. Many engineers still seem to apologize for their use of a "delta function". There is no need to do so since these are well defined proper mathematical entities. Included here also is the associated theory of Fourier transforms that enables one to take Fourier transforms of things like polynomial and trigonometric functions.

Chapter Three contains an introduction to the general theory of orthogonal wavelets. Their construction by a number of different schemes is given as are a number of their properties. These include their multiresolution property in which the terms of the series are naturally grouped at each resolution. The decomposition and reconstruction algorithms of Mallat, which give the coefficients at one resolution in terms of others, are presented here. Some of these properties are extended to tempered distributions in Chapter Five.

In Chapter Four we return to trigonometric Fourier series and discuss more detailed properties such as pointwise convergence and summability. These are fairly well known and many more details may be found in Zygmund's book. A short presentation on expansion of distributions in Fourier series is also presented.

In Chapter Five we also consider orthogonal systems in Sobolev spaces. These can be composed of delta functions as well as ordinary functions. In the former case we obtain an orthonormal series of delta function wavelets.

Chapter Six is devoted to another large class of examples, the orthogonal polynomials. The classical examples are defined and certain of their properties discussed. The Hermite polynomials are naturally associated with tempered distribution; properties of this connection are covered. Other orthogonal series are discussed in Chapter Seven.

Various kinds of convergence of orthogonal series are discussed in Chapter Eight. In particular, pointwise convergence of wavelet series is compared to that of other orthogonal systems. Also, the rate of convergence in Sobolev spaces is determined. Gibbs' phenomenon for wavelet series is compared to that for other series.

Chapter Nine deals with sampling theorems. These arise from many orthogonal systems including the trigonometric and polynomial systems. But the classical Shannon sampling theorem deals with wavelet subspaces for the Shannon wavelet. This can be extended to other wavelet subspaces as well. Both regular and irregular sampling points are considered.

In Chapter Ten we cover the relation between the translation operator and orthogonal systems. Wavelet expansions are not very well behaved with respect to this operator except for certain examples.

Chapter Eleven deals with analytic representation based on both Fourier series and wavelet. These are used to solve boundary value problems for harmonic functions in a half-plane with specified values on the real line.

Chapter Twelve covers probability density estimation with various orthogonal systems. Both Fourier series and Hermite series have been used, but wavelets come out the best.

Finally in the last chapter we cover the Karhunen-Loève theory for representing stochastic processes in terms of orthogonal series. An alternate formulation based on wavelets is developed.

Some of this text material was presented to a graduate course of mixed mathematics and engineering students. While not directly written as a text, it can serve as the basis for a modern course in Special Functions or in mathematics of signal processing. Problems are included at the end of each chapter. For the most part these are designed to aid in the understanding of the text material.

Acknowledgments

Many persons helped in the preparation of the manuscript for this book, but two deserve special mention: Joyce Miezin for her efficient typing and ability to convert my handwriting into the correct symbols, and Bruce O'Neill for catching many of my mathematical misprints.

G. G. Walter

Preface to second edition

In the years since the first edition of this book appeared, the subject of wavelets has continued its phenomenal growth. Much of this growth has been associated with new applications arising out of the multiscale properties of wavelets. Another source has been the widespread use of threshold methods to reduce the data requirements as well as the noise in certain signals. But in the area of wavelets as orthogonal systems, which is the main theme of this book, the growth has not been as marked. The principal new material has been in the area of multiwavelets, which, however, have not found their way into as many applications as the original theory. In addition, there seems to a resurgence of interest in nontensor product higher dimensional wavelets, but this area still needs some time to sort itself out.

In this new edition we have tried to correct many of the misprints and errors in the first edition (and in the process, have probably introduced others). We have reviewed the problems and introduced others in an effort to make their solution possible for average graduate students. We have also introduced a number of illustrations in an attempt to further clarify some of the concepts and examples. The first and fourth chapters remain approximately the same in this edition. The second chapter on distribution theory has been rewritten in order to make it somewhat more readable and self contained. Chapter three on orthogonal wavelet theory has been expanded with some additional examples: the raised cosine wavelets in closed form, and other Daubechies wavelets and their derivation. In Chapter five on wavelets and distributions, a section on impulse trains has been added. Chapter six on orthogonal polynomials remains essentially the same, while in Chapter seven a new section on an alternate approach to periodic wavelets has been added. In Chapter eight on pointwise convergence, an additional section on positive wavelets and their use in avoiding Gibbs' phenomenon is new. Chapter nine has been extensively revised and, in fact, has been split into two chapters, one devoted primarily to the Shannon sampling theorem and its properties and the new Chapter ten which concentrates more on sampling in other wavelet subspaces. New topics include irregular sampling in wavelet subspaces, hybrid wavelet sampling, Gibbs' phenomenon for sampling series in wavelet subspaces, and interpolating multiwavelets.

Chapter eleven on translation and dilation has only minor changes as does most of Chapter twelve except for a few pages on wavelets of entire analytic functions. In Chapter thirteen on statistics a number of new topics have been added. These include positive wavelet density estimators, density estimators with noisy data, and threshold methods. Some additional calculations involving some of these estimators are also included. Chapter fourteen, which deals with stochastic processes, has some new material on cyclostationary processes.

Acknowledgements.

The contributions of many individuals appear in this new edition. In particular the authors wish to acknowledge the work of Youming Liu, Hong-tae Shim, and Luchuan Cai which is covered in more detail here.

Gilbert G. Walter and Xiaoping Shen

Contents

List of Figures

Chapter 1

Orthogonal Series

Orthogonal series play an important part in many areas of mathematics as well as in applications. They constitute an easy way of representing a function in terms of a series and may replace complicated operators on the function by simpler ones on the coefficients of the series. The most familiar orthogonal systems are the trigonometric and the various orthogonal polynomials. Not so familiar but becoming increasingly widely used are the Haar, the Shannon, and wavelet systems.

The basic theory of orthogonal series is deceptively simple, but its detailed study contains many surprisingly difficult questions. In this chapter we skip the latter and present only a few elements of the theory. We first present a little of the general theory and then discuss a few of the principal examples. One, the trigonometric system, will be an important tool in subsequent chapters. The other two, the Haar and the Shannon systems, will serve as prototypes for the construction of wavelets.

1.1 General theory

While there exist many different orthogonal systems, they all have a number of properties in common, which we present here. We shall restrict ourselves to $L^2(a,b)$: the set of square (Lebesgue) integrable functions on (a,b), a real interval. The theory is the same if we introduce a weight function or even if we consider general separable Hilbert spaces. See[R-N].

A nontrivial sequence $\{f_n\}_{n=0}^{\infty}$ of real (or complex) functions in $L^2(a,b)$

is said to be orthogonal if

$$\langle f_n, f_m \rangle = \int_a^b f_n(x)\overline{f_m(x)}\, dx = 0, \qquad n \neq m, \ n, m = 0, 1, 2, \ldots$$

and orthonormal if in addition $\langle f_n, f_n \rangle = 1$, $n = 0, 1, 2, \cdots$. For example, $f_n(x) = \sin(n+1)x$ is orthogonal on $(0, \pi)$. Another example is

$$f_n(x) = \chi_{[n,n+1)}(x) = \begin{cases} 1, & n \leq x < n+1 \\ 0, & 0 \leq x < n, \ n+1 \leq x \end{cases}$$

which in fact is orthonormal on $[0, \infty)$.

The idea is to expand a given function $f(x) \in L^2(a,b)$ in an orthonormal series

$$f(x) = \sum_{n=0}^{\infty} c_n f_n(x). \tag{1.1}$$

This is not always possible (e.g., take $f(x) = \chi_{[0.5,1)}(x)$ in the second example), but if it is, then the c_n's must have a special form. We shall use the usual notation for the L^2 norm, $\|f\| = \langle f, f \rangle^{1/2}$.

Proposition 1.1

Let $\{c_n\}$ be a sequence such that the series in (1.1) converges in the sense of $L^2(a,b)$ to $f(x)$; then $c_n = \langle f, f_n \rangle$.

The proof is immediate. We multiply both sides of (1.1) by $\overline{f_m(x)}$ and then integrate. Because of the orthogonality all the terms in the series drop out except c_m. There is no problem with interchanging the integral and the summation because of the continuity of the inner product with respect to the norm.

Convergence in the sense of $L^2(a,b)$ is also known as mean square convergence, and the error

$$e_N = \left\| f - \sum_{n=0}^{N} c_n f_n \right\|^2$$

is called the mean square error. The coefficients appearing in Proposition 1.1 are called the Fourier coefficients of f with respect to $\{f_n\}$ and have another property that makes them useful.

Proposition 1.2
Let $\{c_n\}$ be the Fourier coefficient of $f \in L^2(a, b)$ and $\{a_n\}$ any other sequence; then we have

$$\left\| f - \sum_{n=0}^{N} c_n f_n \right\|^2 \leq \left\| f - \sum_{n=0}^{N} a_n f_n \right\|^2,$$

i.e., the mean square error is minimized for the series with Fourier coefficients.

The proof is obtained by adding and subtracting the series with the c_n's:

$$\left\| f - \sum_{n=0}^{N} a_n f_n \right\|^2 = \langle f - \sum_{n=0}^{N} a_n f_n, \ f - \sum_{n=0}^{N} a_n f_n \rangle$$

$$= \langle f, f \rangle - \sum_{n=0}^{N} a_n \langle f_n, f \rangle - \sum_{n=0}^{N} \overline{a_n} \langle f, f_n \rangle + \sum_{n=0}^{N} |a_n|^2$$

$$= \langle f, f \rangle - \sum_{n=0}^{N} (a_n \overline{c_n} + \overline{a_n} c_n)$$

$$+ \sum_{n=0}^{N} |a_n|^2 + \sum_{n=0}^{N} |c_n|^2 - \sum_{n=0}^{N} |c_n|^2$$

$$= \langle f, f \rangle + \sum_{n=0}^{N} |a_n - c_n|^2 - \sum_{n=0}^{N} |c_n|^2$$

$$= \left\| f - \sum_{n=0}^{N} c_n f_n \right\|^2 + \sum_{n=0}^{N} |a_n - c_n|^2. \tag{1.2}$$

Since $\sum_{n=0}^{N} |a_n - c_n|^2 \geq 0$, the conclusion follows. $\quad\square$

Another way of thinking of this is that the Fourier coefficients give the orthogonal projection of f onto the subspace V_N spanned by $(f_0, f_1, f_2, \ldots, f_N)$. Indeed by another simple calculation we see that

$$\langle \sum_{n=0}^{N} c_n f_n, \ f - \sum_{n=0}^{N} c_n f_n \rangle = 0.$$

Thus, not only is the best approximation to f in V_N given by this sum, but the error is orthogonal to V_N.

Similar calculations lead to <u>Bessel's inequality</u>

$$\sum_{n=0}^{\infty} |c_n|^2 \leq \|f\|^2 \tag{1.3}$$

since

$$0 \leq \left\| f - \sum_{n=0}^{N} c_n f_n \right\|^2 = \|f\|^2 - \sum_{n=0}^{N} |c_n|^2 ,$$

and therefore $\left\{ \sum\limits_{n=0}^{N} |c_n|^2 \right\}$ is a monotone sequence bounded by $\|f\|^2$. Thus the series of (1.3) converges and has the same bound. A simple consequence of Bessel's inequality is that $\{c_n\} \in \ell^2$ and $c_n \to 0$ as $n \to \infty$.

To round out our theory, we should like to have the series with Fourier coefficients $\sum c_n f_n$ converge to f. By Bessel's inequality the partial sums are a Cauchy sequence in $L^2(\mathbb{R})$, which because of the completeness of this space must converge in the L^2 sense but not to f necessarily. To ensure this we need to add another condition, the <u>completeness</u> of the orthogonal system (not to be confused with the completeness of the space). The orthonormal system $\{f_n\}$ is said to be <u>complete</u> in $L^2(a, b)$ if no nontrivial $f \in L^2(a, b)$ is orthogonal to all the f_n's, i.e., if $\langle f, f_n \rangle = 0$, $n = 0, 1, 2, \ldots$, for $f \in L^2(a, b)$ then $f = 0$, a.e.

THEOREM 1.1

Let $\{f_n\}$ be an orthonormal system in $L^2(a, b)$; let $f \in L^2(a, b)$ with Fourier coefficients $\{c_n\}$; then

$$\left\| f - \sum_{n=0}^{N} c_n f_n \right\| \to 0 \qquad \text{as } N \to \infty$$

if and only if $\{f_n\}$ is complete.

PROOF   By Bessel's inequality we know that the series (1.1) converges to some $g \in L^2(a, b)$, and hence

$$f - \sum_{n=0}^{N} c_n f_n \to f - g .$$

Now the Fourier coefficients of g are given by

$$\langle g, f_m \rangle = \lim_{N \to \infty} \langle \sum_{n=0}^{N} c_n f_n, f_m \rangle = c_m$$

and hence are the same as those of f. Thus $f - g$ has all zero coefficients, and, if the system is complete, $f - g = 0$ a.e. Since the series converges to g it must also converge to f.

On the other hand if the series converges to f, and all the coefficients are zero, then $f = 0$ a.e. as well. $\qquad \square$

The conclusion of the theorems can be restated as Parseval's equality

$$\|f\|^2 = \sum_{n=0}^{\infty} |c_n|^2 \tag{1.4}$$

since

$$\left\| f - \sum_{n=0}^{N} c_n f_n \right\|^2 = \|f\|^2 - \sum_{n=0}^{N} |c_n|^2.$$

An alternate form is given by

$$\sum_{k=0}^{\infty} c_k \overline{d_k} = \langle f, g \rangle$$

where $d_k = \langle g, f_k \rangle$. This is obtained by applying (1.4) to $f + g$ and $f - g$ and then subtracting.

Either of these results can be taken as test for completeness. However, it is sufficient to check (1.4) for a set of functions $\{h_k\}$ whose closed linear span (i.e., the closure in the sense of $L^2(a, b)$ of $\sum_{k=0}^{K} a_k h_k$) is $L^2(a, b)$. In particular we may take h_k to be the set of characteristic functions of subintervals of (a, b).

1.2 Examples

There are many examples of orthonormal systems in the mathematical literature (see [Al], [O], [Sa]). The earliest and most widely studied is the trigonometric system which we consider in more detail below.

We also consider two more recent systems, the Haar and the Shannon, which form prototypes of the newest system, the orthogonal wavelets. In a later chapter we study some aspects of orthogonal polynomials. But there are many others, e.g., Sturm-Liouville systems, which are used to solve partial differential equations; the Walsh functions, which are piecewise constant; and the eigenfunctions of a compact symmetric integral operator, which will not be covered in detail.

1.2.1   Trigonometric system

The trigonometric system is a complete orthogonal system in $L^2(-\pi, \pi)$ given by

$$f_0(x) = 1/2, \quad f_1(x) = \sin x, \quad f_2(x) = \cos x,$$

$$\cdots f_{2n-1}(x) = \sin nx, \quad f_{2n}(x) = \cos nx, \cdots .$$

It is usually not normalized, since $\|f_n\|^2 = \pi, \ n \neq 0$. The series is usually written in the form

$$S(x) = \frac{a_o}{2} + \sum_{n=1}^{\infty} a_n \cos nx + b_n \sin nx . \tag{1.5}$$

If (1.5) is the Fourier series of a function $f \in L^2(-\pi, \pi)$, the coefficients are given by

$$a_n = \frac{1}{\pi} \int_{-\pi}^{\pi} f(x) \cos nx \ dx, \qquad n = 0, 1, \cdots \tag{1.6}$$

$$b_n = \frac{1}{\pi} \int_{-\pi}^{\pi} f(x) \sin nx \ dx, \qquad n = 1, 2, \cdots .$$

The orthogonality of this system is easy to prove by using a few trigonometric identities. However, the completeness, though well known, is not so obvious. In the interest of completeness we present a proof. It involves first showing that the Fourier series converges uniformly for certain functions. It should be remarked that this is not true for all continuous functions; there are examples where the Fourier series fails to converge on a dense set of points [Z, p. 298].

If the series (1.5) is to converge uniformly, then the limit function must be continuous and periodic of period 2π, which we assume f to be. We shall need an expression for the partial sums of the series, obtained by substituting (1.6) into the partial sums of (1.5).

$$S_n(x) = \frac{a_0}{2} + \sum_{k=1}^{n} a_k \cos kx + b_k \sin kx$$

$$= \frac{1}{\pi} \int_{-\pi}^{\pi} f(t) \left\{ \frac{1}{2} + \sum_{k=1}^{n} \cos kt \cos kx + \sin kt \sin kx \right\} dt$$

$$= \frac{1}{\pi} \int_{-\pi}^{\pi} f(t) \left\{ \frac{1}{2} + \sum_{k=1}^{n} \cos k(x - t) \right\} dt$$

$$= \frac{1}{\pi} \int_{-\pi}^{\pi} f(t) \frac{\sin \left(n + \frac{1}{2}\right)(x - t)}{2 \sin(x - t)/2} dt$$

$$= \int_{-\pi}^{\pi} f(x - u) \frac{\sin \left(n + \frac{1}{2}\right) u}{2\pi \sin u/2} du . \tag{1.7}$$

The expression

$$D_n(u) = \frac{1}{\pi} \left[\frac{1}{2} + \sum_{k=1}^{n} \cos ku \right] = \frac{\sin \left(n + \frac{1}{2}\right) u}{2\pi \sin u/2} \tag{1.8}$$

is called the Dirichlet kernel, and it plays a central role in the study of pointwise convergence of Fourier series. It may be shown true by multiplying both sides of

$$\pi D_n(u) = \frac{1}{2} + \sum_{k=1}^{n} \cos ku$$

by $\sin \frac{u}{2}$ and then forming a telescoping sum.

Proposition 1.3
Let f be a 2π periodic function in $C^2(\mathbb{R})$; then
  (i) $\sup_{x \in \mathbb{R}} |S_n(x) - f(x)| \to 0$
  (ii) $\|S_n - f\| \to 0$
  as $n \to \infty$.

PROOF Since $\int_{-\pi}^{\pi} D_n(u) du = 1$, the difference between S_n and f may be expressed as

$$S_n(x) - f(x) = \int_{-\pi}^{\pi} \{f(x - u) - f(x)\} D_n(u) du$$

$$= \frac{1}{\pi} \int_{-\pi}^{\pi} \left\{ \frac{f(x-u)-f(x)}{2\sin u/2} \right\} \sin\left(n+\frac{1}{2}\right) u\, du$$

$$= \frac{1}{\pi} \int_{-\pi}^{\pi} g(x,u) \sin\left(n+\frac{1}{2}\right) u\, du$$

$$= -\frac{1}{\pi} \frac{\cos\left(n+\frac{1}{2}\right)u}{n+\frac{1}{2}} g(x,u) \Big|_{-\pi}^{\pi}$$

$$+ \frac{1}{\pi} \int_{-\pi}^{\pi} \frac{\partial g(x,u)}{\partial u} \frac{\cos\left(n+\frac{1}{2}\right)u}{\left(n+\frac{1}{2}\right)} du \qquad (1.9)$$

where $g(x,u) = \frac{f(x-u)-f(x)}{2\sin u/2}$.

Since both $g(x,u)$ and its derivative are uniformly bounded, the last expression gives us

$$|S_n(x) - f(x)| \le \frac{C}{\left(n+\frac{1}{2}\right)}$$

for some constant C. Hence (i) must hold and by squaring and integration of (i), so must (ii). □

Since twice differentiable functions are dense in $L^2(-\pi,\pi)$, so are trigonometric polynomials by this proposition and therefore it follows that the trigonometric system is complete by Theorem 1.1.

An alternative form for the trigonometric series (1.5) is the exponential form

$$S(x) = \sum_{n=-\infty}^{\infty} c_n e^{inx}, \qquad (1.10)$$

where convergence is with respect to the symmetric partial sums. If (1.10) is a Fourier series the coefficients are

$$c_n = \frac{1}{2\pi} \int_{-\pi}^{\pi} f(x) e^{-inx}\, dx. \qquad (1.11)$$

Of course expression (1.10) is reducible to (1.5) by using

$$e^{\pm inx} = \cos nx \pm i \sin nx.$$

Another way of looking at these expressions is as transforms.

The finite Fourier transform of the periodic function $f(x)$ is given by (1.11) with the inverse transform given by (1.10). It converts a differential operator into a multiplication operator,

$$\frac{1}{2\pi} \int_{-\pi}^{\pi} (Df(x)) e^{-inx}\, dx = in\, c_n,$$

which makes it a useful tool in differential equations. Its absolute value is also shift invariant,

$$\left| \frac{1}{2\pi} \int_{-\pi}^{\pi} f(x-\alpha)e^{-inx}\, dx \right| = |c_n| \, .$$

The infinite Fourier transform of the function $f \in L^1(-\infty, \infty)$ is the expression

$$\hat{f}(w) = \int_{-\infty}^{\infty} f(t)e^{-iwt}\, dt \, , \qquad w \in \mathbb{R}. \tag{1.12}$$

The image of the transform in this case is a continuous function on $\mathbb{R}$, which, if it is also in $L^1(\mathbb{R})$, leads to the inverse [B-C, p. 19]

$$f(t) = \frac{1}{2\pi} \int_{-\infty}^{\infty} \hat{f}(w)e^{iwt}\, dw \, , \qquad t \in \mathbb{R}. \tag{1.13}$$

Versions of Parseval's equality (1.4) also exist for Fourier transforms. They are, for $f, g \in L^2(\mathbb{R})$, [B-C, p. 105]

$$\|f\|^2 = \frac{1}{2\pi}\|\hat{f}\|^2 \qquad \langle f, g \rangle = \frac{1}{2\pi}\langle \hat{f}, \hat{g} \rangle \, .$$

This requires a more general definition of Fourier transform, which together with other properties are found in the next chapter.

One can also go the other way and approximate (1.11) by a discrete sum. This gives us the discrete Fourier transform,

$$\gamma_k = \frac{1}{N} \sum_{j=0}^{N-1} f\left(\frac{2\pi j}{N}\right) e^{-ijk2\pi/N} \, , \qquad k = 0, \cdots, N-1, \tag{1.14}$$

with the inverse given by

$$f\left(\frac{2\pi j}{N}\right) = \sum_{k=0}^{N-1} \gamma_k\, e^{ikj2\pi/N} \, , \qquad j = 0, \cdots, N-1. \tag{1.15}$$

This is the form that leads to the fast Fourier transform which, by grouping terms in (1.14), reduces the computation time considerably. This has made transform methods much more useful in partial differential equations, image processing, time series, and other applied problems [St1, p. 448].

1.2.2 Haar system

The Haar orthogonal system begins with $\phi(t)$, the characteristic function of the unit interval

$$\phi(t) = \chi_{[0,1)}(t).$$

It is clear that $\phi(t)$ and $\phi(t-n)$, $n \neq 0$, $n \in \mathbb{Z}$ are orthogonal since their product is zero. It is also clear that $\{\phi(t-n)\}$ is not a complete orthogonal system in $L^2(\mathbb{R})$ since its closed linear span V_0 consists of piecewise constant functions with possible jumps only at the integers. The characteristic function of $[0, 1/2)$, for example, with a jump at $1/2$, cannot have a convergent expansion.

In order to include more functions we consider the dilated version of $\phi(t)$ as well, $\phi(2^m t)$ where $m \in \mathbb{Z}$. Then by a change of variable we see that $\{2^{m/2}\phi(2^m t - n)\}$ is an orthonormal system. Its closed linear span will be denoted by V_m. Since any function in $L^2(\mathbb{R})$ may be approximated by a piecewise constant function f_m with jumps at binary rationals, it follows that $\bigcup_m V_m$ is dense in $L^2(\mathbb{R})$. Thus the system $\{\phi_{mn}\}$ where

$$\phi_{mn}(t) = 2^{m/2}\phi(2^m t - n)$$

is complete in $L^2(\mathbb{R})$, but, since $\phi(t)$ and $\phi(2t)$ are not orthogonal, it is not an orthogonal system. We must modify it somehow to convert it into an orthogonal system.

Fortunately the cure is simple; we let $\psi(t) = \phi(2t) - \phi(2t-1)$. Then everything works; $\{\psi(t-n)\}$ is an orthonormal system, and $\psi(2t-k)$ and $\psi(t-n)$ are orthogonal for all k and n. This enables us to deduce that $\{\psi_{mn}\}_{m,n\in\mathbb{Z}}$ where

$$\psi_{mn}(t) = 2^{m/2}\psi(2^m t - n)$$

is a complete orthonormal system in $L^2(\mathbb{R})$. This is the Haar system; the expansion of $f \in L^2(\mathbb{R})$ is

$$f(t) = \sum_{m=-\infty}^{\infty} \sum_{n=-\infty}^{\infty} \langle f, \psi_{mn}\rangle \psi_{mn}(t), \qquad (1.16)$$

with convergence in the sense of $L^2(\mathbb{R})$. The standard approximation is the series given by

$$f_m(t) = \sum_{k=-\infty}^{m-1} \sum_{n=-\infty}^{\infty} \langle f, \psi_{kn}\rangle \psi_{kn}(t). \qquad (1.17)$$

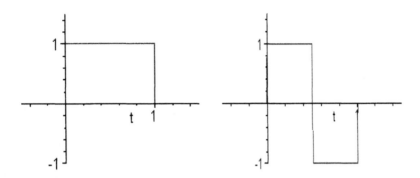

FIGURE 1.1
The scaling function and mother wavelet for the Haar system.

It converges to a piecewise constant function with jumps at $2^{-m}n$, $n \in \mathbb{Z}$, at most.

Hence, $f_m \in V_m$ and since, by Parseval's equality,

$$\langle f_m, \phi_{mn} \rangle = \sum_{k=-\infty}^{m-1} \sum_{j=-\infty}^{\infty} \langle f, \psi_{kj} \rangle \langle \psi_{kj}, \phi_{mn} \rangle = \langle f, \phi_{mn} \rangle,$$

it is the projection of f on V_m, i.e., $f_m = \sum_n \langle f, \phi_{mn} \rangle \phi_{mn}$.

This enables us to get a pointwise convergence theorem.

Proposition 1.4
Let f be continuous on $\mathbb{R}$ and have compact support; then $f_m \to f$ uniformly.

PROOF Since f is uniformly continuous, it follows that for each $\epsilon > 0$ there exists an m such that

$$|f(x) - f(y)| < \epsilon \qquad \text{when } |x - y| \leq 2^{-m}.$$

For $x \in [n2^{-m}, (n+1)2^{-m})$ we have

$$f_m(x) = 2^{m/2} \int_{2^{-m}n}^{2^{-m}(n+1)} f(t)dt \, 2^{m/2} \phi(2^m x - n)$$

since all the other terms in the series are zero in this interval. Therefore by the mean value theorem

$$f_m(x) = f(\zeta_m)2^{-m}2^m\phi(2^m x - n) = f(\zeta_m),$$

for some ζ_m in this interval, and since $|x - \zeta_m| \leq 2^{-m}$, $|f_m(x) - f(x)| < \epsilon$. $\square$

This Haar system is our first prototype of a wavelet system, and we shall return to it several times later. At this point it should be observed that the uniform convergence of f_m to f is a property not shared by the trigonometric system. The uniform convergence of (1.16) also follows since the inner series has only a finite number of terms, and the partial sums of the outer series converge uniformly since they are of the form $f_m(x) - f_{-p}(x)$ and f_{-p} converges to zero uniformly as $p \to \infty$.

The $\phi(t)$ is usually called the <u>scaling function</u> in wavelet terminology while $\psi(t)$ is the <u>mother wavelet</u>.

1.2.3   The Shannon system

A second prototype also begins with the characteristic function of an interval. Now, however, it is the Fourier transform of the scaling function, taken to be

$$\hat{\phi}(w) = \begin{cases} 1 & -\pi \leq w < \pi \\ 0 & \text{o.w.} \end{cases}.$$

Its inverse Fourier transform is

$$\phi(t) = \frac{1}{2\pi}\int_{-\infty}^{\infty}\hat{\phi}(w)e^{iwt}dw = \frac{1}{2\pi}\int_{-\pi}^{\pi}e^{iwt}dw = \frac{\sin\pi t}{\pi t}.$$

The orthogonality of $\phi(t)$ and $\phi(t - n)$ is based on properties of the Fourier transform, of Parseval's equality, and the fact that $(\widehat{\phi(t - \alpha)})(w) = \hat{\phi}(w)e^{-i\alpha w}$:

$$\int_{-\infty}^{\infty}\phi(t)\phi(t - n)dt = \frac{1}{2\pi}\int_{-\infty}^{\infty}\hat{\phi}(w)\overline{\hat{\phi}(w)}e^{iwn}dw$$

$$= \frac{1}{2\pi}\int_{-\pi}^{\pi}e^{iwn}dw = \frac{\sin\pi n}{\pi n} = 0, \ n \neq 0.$$

Let $f(t)$ be a function that is square integrable and whose Fourier transform $\hat{f}(w)$ vanishes for $|w| > \pi$. It has a Fourier series given by

$$\hat{f}(w) = \sum_n c_n e^{iwn}, \qquad |w| \leq \pi \qquad (1.18)$$

where $c_n = \frac{1}{2\pi} \int_{-\pi}^{\pi} \hat{f}(w) e^{-iwn} dw$. By the Fourier integral theorem (1.13) this is just $f(-n)$. This theorem applied to both sides of (1.18) yields

$$f(t) = \frac{1}{2\pi} \int_{-\pi}^{\pi} \hat{f}(w) e^{iwt} dw = \sum_n f(-n) \frac{1}{2\pi} \int_{-\pi}^{\pi} e^{iwn} e^{iwt} dw$$

$$= \sum_n f(-n) \frac{\sin \pi(t+n)}{n(t+n)} . \tag{1.19}$$

We denote by V_0 the set of all such functions. This is a linear space and is closed as well since limits (in the square integrable sense) of the sequences of functions in V_0 are also in V_0.

The formula (1.19) is referred to as the Shannon sampling theorem [Sh]. It enables one to recover a band-limited function in V_0 from its values on the integers. This is used by engineers to convert a digital to an analog signal (as in compact discs). See [B] and [Za] for more properties.

By changing the scale in (1.19) by a factor of 2, we can obtain a sampling theorem on the half integers. (Let $2x = t$ and let $g(x) = f(2x)$). The space with the new scale is V_1 and it will consist of functions whose Fourier transforms vanish outside of $[-2\pi, 2\pi]$. We may repeat this as often as we want and get thereby an increasing sequence of spaces. We can also stretch the scale instead of shrinking it to obtain a sequence $\{V_m\}_{m=-\infty}^{\infty}$ satisfying

$$\cdots \subseteq V_{-m} \subseteq \cdots \subseteq V_{-1} \subseteq V_0 \subseteq V_1 \subseteq \cdots \subseteq V_m \subseteq \cdots .$$

As we go to the left in this sequence the support of the Fourier transform (the set outside of which it vanishes) shrinks to 0. As we go to the right, it expands to all of $\mathbb{R}$. Thus we have

(i) $\bigcap_m V_m = \{0\}$ and

(ii) each $f \in L^2(\mathbb{R})$ can be approximated by a function in V_m for m sufficiently large.

The sequence $\{V_m\}$ is called a "multiresolution analysis" associated with $\phi(t)$.

Just as with the other prototype, we can introduce a function $\psi(t)$ in V_1 which is orthogonal to $\phi(t-n)$, $\psi(t) = 2\phi(2t) - \phi(t)$. Its Fourier transform is given by

$$\hat{\psi}(w) = \hat{\phi}\left(\frac{w}{2}\right) - \hat{\phi}(w)$$

and has support on $[-2\pi, -\pi] \cup [\pi, 2\pi]$. Since the supports of $\hat{\phi}$ and $\hat{\psi}$ are disjoint, the needed orthogonality follows. Thus, the inverse Fourier

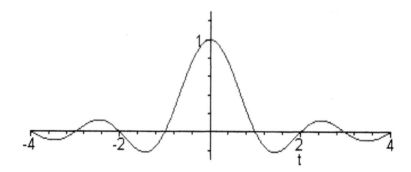

FIGURE 1.2
The scaling function for the Shannon system.

transform $\psi(t)$ is another example of a "mother wavelet", and leads to
another orthonormal sequence $\{\psi_{m,n}(t)\}$. As in the case of the Haar system, it is also complete and the expansion of certain continuous functions
converges uniformly. Rather than require this function to have compact
support, we assume only that it is continuous and that its Fourier transform $\hat{f} \in L^1(\mathbb{R})$.

Proposition 1.5
  *Let $f \in L^2(\mathbb{R})$ be continuous and have a Fourier transform $\hat{f} \in L^1(\mathbb{R})$;
then $f_m \to f$ uniformly on $\mathbb{R}$.*

PROOF   If $f_m = \sum_n \langle f, \phi_{mn}\rangle \phi_{mn}$, then

$$\hat{f}_m(w) = \sum_n \frac{1}{2^{m+1}\pi} \int_{-2^m\pi}^{2^m\pi} \hat{f}(\varsigma)\hat{\phi}(2^{-m}\varsigma)e^{i\varsigma n2^{-m}} d\varsigma \; \hat{\phi}(2^{-m}w)e^{-iwn2^{-m}}.$$

The right hand side is the Fourier series of $\hat{f}(w)$ on $[-2^m\pi, 2^m\pi]$ and
hence is equal to the restriction of $\hat{f}(w)$ to this interval,

$$\hat{f}_m(w) = \hat{f}(w)\chi_{2^m\pi}(w) \tag{1.20}$$

where $\chi_{2^m\pi}(w)$ is the characteristic function of $[-2^m\pi, 2^m\pi]$. The error
then may be given by

$$f(t) - f_m(t) = \frac{1}{2\pi} \int_{-\infty}^{\infty} \left(\hat{f}(w) - \hat{f}_m(w)\right) e^{iwt} dw$$

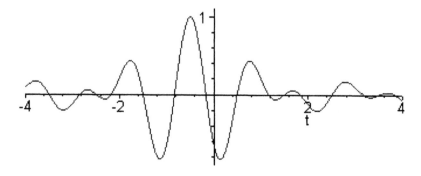

FIGURE 1.3
The mother wavelet for the Shannon system.

$$= \frac{1}{2\pi} \left\{ \int_{2^m \pi}^{\infty} + \int_{-\infty}^{-2^m \pi} \right\} \hat{f}(w) e^{iwt} dw.$$

Hence we obtain

$$|f(t) - f_m(t)| \le \frac{1}{2\pi} \left\{ \int_{2^m \pi}^{\infty} + \int_{-\infty}^{-2^m \pi} \right\} |\hat{f}(w)| \, dw$$

which, since $\hat{f} \in L^1(\mathbb{R})$, must converge to 0 as $m \to \infty$. □

The astute reader will recognize V_0 as the <u>Paley-Wiener</u> space of π band limited functions [P-W]. It is a reproducing kernel Hilbert space with reproducing kernel $q(x,t) = \frac{\sin \pi(x-t)}{\pi(x-t)}$. The expansion of $f \in V_0$ with respect to $\phi(x - n)$ given by (1.19) not only converges in $L^2(\mathbb{R})$ but converges uniformly on all of $\mathbb{R}$. The proof is implicit in (1.19).

1.3   Problems

1. The following are orthogonal systems in $L^2(0, \pi)$:

   (a) $\{\cos nx\}_{n=0}^{\infty}$,
   (b) $\{\sin nx\}_{n=1}^{\infty}$,
   (c) $\{\sin(n + \frac{1}{2})x\}_{n=0}^{\infty}$.

Show they are also complete in $L^2(0,\pi)$ by using the symmetry of these functions and the completeness of the full trigonometric system in $L^2(-\pi,\pi)$.

2. Show that any orthogonal system $\{f_n\}$ is also linearly independent.

3. Let $\{f_n\}_{n=0}^{\infty}$ be a linearly independent sequence in $L^2(a,b)$; define
$$\phi_0 = f_0, \quad \phi_1 = f_1 - \frac{\langle f_1,\phi_0\rangle\phi_0}{\|\phi_0\|^2}, \quad \phi_2 = f_2 - \frac{\langle f_2,\phi_1\rangle\phi_1}{\|\phi_1\|^2} - \frac{\langle f_2,\phi_0\rangle\phi_0}{\|\phi_0\|^2}, \cdots.$$
Show that $\{\phi_n\}$ is an orthogonal system.

4. Let $\{f_n\}$ be linearly independent and complete in $L^2(a,b)$ (i.e., the closed linear span of $\{f_n\}$ is $L^2(a,b)$). Then $\{f_n\}$ is said to be a <u>Riesz basis</u> if there are positive constants A and B such that

$$A \sum_{i=1}^{n} |c_i|^2 \leq \|\sum_{i=1}^{n} c_i f_i\|^2 \leq B \sum_{i=1}^{n} |c_i|^2$$

for each sequence $\{c_i\}$ of complex numbers. Take $\{g_n\}$ to be a biorthogonal sequence to $\{f_n\}$ ($\langle g_n, f_k\rangle = \delta_{nk}$) in $L^2(a,b)$. Show that

(a) $\{g_n\}$ is the unique biorthogonal sequence to $\{f_n\}$,
(b) If $\{c_n\} \in \ell^2$ then $\sum_n c_n f_n$ converges in $L^2(a,b)$,
(c) For each $f \in L^2(a,b)$, $\{\langle f, g_n\rangle\} \in \ell^2$,
(d) For each $f \in L^2(a,b)$, $f = \sum_{n=0}^{\infty} \langle f, g_n\rangle f_n$.

5. Let $f_1(x) = x$ and $f_2(x) = \pi^2 - 3x^2$, $-\pi < x < \pi$. Find the Fourier series of f_1 and f_2 on $(-\pi,\pi)$ and use them to sum the series

(a) $\sum_{n=1}^{\infty} \frac{(-1)^n}{n}$

(b) $\sum_{n=1}^{\infty} \frac{1}{n^2}$

(c) $\sum_{n=1}^{\infty} \frac{(-1)^n}{n^2}$

(d) $\sum_{n=1}^{\infty} \frac{1}{n^4}$

6. Show that the Haar system

$$\{\phi(t),\, 2^{m/2}\psi(2^m t - n)\}, \quad m = 0,1,2,\cdots, \quad n = 0,1,\cdots,2^m - 1,$$

is an orthonormal basis of $L^2(0,1)$.

7. Find the expansion with respect to $\{\phi(t-n)\}$, the Shannon system in V_0, of the functions given by

(a) $f_1(x) = \frac{1}{2\pi} \int\limits_{-\pi}^{\pi} w\, e^{ixw}\, dw$

(b) $f_2(x) = \frac{1}{2\pi} \int\limits_{-\pi}^{\pi} (\pi^2 - 3w^2)e^{ixw}\, dw.$

Chapter 2

A Primer on Tempered Distributions

For many applications it is necessary to work with a larger class of functions than $L^1(\mathbb{R})$. It is also desirable to have a more general theory of the Fourier transform, which would enable us to calculate Fourier transforms of periodic functions or polynomials. This is impossible with the L^1 (or the usual L^2) theory, which does not even work for constant functions.

The simplest theory that extends the Fourier transform to all functions of polynomial growth is the theory of "tempered distributions". These include, in addition to functions, certain objects such as the "delta function," which is not strictly a function. Nonetheless, it is widely used and is central to some applications such as the study of transfer functions of linear filters in engineering.

These tempered distributions may be considered limits in some sense of sequences of ordinary $L^1(\mathbb{R})$ functions. In fact, one approach is to define them as limits of Cauchy sequences of such functions. In this case the familiar process of extending the rational numbers to obtain the real numbers is imitated (see [K]). However, there are other approaches that can be used for the formal definitions, but they will still be limits of sequences of functions and are best thought of as such.

In our approach, which is due to L. Schwartz [S], these ideal elements are defined as continuous linear functionals on a space of "test functions". The resulting objects will include all locally integrable functions of polynomial growth and all tempered measures as well as their derivatives of every order. The set of such objects is closed under the Fourier transform and provides an appropriate setting for much of signal processing.

We shall omit many details. The reader completely unfamiliar with the subject should consult one of the many excellent texts devoted to it, e.g., [K1], [S], [Br], [Ze], [G-S].

2.1   Intuitive introduction

One of the difficulties with the L^1 theory of functions is that we cannot always perform operations that we would like. We cannot, in general, differentiate such a function, nor can we even multiply it by a polynomial and get another function in this class. We should like to extend our class of functions to a larger class of objects (no longer necessarily functions) for which these operations are always possible. These will be our "tempered distributions".

We begin with the "tempered functions". These are functions which are locally in L^1 and are of at most polynomial growth. This may be expressed by requiring that

$$\left|\int_a^x f(t)dt\right| \leq C(x^2+1)^p, \ x \in \mathbb{R}$$

for any a and some positive constants C and p. The class of such tempered functions includes

- all polynomials
- all piecewise continuous functions of polynomial growth
- all classical orthogonal functions of Chapter 1
- all piecewise continuous periodic functions
- all scaling functions and wavelets.

However, functions of exponential growth are excluded, and hence so are solutions of certain differential equations (e.g., $y' = y$) on the real line.

Our goal is to extend the class of these tempered functions in such a way that the following operations hold for f in the extended class:

- differentiation $\frac{df}{dx}$,
- dilation $f(ax), a \in \mathbb{R}^+$,

- translation $f(x - b), b \in \mathbb{R}$,

- multiplication by $\theta(x)$, a C^∞ function of polynomial growth, $f(x)\theta(x)$,

- convolution with a function g of compact support $(f * g)(x)$.

One example that will recur is the Heaviside function given by

$$H(x) = \begin{cases} 0, & x \leq 0 \\ 1, & x > 0 \end{cases}.$$

Its derivative does not exist in the usual sense at $x = 0$, but is equal to 0 for all $x \neq 0$. We denote by δ the formal derivative of the Heaviside function. If it is to behave as a usual derivative, then since $\delta = 0$ a.e., the integral

$$\int_{-\varepsilon}^{\varepsilon} \delta(t)dt = 0,$$

but on the other hand

$$\int_{-\varepsilon}^{\varepsilon} \delta(t)dt = H(\varepsilon) - H(-\varepsilon) = 1 - 0 = 1,$$

for any positive ε. Therefore δ cannot be an ordinary function, but is some different sort of object.

Even though it is not a function, we can define all the operations listed above for δ. The intuitive way to think of δ is as approximately a spike, a function δ_λ given by

$$\delta_\lambda(x) = \begin{cases} 0, & |x| > 1/\lambda \\ \lambda(1 - \lambda|x|), & |x| \leq 1/\lambda \end{cases},$$

where λ is some large positive number. Then some of the properties of δ can be found by performing operations on δ_λ to determine the expected behavior of the same operations on δ. For example, $\delta(ax)$ is given approximately by

$$\delta_\lambda(ax) = \begin{cases} 0, & |ax| > 1/\lambda \\ \lambda(1 - \lambda|ax|), & |ax| \leq 1/\lambda \end{cases} = \frac{1}{a}\delta_{a\lambda}(x).$$

Thus we might expect δ to satisfy $\delta(ax) = \frac{1}{a}\delta(x)$ and we would be right. Similarly the integral of δ_λ

$$\int_{-\varepsilon}^{\varepsilon} \delta_\lambda(t)dt = 1 \text{ for } 1/\lambda \leq \varepsilon,$$

as expected. Other properties follow similarly, in particular, the fact that

$$\theta(x)\delta(x) = \theta(0)\delta(x)$$

for any C^∞ function of polynomial growth.

Other tempered distributions are obtained by differentiating different tempered functions. One such example is the function $f(x) = |x|^\alpha$ where $\alpha > -1$. Its derivative is $\alpha|x|^{\alpha-1}sgn(x)$, which is an ordinary function for $|x| \neq 0$, but is not a tempered function when $\alpha < 0$. Similarly the function $g(x) = |x|^\alpha sgn(x)$ has derivative $\alpha|x|^{\alpha-1}$. These derivatives and others lead to various principal value distributions.

If we treat these tempered distributions as ordinary functions as far as the operations are concerned, we will not usually be far wrong. The exception is multiplication. The product of two tempered functions is not a tempered function necessarily, nor is the product of two tempered distributions a tempered distribution. Even the product of δ with itself does not exist.

The next sections will present a more rigorous approach in which the tempered distributions are defined as continuous linear functionals on a space of *test functions*.

2.2 Test functions

The device for defining our objects (the tempered distributions) is in terms of a certain space of test functions. These functions test our objects by averaging them or "smearing" them. Since our objects are not functions for which we know the exact values at points, the next best thing is to know the average or smeared values which we do know. These smeared values correspond to integrals or more generally to continuous linear functionals, i.e., linear functions from this space of test functions to the complex numbers which are continuous. The properties of the tempered distributions are based on the properties of these test functions which we must first define and study.

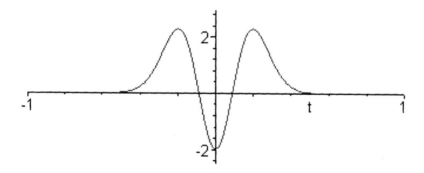

FIGURE 2.1
A test function in the space S (the Hermite function $h_4(x)$).

Our test functions will belong to the space S of rapidly decreasing C^∞ functions on $\mathbb{R}$, i.e., functions that satisfy

$$\left|\theta^{(k)}(t)\right| \leq C_{pk}(1+|t|)^{-p}, \quad p, k = 0, 1, 2, \ldots, t \in \mathbb{R}. \qquad (2.1)$$

S is clearly a linear space, that is, linear combinations of elements of S also belong to S. But in order to define continuity of the linear functionals, we need a notion of convergence in S. This is given by the semi-norms

$$\gamma_{pk} = \sup_t (1+|t|)^p \left|\theta^{(k)}(t)\right|, \quad p, k = 0, 1, 2, \ldots. \qquad (2.2)$$

This is used to define that $\theta_\nu \to \theta$ in S whenever

$$(1+|t|)^p D^k(\theta_\nu(t) - \theta(t)) \to 0$$

uniformly in t for each p and k as $\nu \to \infty$. (D is the derivative operator.)

The space S is dense in $L^2(\mathbb{R})$ in the sense that each $f \in L^2(\mathbb{R})$ may be approximated by some $\theta \in S$ in the norm of $L^2(\mathbb{R})$. This may be shown by observing that S contains the Hermite functions $\{h_n\}$ given by

$$h_0(x) = \frac{e^{-x^2/2}}{\pi^{1/4}}, \qquad x \in \mathbb{R}$$

and recursively

$$(x - D)h_n(x) = \sqrt{2n+2}h_{n+1}(x), \quad n = 0, 1, \ldots, x \in \mathbb{R}.$$

These constitute an orthonormal basis of $L^2(\mathbb{R})$ (see Chapter 6). Hence linear combinations of these $\{h_n\}$ also belong to S and may be used to approximate a given $f \in L^2(\mathbb{R})$ in the L^2 norm.

The C^∞ functions of compact support are also contained in S since they trivially satisfy the decay condition (2.1). But polynomials and trigonometric functions do not belong to S since they do not converge to 0 as $t \to \infty$.

Some of the properties of S are as follows:

1. S is complete with respect to the convergence consistent with (2.2), i.e., every Cauchy sequence in this sense converges to an element of S.

2. Differentiation is a continuous operation in S, i.e., if $\theta_\nu \to \theta$ in the sense of (2.2) then $D\theta_\nu \to D\theta$ in the same sense.

3. Multipliction by a polynomial is a continuous operation in S.

4. S is closed under dilations and translations.

5. The Fourier transform given by (1.12) is a 1-1 mapping of S onto itself.

The two operators of property 4 are important for wavelets. These are the dilation operator D_a, $\alpha > 0$ $(D_a f(t) = f(\alpha t))$ and translation operator T_β $(T_\beta f(t) = f(t - \beta))$. Our space S is not only closed under these operators but both are continuous with respect to (2.2).

The Fourier transform $\mathcal{F}$ is given by

$$(\mathcal{F}\theta)(w) = \hat{\theta}(w) = \int_{-\infty}^{\infty} e^{-iwt}\theta(t)dt. \tag{2.3}$$

That it maps S into S follows from the fact that $\mathcal{F}$ converts differentiation into multiplication by a multiple of w and vice versa. In fact by the Fourier integral theorem [B-C, p.10],

$$\theta(t) = \frac{1}{2\pi}\int_{-\infty}^{\infty} e^{iwt}\hat{\theta}(w)dw \tag{2.4}$$

and hence $\mathcal{F}$ can be seen to be one to one and onto as well.

For some of our examples we will need a related smaller space than the space of all tempered distributions. This makes it necessary to define a corresponding test function space which will be larger. We denote by

S_r the space containing S consisting of all C^r functions such that (2.1) is satisfied for $k \leq r$ but for all p. The convergence is the same but with this restriction on k. Its Fourier transform will be denoted $\hat{S}_r$ and will consist of all C^∞ functions such that (2.1) is satisfied for all $p \leq r$ and all k.

2.3   Tempered distributions

A tempered distribution is an element of the dual space S' of S. This space is composed of all continuous linear functionals on S; functions from S to $\mathbb{C}$ (denoted by (T, θ), the value of T on θ), which are linear

$$(T, \alpha_1\theta_1 + \alpha_2\theta_2) = \alpha_2(T, \theta_1) + \alpha_2(T, \theta_2)$$

and continuous

$$\theta_n \to 0 \text{ in } S \Rightarrow (T, \theta_n) \to 0 \text{ in } \mathbb{C}.$$

These are the ideal elements mentioned at the start of the chapter; they inherit many of the properties of S.

The limit of any bounded sequence of L^1 functions convergent in S' (i.e., convergent for each $\theta \in S$) defines such an element. Indeed, let $\{f_n\}$ be a sequence in $L^1(\mathbb{R})$ such that

$$\int_{-\infty}^{\infty} |f_n| \leq C$$

and $\{(f_n, \theta)\}$ $(= \{\int f_n \theta\})$ is a Cauchy sequence of complex numbers for each $\theta \in S$. Then we may define T as

$$(T, \theta) = \lim_{n \to \infty} (f_n, \theta).$$

This will be a functional since the complex numbers are complete; it will be linear since each term in the sequence is linear; it will be continuous as well since if $\theta_m \to \theta$ in S then $\theta_m \to \theta$ uniformly on $\mathbb{R}$ and

$$|(f_n, \theta_m - \theta)| = |\int f_n(\theta_m - \theta)|$$

$$\leq \int |f_n| \sup_x |\theta_m(x) - \theta(x)| \leq C \sup_x |\theta_m(x) - \theta(x)|$$

which converges to 0. Hence we can interchange the limits in

$$\lim_{m\to\infty} (T,\theta_m) = \lim_{m\to\infty} \lim_{n\to\infty} (f_n,\theta_m) = \lim_{n\to\infty} (f_n,\theta) = (T,\theta),$$

thereby proving continuity and membership in $\mathcal{S}'$. This last equality corresponds to convergence in $\mathcal{S}'$. We say that $f_n \to f$ in the sense of $\mathcal{S}'$ if it converges in the sense that

$$\lim_{n\to\infty} (f_n,\theta) = (f,\theta)$$

for each $\theta \in \mathcal{S}$.

We could go further than this to prove that each Cauchy sequence converges in $\mathcal{S}'$ so that this space is complete. We shall not do so (see, however [S, II, p.94]).

Each locally integrable function f of polynomial growth belongs to $\mathcal{S}'$. The functional corresponding to it is

$$(f,\theta) = \int_{-\infty}^{\infty} f(t)\theta(t)dt$$

where we have used the same symbol f for the functional in $\mathcal{S}'$ and the function $f(t)$. This functional clearly exists and is linear. Furthemore, since

$$f(t)/(t^2+1)^r \in L^1(\mathbb{R})$$

for some integer r, we see that

$$(f,\theta_n) = \int_{-\infty}^{\infty} \frac{f(t)}{(t^2+1)^r}(t^2+1)^r\theta_n(t)dt.$$

This is continuous since if $\theta_n \to 0$ in the sense of $\mathcal{S}$, then $(t^2+1)^r\theta_n(t) \to 0$ uniformly on $\mathbb{R}$ and hence $(f,\theta_n) \to 0$.

Each $T \in \mathcal{S}'$ has a derivative defined by

$$(DT,\theta) = -(T,\theta'), \qquad \theta \in \mathcal{S}.$$

Thus even if T corresponds to a nowhere differentiable function, its derivative exists as a tempered distribution. By combining this with the previous example we see that each derivative of any order of a locally integrable function of polynomial growth belongs to $\mathcal{S}'$. In fact this characterizes $\mathcal{S}'$.

Proposition 2.1
Let $T \in S'$; then there exists a locally integrable function of polynomial growth $F(x)$ and an integer p such that

$$T = D^p F$$

(see [Ze, p.111]).

Thus, we may interpret elements of S' as generalizations of functions (in fact, they constitute examples of "generalized functions"), and shall adopt the same notation as for functions. The symbol (f, θ) will stand for either the value of the functional or, in the case of tempered distributions given by functions, the integral.

We already have the notation $\langle f, \bar{\theta} \rangle$ for the latter and hence $(f, \bar{\theta}) = \langle f, \theta \rangle$.

2.3.1 Simple properties based on duality

In addition to differentiation, we have a number of other operations defined on S'. Each is defined by the same device. Any operation in S is translated to a corresponding operation in S' by first observing what happens to (f, θ) when f is an ordinary function and (f, θ) is an integral. Then this behavior is extended to all of S'. For example, translation T_α is defined on S' as

$$(T_\alpha f, \theta) := (f, T_{-\alpha} \theta)$$

since for ordinary functions

$$(T_\alpha f, \theta) = \int f(t - \alpha) \theta(t) dt = \int f(t) \theta(t + \alpha) dt = (f, T_{-\alpha} \theta).$$

Similarly for dilation by a positive quantity α, we have

$$(D_\alpha f, \theta) := \frac{1}{\alpha} (f, D_{\frac{1}{\alpha}} \theta)$$

and for multiplication by a C^∞ function of polynomial growth $F(t)$,

$$(Ff, \theta) := (f, F\theta).$$

Of course we need to check for each definition that the result is in fact in S'. But that again is clear in each of the cases.

The "delta function", δ_α, which we discussed in the first section, is properly defined simply as

$$(\delta_\alpha, \theta) := \theta(\alpha),$$

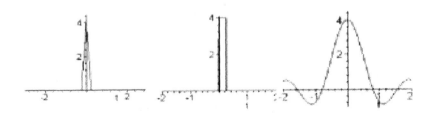

FIGURE 2.2
Some approximations to the delta function in $\mathcal{S}'$.

i.e., the unit point mass at α. It can also be given as the derivative of
the Heaviside function

$$\delta_\alpha = DH_\alpha$$

where

$$H_\alpha(t) = \begin{cases} 1, & t > \alpha \\ 0, & t \le \alpha \end{cases}.$$

Indeed, if $\theta \in \mathcal{S}$,

$$-(DH_\alpha, \theta) = (H_\alpha, \theta') = \int_\alpha^\infty \theta'(t)dt = -\theta(\alpha) = -(\delta_\alpha, \theta).$$

Still a third way of defining δ_α is as the limit of a sequence of functions

$$\delta = \delta_0 = \lim_{n \to \infty} n\chi_{[0,n^{-1})}$$

where $\chi_{[0,n^{-1})}(t)$ is the characteristic function of $[0, n^{-1})$. This limit
must be taken in the sense of $\mathcal{S}'$ and not pointwise. Then $\delta_\alpha = T_\alpha \delta$. It
is easy to check that this is consistent with the other definitions.
    Other elements of $\mathcal{S}'$ include the pseudofunctions such as $pv(\frac{1}{t})$ defined
by

$$\left(pv\frac{1}{t}, \theta\right) = cpv \int_{-\infty}^\infty \frac{\theta(t)}{t}dt$$

where *cpv* stands for Cauchy principal value.

2.3.2 Further properties

In addition to having a derivative, each $f \in \mathcal{S}'$ has an antiderivative $f^{(-1)} \in \mathcal{S}'$. Let $\theta \in \mathcal{S}$ and define $\theta_1(t) = \theta(t) - \left(\int_{-\infty}^{\infty} \theta \right) h_0^2(t)$ where h_0 is the Hermite function of order 0. Then $\int_{-\infty}^{\infty} \theta_1 = 0$ and hence $\int_{-\infty}^{t} \theta_1 \in \mathcal{S}$. We now define $f^{(-1)}$ by

$$(f^{(-1)}, \theta) := -(f, \int_{-\infty}^{t} \theta_1) + C(1, \theta)$$

where C is arbitrary. Then

$$(Df^{(-1)}, \theta) = -(f^{(-1)}, \theta')$$
$$= (f, \int_{-\infty}^{t} \theta_1(\theta)) + C(1, \theta')$$
$$= (f, \int_{-\infty}^{t} \theta') = (f, \theta)$$

since $(1, \theta') = 0$ and the θ_1 corresponding to θ' equals θ'.

The <u>support</u> of a function $\theta \in \mathcal{S}$ is the smallest closed set outside of which θ is identically zero. Two distributions f and $g \in \mathcal{S}'$ are said to be equal in an open set Ω if their difference is zero on each θ with support in Ω, i.e., if

$$(f, \theta) = (g, \theta), \quad \text{supp } \theta \subset \Omega.$$

In particular f could be zero on some such Ω. The <u>support</u> of f then is the smallest closed set K such that $f = 0$ on the complement of K. Clearly δ has support on the single point $\{0\}$, as do all of its derivatives. In fact, the only $f \in \mathcal{S}'$ with point support are linear combinations of δ and its derivatives ([Ze, p.81]).

The delta function also satisfies the useful <u>sifting</u> property. For $F(t)$ a continuous function at $t = \alpha$, we have

$$F(t)\delta_\alpha(t) = F(\alpha)\delta_\alpha(t).$$

This is clear for $F \in C^\infty$ and of polynomial growth, but may be shown for other functions by using the sequence approach to δ.

There is also an intimate connection between δ and convolution. For a tempered distribution g with compact support, the convolution with any $f \in \mathcal{S}'$ can be defined. (The convolution of two $L'(\mathbb{R})$ functions is defined to be $f * g(t) := \int_{-\infty}^{\infty} f(t - s)g(s)ds$.)

We first take the convolution of g with $\theta \in \mathcal{S}$ defined as

$$(g * \theta)(t) = (g, T_t \tilde{\theta})$$

where $\tilde{\theta}(t) := \theta(-t)$. Then $g * \theta$ is a continuously differentiable function which belongs to $\mathcal{S}$. The convolution of f with g is then defined as

$$(f * g, \theta) := (f, \widetilde{g * \tilde{\theta}}).$$

There is a lot of checking to do here but it all works if g has compact support.

Since δ has compact support, its convolution exists for every $f \in \mathcal{S}'$ and in fact

$$f * \delta = f.$$

These distributions of compact support have a simple characterization. If g has support on a compact set K contained in an interval $[a, b]$ such that $a < 0 < b$, then there is a continuous function G with support on $[a, b]$ and an integer p such that

$$g = D^p G + \sum_{i=0}^{p-1} c_i \delta^{(i)}.$$

This is proved by observing that $g = D^p F$ for some continuous function F of polynomial growth (Proposition 2.1). Since $K \subseteq [a, b]$, it follows that $D^p F$ is zero on the complement of $[a, b]$, and therefore F is a polynomial P_+ of degree $\leq p - 1$ for $t > b$, and a polynomial P_- of degree $\leq p - 1$ for $t < a$. We take

$$G(t) := \begin{cases} F(t) - P_+(t) & \text{for } t > 0 \\ F(t) - P_-(t) & \text{for } t \leq 0 \end{cases}.$$

Then G is a tempered function and $D^p G - g$ has support on $\{0\}$ and must be a linear combination of δ and its derivatives.

2.4   Fourier transforms

The definition of Fourier transform may be extended from $\mathcal{S}$ to $\mathcal{S}'$ by the same device as the other operations. For $f \in \mathcal{S}'$ we define $\hat{f} = \mathcal{F}f$ to be that element of $\mathcal{S}'$ given by

$$(\hat{f}, \theta) := (f, \hat{\theta}).$$

It is easy to check that this is a linear functional and that it is continuous. Moreover the map $f \mapsto \hat{f}$ is a continuous linear transformation from $\mathcal{S}'$ onto $\mathcal{S}'$ whose inverse exists and is continuous as well.

Some examples of Fourier transforms on $\mathcal{S}'$ are:

$$\mathcal{F}\delta_\alpha = e^{-iw\alpha},$$

$$\mathcal{F}\sum_k \delta(t - 2\pi k) = \sum_n \delta(w - n),$$

$$\mathcal{F}(t^k e^{-it\tau}) = i^k 2\pi \delta^{(k)}(w + \tau),$$

$$\mathcal{F}\left(pv\frac{1}{t}\right) = -\pi i \, sgn(w),$$

$$\mathcal{F}(\sum_n a_n e^{int}) = 2\pi \sum_n a_n \delta(w - n),$$

$$\mathcal{F}\left(\sum_n a_n e^{int} \chi_{[-\pi,\pi]}(t)\right) = 2\pi \sum_n a_n \frac{\sin \pi(w - n)}{\pi(w - n)},$$

where $\chi_{[-\pi,\pi]}$ is the characteristic function of $[-\pi, \pi]$, $\{a_n\} \in \ell^2$.

Each of these may be found by using the definition. For example

$$(\mathcal{F}\delta_\alpha, \theta) = (\delta_\alpha, \hat{\theta}) = \hat{\theta}(\alpha) = \int_{-\infty}^{\infty} e^{-iw\alpha} \theta(w)dw = (e^{-iw\alpha}, \theta).$$

Furthermore, this definition of Fourier transform is consistent with the usual definition for functions in $L^1(\mathbb{R})$. Since elements of $\mathcal{S}'$ may be approximated by such functions, the usual definition may be used for $f \in \mathcal{S}'$ provided it makes sense. Again $\mathcal{F}\delta_\alpha$ may be found by using the sifting property $\theta(t)\delta_\alpha = \theta(\alpha)\delta_\alpha$ to again get

$$\mathcal{F}\delta_\alpha = (\delta_\alpha, e^{-iw\cdot}) = e^{-iw\alpha}.$$

Still another approach involves using Hermite functions $h_n(x)$, $n = 0, 1, \cdots$. It is well known that $\hat{h}_0(w) = \sqrt{2\pi}h_0(w)$, i.e., $h_0(w) = \pi^{-1/4}e^{-w^2/2}$ is an eigenfunction of the Fourier transform. Since $\mathcal{F}(xf)(w) = i\hat{f}'(w)$ and $\mathcal{F}(f')(w) = iw\hat{f}(w)$, the equation

$$(x - D)h_n = \sqrt{2n + 2}h_{n+1}$$

can be used to find the Fourier transform of all h_n. For example, $h_1 = \frac{1}{\sqrt{2}}(x - D)h_0$ and therefore

$$\hat{h}_1(w) = \frac{1}{\sqrt{2}}i(\hat{h}_0'(w) - w\hat{h}_0(w))$$

$$= -\frac{i}{\sqrt{2}}\left(w - \frac{d}{dw}\right)\sqrt{2\pi}h_0(w) = -\sqrt{2\pi}i\, h_1(w).$$

Similarly for all values of n we find

$$\hat{h}_n(w) = \sqrt{2\pi}(-i)^n h_n(w),$$

i.e., all h_n are eigenfunctions of the Fourier transform. Thus if f has an expansion

$$f = \Sigma a_n h_n$$

then

$$\hat{f} = \sqrt{2\pi}\Sigma a_n (-i)^n h_n. \tag{2.5}$$

It's fortunate that all $f \in S'$ have such an expansion (see Chapter 6).

THEOREM 2.1
Let $f \in S'$; then there is a $p \in \mathbb{N}$ such that

$$a_n = (f, h_n) = O(n^p),$$

and

$$f = \sum_n a_n h_n$$

with convergence in the sense of S'.

Again we may find $\mathcal{F}\delta_\alpha$ by this device.

$$\delta_\alpha = \sum_n (\delta_\alpha, h_n)h_n = \sum_n h_n(\alpha)h_n$$

and

$$(\mathcal{F}\delta_\alpha)(w) = \sqrt{2\pi}\sum_n (-i)^n h_n(\alpha)h_n(w) = \sum_n (e^{-i\cdot\alpha}, h_n)h_n(w) = e^{-iw\alpha}.$$

2.5   Periodic distributions

Some tempered distributions are periodic of period 2π, i.e., $T_{2\pi}f = f$ where $T_{2\pi}$ is the operator of translation by an amount 2π. Such tempered distributions have a Fourier series with coefficient c_n given by

$$c_n = \frac{1}{2\pi}\left[T_{2n}\left(fe^{int}\right)^{(-1)} - \left(fe^{int}\right)^{(-1)}\right]. \tag{2.6}$$

While this seems a rather awkward definition, it reduces to

$$c_n = \frac{1}{2\pi} \int_t^{t+2\pi} f(x)e^{-inx}\,dx$$

when f is an ordinary function. In fact, this integral may be taken as the definition of Fourier coefficients for $f \in \mathcal{S}'$ provided it is interpreted as (2.6).

The resulting c_n is a constant since

$$\frac{dc_n}{dt} = \frac{1}{2\pi}\left[T_{2\pi}(fe^{-int}) - (fe^{-int})\right] = 0$$

by the periodicity of fe^{-int}. We then associate with f the Fourier series

$$\sum_n c_n e^{int}.$$

THEOREM 2.2
Let $f \in \mathcal{S}'$ be periodic of period 2π; let c_n be given by (2.6); then there is a $p \in \mathbb{N}$ such that

$$c_n = O(|n|^p)$$

and

$$f = \sum c_n e^{int}$$

with convergence in the sense of $\mathcal{S}'$.

See Chapter 4 for more details and a proof.

2.6   Analytic representations

The elements of $\mathcal{S}'$ are not necessarily ordinary functions but can be represented by pairs of ordinary functions in the complex plane. This pair is the analytic representation given for ordinary functions by

$$f^{\pm}(z) = \frac{1}{2\pi i} \int_{-\infty}^{\infty} \frac{f(x)}{x-z}\,dx, \qquad \text{Im } z \neq 0$$

for $f \in L^2(\mathbb{R})$. The functions $f^{\pm}(z)$ are analytic in the upper half-plane (+) and in the lower half-plane (−) and

$$\lim_{\epsilon \to 0}(f^+(x + i\epsilon) - f^-(x - i\epsilon)) = f(x) \qquad \text{a.e.}$$

The definition may be first extended to elements of $\mathcal{S}'$ whose Fourier transforms are continuous functions of polynomial growth by

$$f^+(z) = \frac{1}{2\pi} \int_0^\infty e^{iwz} \hat{f}(w)dw, \ \mathrm{Im}\, z > 0, \qquad (2.7)$$

$$f^-(z) = \frac{1}{2\pi} \int_{-\infty}^0 e^{iwz} \hat{f}(w)dw, \ \mathrm{Im}\, z < 0.$$

The integrals exist as ordinary Lebesgue integrals since the integrand in the first, for example, is dominated by

$$e^{-w\, \mathrm{Im}\, z} \hat{f}(w).$$

Furthermore, the complex derivative of the integrand is also integrable and hence $f^+(z)$ is analytic in the upper and $f^-(z)$ in the lower half-plane. By the inversion formula for Fourier transforms we get

$$(f^+(x + i\epsilon) - f^-(x - i\epsilon), \theta(x)) \to (f, \theta) \qquad \text{as } \epsilon \to 0 \qquad (2.8)$$

for all $\theta \in \mathcal{S}$, [Br, p.87]. Since all $f \in \mathcal{S}'$ are finite order derivatives of continuous functions of polynomial growth, we may represent $\hat{f}$ by

$$\hat{f}(w) = D^p \hat{g}(w)$$

where $\hat{g}$ is such a function. This defines a pair of analytic functions

$$f^{\pm}(z) = (-iz)^p g^{\pm}(z),$$

which also satisfy (2.8) [Br, p.90].

Some examples of analytic representations are

$$\delta^{\pm}(z) = -\frac{1}{2\pi i z}, \qquad \mathrm{Im}\, z \neq 0,$$

$$\left(pv\frac{1}{t}\right)^{\pm}(z) = \frac{\pm 1}{2z}, \qquad \mathrm{Im}\, z \neq 0,$$

$$p^{\pm}(z) = \pm p(z)/2, \quad \mathit{Im}\, z \neq 0, \qquad p(t) \text{ a polynomial,}$$

$$H^{\pm}(z) = -\frac{1}{2\pi i} \log(-z), \qquad \mathrm{Im}\, z \neq 0, \qquad H(t) = \text{Heaviside function}$$

2.7 Sobolev spaces

A convenient Hilbert space structure to use on certain subsets of $\mathcal{S}'$ is that of Sobolev spaces [R]. The Sobolev space H^α, $\alpha \in \mathbb{R}$ consists of all $f \in \mathcal{S}'$ such that

$$\int_{-\infty}^{\infty} |\hat{f}(w)|^2 (w^2 + 1)^\alpha dw < \infty.$$

For $\alpha = 0$, this is just $L^2(\mathbb{R})$; for $\alpha = 1, 2, \ldots$, H^α is composed of ordinary $L^2(\mathbb{R})$ functions that are $(\alpha - 1)$ times differentiable and whose α-th derivative is in $L^2(\mathbb{R})$; for $\alpha = -1, -2, \ldots$, H^α contains the $-\alpha$-th derivatives of $L^2(\mathbb{R})$ and in particular all distributions with point support of order $\leq \alpha$ (i.e., $\delta, \delta', \ldots, \delta^{(\alpha-1)}$). Clearly $H^\alpha \supset H^\beta$ when $\alpha < \beta$.

The inner product of f and $g \in H^\alpha$ is defined as

$$\langle f, g \rangle_\alpha = \frac{1}{2\pi} \int_{-\infty}^{\infty} \hat{f}(w)\overline{\hat{g}(w)}(w^2 + 1)^\alpha dw.$$

It is complete with respect to this inner product and hence is a Hilbert space.

Concluding remark

This has been a somewhat abbreviated introduction to tempered distributions. A more detailed but still simple introduction is given by Zemanian [Ze]. A much more complete treatment is found in the five volumes of Gelfand, Shilov [G-S].

These tempered distributions constitute a subspace of the space of distributions $\mathcal{D}'$ of L. Schwartz [S], which we have not discussed. However we shall often refer to particular elements of $\mathcal{S}'$ as "distributions," which is proper because of this containment relation.

2.8 Problems

1. Check that the definition of convolution of $f \in \mathcal{S}'$ with a $g \in \mathcal{S}'$ which has compact support is linear and continuous on $\mathcal{S}$.

2. Check that the approximation to δ given in Section 2.1 is in fact an approximation in the sense of $\mathcal{S}'$.

3. Show that $f(x)\delta(x-a) = f(a)\delta(x-a)$ for a continuous function f of polynomial growth by using the sequence approach in Section 2.3.1 for defining δ.

4. Leibniz rule for the derivative of a product is given by the formula

$$D^p[f(x)g(x)] = \sum_{k=0}^{p} \binom{p}{k} f^{(k)}(x)g^{(p-k)}(x).$$

Use it to find $D^p[f(x)\delta(x)]$ and then solve for the term

$$f(x)\delta^{(p)}(x)$$

for $f \in C^p$ and of polynomial growth.

5. Use the result in Problem 4 to show that

$$x^n \delta^{(m)}(x) = (-1)^n \frac{m!}{(m-n)!} \delta^{(m-n)}(x), \quad m \geq n.$$

6. Show that

$$\delta^{(k)}(2^m x) = 2^{-m(k+1)} \delta^{(k)}(x).$$

7. Show that if $\theta_n \to 0$ in $\mathcal{S}$ (i.e., $x^p \theta_n^{(q)}(x) \to 0$ converges uniformly to zero for non-negative integers p, q) so does its Fourier transform $\hat{\theta}_n$.

8. Show that the Fourier transform of the periodic distribution

$$\delta^*(x) = \sum_{k=-\infty}^{\infty} \delta(x - 2\pi k)$$

is as given in Section 2.4.

9. Show that the other examples in Section 2.4 have the Fourier transforms indicated. Sometimes it will be easier to find the inverse Fourier transform and then use (2.4).

10. Find the Fourier series of $\delta^*(x)$ given in Problem 8.

11. Show that the analytic representation $\delta^{\pm}(z)$ of δ satisfies $\lim_{\epsilon \to 0} \delta^+(x + i\epsilon) - \delta^-(x - i\epsilon) = \delta(x)$.

12. Show that the Sobolev space H^{-1} contains $\delta(x)$ but not $\delta'(x)$.

Chapter 3

An Introduction to Orthogonal Wavelet Theory

The structure of systems of orthogonal wavelets is similar to that of the prototypes in Chapter One. However, both they and their Fourier transforms have a certain degree of smoothness that the prototypes do not have. It was the discovery of orthogonal wavelets of analytic functions with smooth Fourier transforms by Meyer [M], and of smooth functions with compact support by Daubechies [D1], that led to the explosion of activity in the subject.

Wavelets began not as an orthogonal system but, rather, as an integral transform involving dilations and translations of a fixed function. These were found to give a better representation for certain seismic signals than Fourier methods [Mo]. The original theory was soon put into a rigorous framework by a number of mathematicians [G-M], [M]. Next, a discrete nonorthogonal representation based on frames evolved [D1]. Finally, the orthogonal theory was discovered [D2], [M1]. Many of the techniques, based on harmonic analysis, had been around for a long time, but had not been directed toward the particular problems associated with wavelets.

The wavelet basis is composed of functions $\psi_{mn}(t)$ given by

$$\psi_{mn}(t) = 2^{m/2}\psi(2^m t - n), \qquad m, n \in \mathbb{Z},$$

where ψ is a fixed function in $L^2(\mathbb{R})$, the so-called "mother wavelet". It's not clear that such an orthonormal basis consisting of dilations and translations of a smooth function exists. Much of the initial theory has been devoted to constructing $\psi(t)$ for which such a basis exists. This construction is usually based on another function $\phi(t)$, the "scaling function".

3.1　Multiresolution analysis

We begin with a scaling function ϕ, a real valued function on $\mathbb{R}$ which is r times differentiable and whose derivatives are continuous and rapidly decreasing. That is, ϕ satisfies

$$|\phi^{(k)}(t)| \leq C_{pk}(1+|t|)^{-p}, \quad k = 0, 1, \ldots, r, \quad p \in \mathbb{Z}, \quad t \in \mathbb{R}, \quad (3.1)$$

and hence $\phi \in \mathcal{S}_r$. In order for it to qualify as a scaling function, there must be associated with ϕ a multiresolution analysis of $L^2(\mathbb{R})$, i.e., a nested sequence of closed subspaces $\{V_m\}_{m \in \mathbb{Z}}$ such that

$$\begin{cases} (i) & \{\phi(t-n)\} \text{ is an orthonormal basis of } V_0 \\[2mm] (ii) & \cdots \subset V_{-1} \subset V_0 \subset V_1 \subset \cdots \subset L^2(\mathbb{R}) \\[2mm] (iii) & f \in V_m \Leftrightarrow f(2\cdot) \in V_{m+1} \\[2mm] (iv) & \bigcap_m V_m = \{0\}, \quad \overline{\bigcup_m V_m} = L^2(\mathbb{R}). \end{cases} \quad (3.2)$$

Clearly $\{\sqrt{2}\phi(2t-n)\}$ is an orthonormal basis for V_1 since the map $f \vdash \sqrt{2}f(2\cdot)$ is an isometry from V_0 onto V_1. Since $\phi \in V_1$, it must have an expansion, the dilation equation,

$$\phi(t) = \sum_k c_k \sqrt{2}\phi(2t-k), \qquad \{c_k\} \in \ell^2, \qquad t \in \mathbb{R}. \quad (3.3)$$

Example 1
A simple example, which however does not satisfy (3.1), is given by $\phi(t) = \chi_{[0,1]}(t)$, the characteristic function of the unit interval. Its translates $\{\phi(t-n)\}$ are orthonormal and V_0 is composed of piecewise constant functions. This is prototype I of Chapter 1.

It is not obvious that a multiresolution analysis exists for ϕ other than this example and prototype II. In order to find other ϕ's, three approaches have been used. One begins with an existing multiresolution analysis of subspaces $\{V_m\}$ defined in some other way and then tries to find an orthogonal basis. In particular, V_0 may be defined by means of a Riesz basis of translates of a fixed function $\{\theta(t-n)\}$. Then an orthogonalization procedure due to Lemarié and Meyer [L-M] is used to find an orthogonal $\{\phi(t-n)\}$. This is not the familiar Gram-Schmidt

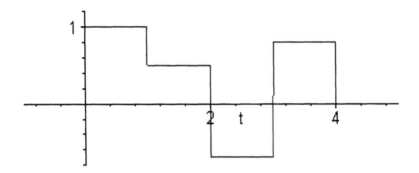

FIGURE 3.1
Typical functions in the subspaces V_0 of the multiresolution analysis for the Haar scaling function.

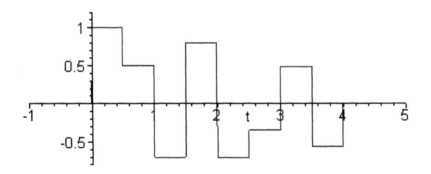

FIGURE 3.2
Typical functions in the subspaces V_1 of the multiresolution analysis for the Haar scaling function.

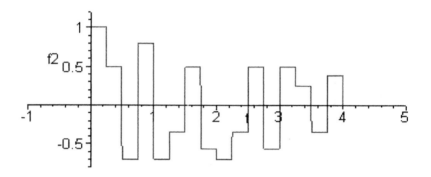

FIGURE 3.3
Typical functions in the subspaces V_2 of the multiresolution analysis for the Haar scaling function.

orthogonalization, which, while it would give an orthogonal basis, would not guarantee that it be composed of translates of a fixed function.

The second method, due to Daubechies [D2], consists of choosing the dilation coefficients $\{c_k\}$ in (3.3) such that all the required conditions are satisfied. The third approach is similar; it consists of choosing the Fourier transform of $\phi(t)$ in such a way that it has compact support, and the transformed versions of (3.2) and (3.3) are satisfied.

Example 2
The simplest example of the orthogonalization approach leads to the Franklin wavelets. It starts with the "hat" function

$$\theta(t) = (1 - |t - 1|)\chi_{[0,2]}(t).$$

This corresponds to a V_0 composed of continuous piecewise linear functions with breaks at the integers. Its Fourier transform is

$$\hat{\theta}(w) = \int_{-\infty}^{\infty} e^{-iwt}\theta(t)dt = \left[\frac{1 - e^{-iw}}{iw}\right]^2 = \frac{e^{-iw}\sin^2 w/2}{(w/2)^2}. \qquad (3.4)$$

To find the orthogonal ϕ we use the isometry property of the Fourier transform:

$$\delta_{on} = \int_{-\infty}^{\infty} \phi(t - n)\phi(t)dt = \frac{1}{2\pi}\int_{-\infty}^{\infty} \hat{\phi}(w)e^{-iwn}\overline{\hat{\phi}(w)}dw$$

$$= \frac{1}{2\pi}\sum_{k=-\infty}^{\infty}\int_{0}^{2\pi} \left|\hat{\phi}(w + 2\pi k)\right|^2 e^{-iwn}dw$$

$$= \frac{1}{2\pi} \int_0^{2\pi} \sum_k \left| \hat{\phi}(w + 2\pi k) \right|^2 e^{-iwn} dw.$$

This is the integral that gives the Fourier coefficient of the periodic function $\left| \hat{\phi}_2^*(w) \right|^2 := \sum_k \left| \hat{\phi}(w + 2\pi k) \right|^2$, which by orthogonality must be δ_{on}. Hence $\left| \hat{\phi}^*(w) \right|^2$, which is equal to its Fourier series, is identically equal to 1, i.e.,

$$\sum_k |\hat{\phi}(w + 2\pi k)|^2 = 1, \tag{3.5}$$

a necessary condition for orthogonality.

Since $\phi \in V_0$, we have $\phi(t) = \sum a_n \theta(t - n)$ for some $\{a_n\} \in \ell^2$. Again by taking Fourier transforms we find

$$\hat{\phi}(w) = \sum_n a_n e^{-iwn} \hat{\theta}(w) = \alpha(w) \hat{\theta}(w) \tag{3.6}$$

and hence

$$\left| \hat{\phi}_2^*(w) \right|^2 = |\alpha(w)|^2 \left| \hat{\theta}_2^*(w) \right|^2 = 1. \tag{3.7}$$

Now $\left| \hat{\theta}_2^*(w) \right|^2$ can be found in closed form by summing the series obtained from (3.4)

$$\left| \hat{\theta}_2^*(w) \right|^2 = \sum_k \left| \frac{\sin(w + 2\pi k)/2}{(w + 2\pi k)/2} \right|^4 = (\sin w/2)^4 \sum_k \frac{1}{\left(\frac{w}{2} + \pi k \right)^4}.$$

This may be done by using the Fourier series of e^{-ixt}

$$e^{-ixt} = \sum_k \frac{\sin \pi(x + k)}{\pi(x + k)} e^{ikt}, \quad -\pi < t < \pi.$$

Then by Parseval's equality we have

$$1 = \sum_k \frac{\sin^2 \pi x}{\pi^2 (x + k)^2}$$

or

$$\frac{1}{\sin^2 \pi x} = \sum_k \frac{1}{(\pi x + \pi k)^2}.$$

By differentiating both sides twice we obtain

$$\frac{2\pi^2 (\sin^2 \pi x + 3 \cos^2 \pi x)}{\sin^4 \pi x} = \sum_k \frac{6\pi^2}{(\pi x + \pi k)^4}.$$

We then replace πx by $\frac{w}{2}$ to find

$$\sum_k \frac{1}{\left(\frac{w}{2} + \pi k\right)^4} = \frac{1 - \frac{2}{3}\sin^2 \frac{w}{2}}{\sin^4 \frac{w}{2}}.$$

Hence by multiplying this by the periodic factor $\sin^4 w/2$ we find

$$|\hat{\theta}_2^*(w)|^2 = 1 - \frac{2}{3}\sin^2 \frac{w}{2}.$$

We then substitute this result in (3.7) and solve for $|\alpha(w)|^2$. We then choose an appropriate square root, and, by substituting it in (3.6), find ϕ to be

$$\hat{\phi}(w) = \frac{\sin^2 w/2}{(w/2)^2}\left\{1 - \frac{2}{3}\sin^2 w/2\right\}^{-\frac{1}{2}}. \tag{3.8}$$

This times any unimodular function gives another square, but (3.8) is real and symmetric.

Since $\left\{1 - \frac{2}{3}\sin^2 w/2\right\}^{-1/2}$ is an analytic function in a strip about the real axis and is of period 2π, its Fourier series has coefficients that are exponentially decreasing and

$$\hat{\phi}(w) = e^{iw}\left[\frac{1 - e^{-iw}}{iw}\right]^2 \sum_k a_k e^{iwk}.$$

The inverse Fourier transform is

$$\phi(t) = \sum_k a_k \theta(t + k + 1), \qquad a_k = O(e^{-a|k|}).$$

The coefficients a_k may be found explicitly by expanding

$$\left(1 - \frac{2}{3}\sin^2 w/2\right)^{-\frac{1}{2}} = \sqrt{\frac{3}{2}}\left(1 + \frac{\cos w}{2}\right)^{-\frac{1}{2}}$$

in powers of $\frac{\cos w}{2}$ and then expressing these powers in exponential form.

Example 3

The Daubechies wavelets are based on the other approach. We look for $\{c_k\}$ in (3.3) such that the orthogonality condition is satisfied. This becomes

$$\int \phi(t)\phi(t - n)dt = \sum_k c_k \int \sqrt{2}\phi(2t - k)\sum_j c_j\sqrt{2}\phi(2t - 2n - j)dt$$

$$= \sum c_k c_{k-2n} = \delta_{on}, \qquad n \in \mathbb{Z}. \tag{3.9}$$

The conditions $\hat{\phi}(0) = 1$, $\hat{\phi}(2\pi) = 0$ implied by orthogonality (see Section 3.4) become

$$\sum_k \frac{c_k}{\sqrt{2}} = 1, \qquad \sum_k \frac{c_k}{\sqrt{2}}(-1)^k = 0 \tag{3.10}$$

if $\hat{\phi}(\pi) \neq 0$. These are redundant since both are based on orthogonality. However, there are three independent equations given by (3.9) for $n = 0$ and (3.10). If c_k is chosen by

$$c_0 = \nu(\nu - 1)/(\nu^2 + 1)\sqrt{2},$$
$$c_1 = (1 - \nu)/(\nu^2 + 1)\sqrt{2},$$
$$c_2 = (\nu + 1)/(\nu^2 + 1)\sqrt{2},$$
$$c_3 = \nu(\nu + 1)/(\nu^2 + 1)\sqrt{2},$$

where $\nu \in \mathbb{R}$ (and all the other $c_k = 0$), then (3.9) and (3.10) will be satisfied.

A greater degree of smoothness than continuity, as we shall see later, leads to vanishing moments for the mother wavelet. We can turn this around and hope that a vanishing moment will give us greater smoothness. So we add the moment condition

$$\int_{-\infty}^{\infty} t\,\psi(t)dt = 0,$$

which can be shown (see Section 3.4) to lead to the equation

$$\sum_k c_k(-1)^k k = 0.$$

There are then two solutions to these four equations given by $\nu = \pm 1/\sqrt{3}$.

This does not directly give $\phi(t)$, which rather must be found recursively by starting with, say, $\phi_0(t) = \chi_{[0,1)}(t)$ and then defining

$$\phi_n(t) = \sum_k c_k \sqrt{2}\phi_{n-1}(2t - k). \tag{3.11}$$

This sequence converges to a continuous function when $\nu = \pm 1/\sqrt{3}$ but, in general, does not, although it does converge weakly [V], whatever the choice of $\{c_n\}$.

The dilation equation (3.3) may also be expressed in terms of Fourier transforms. It is

$$\hat{\phi}(w) = \sum_k \frac{c_k}{\sqrt{2}} \hat{\phi}\left(\frac{w}{2}\right) e^{-iwk/2}$$

and if we denote by

$$m_0\left(\frac{w}{2}\right) = \sum_{k=-\infty}^{\infty} \frac{c_k}{\sqrt{2}} e^{-iwk/2}, \tag{3.12}$$

the orthogonality condition (3.5) becomes

$$\left|\hat{\phi}_2^*(w)\right|^2 = \sum_k \left|m_0\left(\frac{w+2\pi k}{2}\right)\right|^2 \left|\hat{\phi}\left(\frac{w+2\pi k}{2}\right)\right|^2$$

$$= \sum_j \left|m_0\left(\frac{w}{2} + 2\pi j\right)\right|^2 \left|\hat{\phi}\left(\frac{w}{2} + 2\pi j\right)\right|^2$$

$$+ \left|m_0\left(\frac{w}{2} + (2j+1)\pi\right)\right|^2 \left|\hat{\phi}\left(\frac{w}{2} + (2j+1)\pi\right)\right|^2$$

$$= \left|m_0\left(\frac{w}{2}\right)\right|^2 \left|\hat{\phi}^*\left(\frac{w}{2}\right)\right|^2 + \left|m_0\left(\frac{w}{2} + \pi\right)\right|^2 \left|\hat{\phi}^*\left(\frac{w}{2} + \pi\right)\right|^2$$

or

$$\left|m_0\left(\frac{w}{2}\right)\right|^2 + \left|m_0\left(\frac{w}{2} + \pi\right)\right|^2 = 1. \tag{3.13}$$

This, then, is a necessary condition for orthogonality, which we use later to construct more examples of Daubechies wavelets. But first we need to introduce some additional concepts.

3.2   Mother wavelet

Once we have the scaling function $\phi(t)$, we may use it to construct the "mother wavelet" $\psi(t)$. It must be chosen such that $\{\psi(t-n)\}$ is an orthonormal basis of the space W_0, given by the orthogonal complement of V_0 in V_1. Then

$$V_1 = V_0 \oplus W_0.$$

If such a $\psi(t)$ can be found, then $2^{m/2}\psi\left(2^{m/2}t - n\right) = \psi_{nm}(t)$ is an orthonormal basis of W_m, the dilation space of W_0. Furthermore W_m will be the orthogonal complement of V_m in V_{m+1}. Hence we have

$$V_{m+1} = V_m \oplus W_m = \cdots = V_0 \oplus W_0 \oplus W_1 \oplus \cdots \oplus W_m,$$

and since $\bigcup_m V_m$ is dense in $L^2(\mathbb{R})$, we may take the limit as $m \to \infty$ to obtain

$$V_0 \oplus \left(\bigoplus_{m=0}^{\infty} W_m\right) = L^2(\mathbb{R}).$$

We may also go in the other direction to write

$$V_0 = V_{-1} \oplus W_{-1} = V_{-k} \oplus W_{-k} \oplus \cdots \oplus W_{-1}.$$

We again take the limit as $k \to \infty$. Since $\bigcap_m V_m = \{0\}$, it follows that $V_{-k} \to \{0\}$; from this we see that

$$\bigoplus_{m \in Z} W_m = L^2(\mathbb{R}),$$

and hence $\{\psi_{nm}\}_{n,m \in Z}$ is an orthonormal basis of $L^2(\mathbb{R})$.

Again there are two methods of finding $\psi(t)$. One uses the orthogonalization procedure similar to that illustrated in Example 2. But $\psi(t)$ must also be orthogonal to $\phi(t - n)$. Hence the two conditions

$$
\begin{aligned}
&(a)\ \sum_k \hat{\psi}(w + \pi k)\overline{\hat{\phi}(w + 2\pi k)} = 0 \\
&(b)\ \sum_k \left|\hat{\psi}(w + 2\pi k)\right|^2 = 1
\end{aligned}
\tag{3.14}
$$

must be satisfied. Since $\psi(t) \in V_1$, it has an expansion similar to (3.3)

$$\psi(t) = \sum_k d_k \sqrt{2}\phi(2t - k) \tag{3.15}$$

and hence

$$\hat{\psi}(w) = m_1\left(\frac{w}{2}\right)\hat{\phi}\left(\frac{w}{2}\right). \tag{3.16}$$

This may be substituted into (3.14) and the two equations solved for $m_1\left(\frac{w}{2}\right)$, which is then substituted back in (3.16) to find $\hat{\psi}(w)$, i.e., we find

$$
\begin{aligned}
&(a)\ m_1\left(\frac{w}{2}\right)\overline{m_0\left(\frac{w}{2}\right)} + m_1\left(\frac{w}{2} + \pi\right)\overline{m_0\left(\frac{w}{2} + \pi\right)} = 0 \quad (3.17) \\
&(b)\ m_1\left(\frac{w}{2}\right)\overline{m_0\left(\frac{w}{2}\right)} + m_1\left(\frac{w}{2} + \pi\right)\overline{m_0\left(\frac{w}{2} + \pi\right)} = 1.
\end{aligned}
$$

These two equations may then be solved simultaneously for $\left|m_1\left(\frac{w}{2}\right)\right|^2$ and then a square root taken. However, the positive square root doesn't work, but

$$m_1\left(\frac{w}{2}\right) = e^{-iw/2}\overline{m_0\left(\frac{w}{2} + \pi\right)}$$

does. Of course, there are other solutions and we shall sometimes use a slightly different one.

The other approach involving the dilation equation is straightforward; $\psi(t)$ is defined as

$$\psi(t) := \sqrt{2}\sum_k c_{1-k}(-1)^{k-1}\phi(2t - k). \tag{3.18}$$

Then it is merely a matter of checking that the two orthogonality conditions are satisfied [D2, p.944].

By taking the Fourier transform of (3.18) we obtain the expression

$$\hat{\psi}(w) = -1/\sqrt{2}\sum_k c_{1-k}e^{-i(\frac{w}{2}+\pi)k}\hat{\phi}(\frac{w}{2})$$

$$= 1/\sqrt{2}\,e^{-\frac{iw}{2}}\sum_j c_j e^{i(\frac{w}{2}+\pi)j}\hat{\phi}(\frac{w}{2}) = e^{-i\frac{w}{2}}\overline{m_0(\frac{w}{2} + \pi)}\hat{\phi}(\frac{w}{2}),$$

which agrees with the calculation based on (3.17). Other solutions are obtained by multiplying $\hat{\phi}(w)$ by any unimodular functions, in particular (-1) and e^{iwn} which sometimes lead to other standard mother wavelets.

Examples 1-3

In Example 1, $\psi(t) = \phi(2t) - \phi(2t - 1) = \begin{cases} 1, & 0 \le t < 1/2 \\ -1, & 1/2 \le t < 1 \\ 0, & \text{otherwise.} \end{cases}$

In Example 2, we find from (3.8) that

$$m_0(\frac{w}{2}) = \frac{2 + 3\cos w/2 + \cos^2 w/2}{1 + 2\cos^2 w/2}$$

and hence that

$$\hat{\psi}(w) = e^{-i\frac{w}{2}}\frac{2 - 3\cos w/2 + \cos^2 w/2}{1 + 2\cos^2 w/2}\hat{\phi}(\frac{w}{2}).$$

In Example 3, the best we can do is use the coefficients c_0, c_1, c_2, and c_3 in Equation (3.18). Again as in Example 2, there is no formula, but the mother wavelet can be based on ϕ in Equation (3.18).

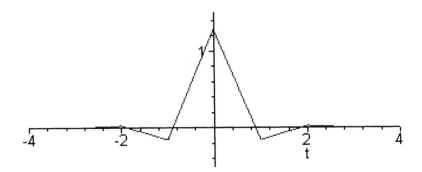

FIGURE 3.4
The scaling function of the Franklin wavelet arising from the
hat function.

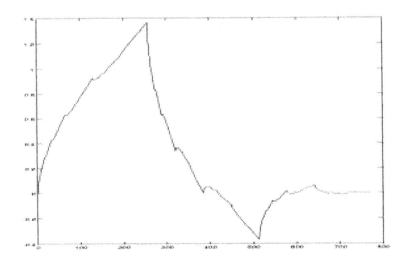

FIGURE 3.5
The Daubechies scaling function of example 3.

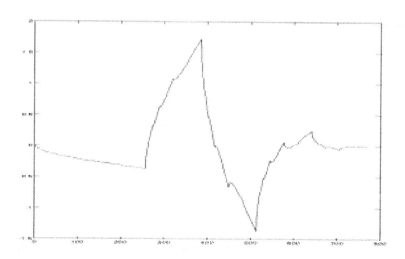

FIGURE 3.6
The mother wavelet of example 3.

Example 4
The B-spline of order n is the function $\theta_n(t)$ obtained by convolving the
Haar scaling function with itself n-times. Its Fourier transform is

$$\hat{\theta}_n(w) = \left[\frac{1 - e^{-iw}}{iw}\right]^{n+1}.$$

For $n = 1$, these give the Franklin wavelets of Example 2. For other
values of n, $\theta_n(t) \in C^n$ has support on $[0, n+1]$ and is a polynomial of
degree $n+1$ in each interval $(k, k+1)$ where $k \in \mathbb{Z}$ [C], [M]. They may
be orthogonalized by the same trick as in Example 2 to obtain

$$\hat{\phi}(w) = \left(\frac{\sin w/2}{w/2}\right)^{n+1} [P(\cos w/2)]^{-1/2}$$

where P is a polynomial of degree $2n$, strictly positive on $[0, 1]$, when n
is odd. When n is even the expression becomes

$$\hat{\phi}(w) = \left(\frac{\sin w/2}{w/2}\right)^{n+1} e^{-i\frac{w}{2}} [P(\cos w/2)]^{-1/2}.$$

The mother wavelet is obtained from (3.18). These are the <u>Battle-Lemarie wavele</u>

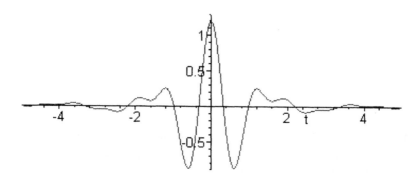

FIGURE 3.7
The mother wavelet of Example 5 in the time domain.

Example 5

For this example the role of ϕ and $\hat{\phi}$ in Example 1 is reversed. The Shannon scaling function, prototype II of Chapter 1, is

$$\phi(t) = \frac{\sin \pi t}{\pi t}$$

or

$$\hat{\phi}(w) = \chi_{[-\pi, \pi)}(w).$$

In this case ϕ is not rapidly decreasing, but ϕ has other nice properties. In particular for $f \in V_0$, which in this case is composed of π band limited functions, we have the Shannon Sampling Theorem

$$f(t) = \sum_{n=-\infty}^{\infty} f(n)\phi(t - n).$$

A possible mother wavelet is given by

$$\psi(t) = \frac{\sin \pi(t - \frac{1}{2}) - \sin 2\pi(t - \frac{1}{2})}{\pi(t - \frac{1}{2})};$$

which differs from the one given in Section 1.2.3, but corresponds more closely to the ones given in other cases.

Example 6

A smoothed version of Example 5 is the Meyer wavelet. Here $\hat{\phi} \in C^\infty$, $\hat{\phi}(w) = 1$ for $\frac{-2\pi}{3} \le w \le \frac{2\pi}{3}$, $\hat{\phi}(w) = 0$ for $|w| \ge \frac{4\pi}{3}$, and in between it satisfies the condition

$$[\hat{\phi}(w - \pi)]^2 + [\hat{\phi}(w + \pi)]^2 = 1.$$

(See [M, p.60].) In this case $\phi \in \mathcal{S}$.

Example 7

Examples 5 and 6 can both be generalized as follows: Let P be a probability measure with support in $[-\varepsilon, \varepsilon] \subset \left[-\frac{\pi}{3}, \frac{\pi}{3}\right]$ and define $\phi(t)$ as the function whose Fourier transform is

$$\hat{\phi}(w) = \left[\int_{w-\pi}^{w+\pi} dP\right]^{\frac{1}{2}},$$

the nonnegative square root of the integral. Then $\hat{\phi}$ has support in $[-\pi - \varepsilon, \pi + \varepsilon] \subseteq \left[-\frac{4\pi}{3}, \frac{4\pi}{3}\right]$, and $\hat{\phi}(w) = 1$ for $|w| < \pi - \varepsilon$. The orthogonality condition is satisfied since

$$\sum_k |\hat{\phi}(w + 2\pi k)|^2 = \sum_k \int_{w+(2k-1)\pi}^{w+(2k+1)\pi} dP = \int_{-\infty}^{\infty} dP = 1$$

as is the dilation condition

$$\hat{\phi}(w) = m_0\left(\frac{w}{2}\right) \hat{\phi}\left(\frac{w}{2}\right)$$

with $m_0(w/2)$ defined on $[-2\pi, 2\pi]$ as

$$m_0\left(\frac{w}{2}\right) = \begin{cases} \hat{\phi}(w), & |w| \le \frac{4\pi}{3} \\ 0, & \frac{4\pi}{3} < |w| \le 2\pi \end{cases}$$

and extended (4π) periodically to all $w \in \mathbb{R}$. The mother wavelet is given by

$$\hat{\psi}(w) = e^{-iw/2} \left[\int_{\left|\frac{w}{2}\right|-\pi}^{|w|-\pi} dP\right]^{\frac{1}{2}}.$$

Example 5 results from taking $dP(w) = \delta(w)dw$, while Example 6 results from taking P to be a C^∞ function with compact support in $\left[-\frac{\pi}{3}, \frac{\pi}{3}\right]$. Notice that the coefficients of the dilation equation are given by the Fourier series of $\hat{\phi}(2w)$

$$m_0(w) \sim \sum_k c_k e^{-iwk}, \quad c_k = \frac{1}{2\pi} \int_{-\pi}^{\pi} \hat{\phi}(2w) e^{iwk} dw.$$

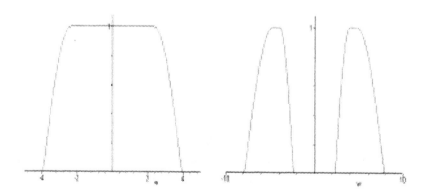

FIGURE 3.8
The scaling function and absolute value of the mother wavelet of Example 7 in frequency domain.

As additional examples, we can use the rescaled B-splines of order n (Example 4) as our probability,

$$dP_n(w) = \theta_n \left((\frac{3}{\pi}w + 1)\frac{n+1}{2} \right) \frac{3}{2\pi}(n+1)dw, \quad n = 0, 1, 2, \cdots.$$

Since θ_n has support on $[0, n+1]$, we have the desired support for P_n.

Example 8
Most scaling functions and wavelets cannot be given in closed form but must be given in terms of integrals or recurrence formulae. There are two notable exceptions, the raised cosine wavelets, which arise from a particular case of example 7 [W-Z1]. They both involve the same probability measure given by the function $dP(w) = \cos(w/2\beta)\chi_{[-\beta\pi,\beta\pi]}(w)dw$, where $\beta < 1/3$. In this case, we can actually integrate the expression for $|\widehat{\phi}(w)|^2$ to obtain

$$|\widehat{\phi}(w)|^2 = \begin{cases} 0, & |w| \geq \pi(1+\beta) \\ \frac{1}{2}(1 + \cos\frac{|w| - \pi(1-\beta)}{2\beta}), & \pi(1-\beta) < |w| < \pi(1+\beta). \\ 1, & |w| \leq \pi(1-\beta) \end{cases}$$

One example involves the positive square root and the other a complex square root:

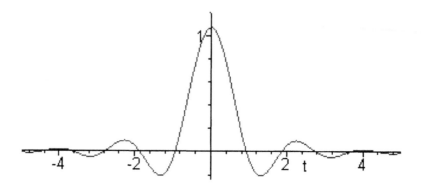

FIGURE 3.9
The scaling function of Figure 3.7 in the time domain.

$$\widehat{\phi}_1(w) = \begin{cases} 0, & |w| \geq \pi(1+\beta) \\ \cos \frac{|w| - \pi(1-\beta)}{4\beta}, & \pi(1-\beta) < |w| < \pi(1+\beta) \\ 1, & |w| \leq \pi(1-\beta) \end{cases}$$

$$\widehat{\phi}_2(w) = \begin{cases} 0, & |w| \geq \pi(1+\beta) \\ \frac{1}{2}(1 + \exp i(|w| - \frac{\pi(1-\beta)}{2\beta})), & \pi(1-\beta) < |w| < \pi(1+\beta) \\ 1, & |w| \leq \pi(1-\beta) \end{cases} .$$

The inverse Fourier transform can also be found in closed form in both cases. The calculations are straightforward, but tedious, and result in the expressions

$$\phi_1(t) = \frac{\sin \pi(1-\beta)t + 4\beta t \cos \pi(1+\beta)t}{\pi t(1 - (4\beta t)^2)}$$

and

$$\phi_2(t) = \frac{\sin \pi(1-\beta)t + \sin \pi(1+\beta)t}{2\pi t(1 + 2\beta t)}.$$

Notice that ϕ_2 is a real function even though its Fourier transform is not. However it is not symmetric whereas ϕ_1 is, but, in common with the sinc function, is a sampling whereas ϕ_1 is not. In both cases the mother wavelet can also be found in closed form by using the usual formulae or alternatively, by using the integral formula in example 7. The two are:

$$\psi_1(t + 1/2) = \frac{\sin \pi(1 + \beta)t - 4\beta t \cos \pi(1 - \beta)t}{\pi t((4\beta t)^2 - 1)}$$

$$- \frac{\sin 2\pi(1 - \beta)t + 8\beta t \cos \pi(1 - \beta)t}{\pi t((8\beta t)^2 - 1)},$$

$$\psi_2(t + 1/2) = \frac{\sin 2\pi(1 - \beta)t + \sin 2\pi(1 + \beta)t}{2\pi t(1 + 4\beta t)}$$

$$- \frac{\sin \pi(1 - \beta)t + \sin \pi(1 + \beta)t}{2\pi t(1 - 2\beta t)}.$$

3.3   Reproducing kernels and a moment condition

Each of the spaces V_m in the multiresolution analysis is a reproducing kernel Hilbert space (RKHS) [Ar]. Such a space consists of a Hilbert space H of functions f on an interval T in which all evaluation functions $\xi_t(f) := f(t)$, $f \in H$, $t \in T$ are continuous on H. Then by the Riesz representation theorem for each $t \in T$ there is a unique element $k_t \in H$ such that for each $f \in H$

$$f(t) = \langle f, k_t \rangle.$$

The function defined by $k(t, u) = \langle k_t, k_u \rangle$ for $t, w \in T$ is the reproducing kernel. Note that $L^2(T)$ is not a RKHS but has subspaces that are. The reproducing kernel of V_0 is given by

$$q(x, t) = \sum_n \overline{\phi(x - n)} \phi(t - n), \tag{3.19}$$

where $\phi(x)$ is the scaling function. The series and its derivatives with respect to t of order $\leq r$ converge uniformly for $x \in \mathbb{R}$, since, by the regularity of ϕ,

$$|\phi^{(k)}(x)| \leq \frac{C_{kn}}{(1 + |x|)^n}, \quad k = 0, 1, \cdots, r; \ n = 0, 1, 2, \cdots.$$

We then use the inequality

$$(1 + |b|)(1 + |a|) \geq 1 + |b - a| \quad a, b \in \mathbb{R}$$

to find that

$$|q(x, t)| \leq \sum_j |\phi(x - j)||\phi(t - j)|$$

$$\leq C_{0,n+2}^2 \sum_j \frac{1}{(1 + |x - j|)^{n+2}} \frac{1}{(1 + |t - j|)^{n+2}}$$

$$\leq C_{0,n+2}^2 \sum_j \frac{1}{(1 + |x - j|)^2} \frac{1}{(1 + |t - j|)^2} \frac{1}{(1 + |x - t|)^n}$$

and hence that

$$|q(x, t)| \leq \frac{C_n}{(1 + |x - t|)^n}, \quad n \in \mathbb{N}.$$

The RK for V_m is obtained from that of V_0 by

$$q_m(x, t) = 2^m q(2^m x, 2^m t).$$

Similarly W_m, the orthogonal complements of V_m in V_{m+1}, is a RKHS with RK

$$r_m(x, t) = 2^m \sum_n \overline{\psi(2^m x - n)} \psi(2^m t - n)$$

where $\psi(t)$ is the mother wavelet.

In Example 1, $q(x, t)$ is easy to find in closed form. It is just

$$q(x, t) = \phi(x - [t])$$

where $[t]$ is the largest integer $\leq t$. In Example 5, it is

$$q(x, t) = \phi(x - t)$$

but in other cases has no obvious closed form.

The sequence $\{q_m(x, t)\}$ is a delta sequence in S'_r, i.e., $q_m(x, t) \rightarrow \delta(x - t)$. This follows from the fact that

$$\int_{-\infty}^{\infty} q_m(x, t)\theta(t)dt \rightarrow \theta(x) \quad \text{as} \quad m \rightarrow \infty$$

for each $\theta \in S_r$ where convergence is in the L^2 sense. As we shall see in a later chapter, this convergence is uniform on bounded sets. These kernels have a number of interesting properties, some of which come out of the wavelet moment theorem.

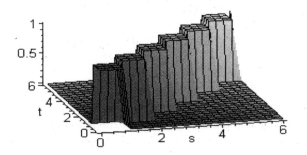

FIGURE 3.10
The reproducing kernel $q(x,t)$ for V_0 in the case of Haar wavelets.

3.4 Regularity of wavelets as a moment condition

We have assumed that the scaling function $\phi \in S_r$ and, hence, the mother wavelet $\psi \in S_r$. This condition leads to another condition in which ψ is orthogonal to all polynomials of degree $\le r$.

THEOREM 3.1
Let $\psi \in S_r$ with $\psi_{mn}(x) = 2^{m/2}\psi(2^m x - n)$ an orthonormal system in $L^2(\mathbb{R})$. Then

$$\int_{-\infty}^{\infty} x^k \psi(x)dx = 0, \qquad k = 0, 1, \cdots, r.$$

The proof is a straightforward application of the Lebesgue dominated convergence theorem and is similar to the one in [D]. We proceed by induction. Let N be a dyadic rational number $(= 2^{-j_o}k_o)$ such that $\psi(N) \neq 0$. This is possible since $\psi \neq 0$ and is continuous, and since such numbers are dense in $\mathbb{R}$. Let $j > 1$ be sufficiently large so that $2^j N$ is an integer. Then by orthogonality

$$0 = 2^j \int_{-\infty}^{\infty} \psi(x)\psi(2^j x - 2^j N)dx = \int_{-\infty}^{\infty} \psi(2^{-j}y + N)\psi(y)dy. \quad (3.20)$$

Since $\psi(2^{-j}y + N)$ is uniformly bounded, the integrand is dominated by a multiple of $|\psi(y)|$ and hence by the dominated convergence theorem

$$
0 = \lim_{j \to \infty} \int_{-\infty}^{\infty} \psi(2^{-j}y + N)\psi(y)dy = \int_{-\infty}^{\infty} \lim_{j \to \infty} \psi(2^{-j}y + N)\psi(y)dy
$$

$$
= \int_{-\infty}^{\infty} \psi(N)\psi(y)dy = \psi(N) \int_{-\infty}^{\infty} \psi(y)dy.
$$

Hence the conclusion holds for $k = 0$. By the induction hypothesis we assume it holds for $k = 0, 1, \cdots, n-1$; $n \leq r$, and deduce, by Taylor's theorem,

$$
\psi(x) = \sum_{k=0}^{n} \psi^{(k)}(N)\frac{(X-N)^k}{k!} + r_n(x)\frac{(X-N)^n}{n!},
$$

where $r_n(x)$ is uniformly bounded and $\lim_{x \to N} r_n(x) = 0$. We now assume that $\psi^{(n)}(N) \neq 0$, and find by substituting this in (3.20) that

$$
0 = \int_{-\infty}^{\infty} \left\{ \sum_{k=0}^{n} \psi^{(k)}(N)\frac{(2^{-j}y)^k}{k!} + r_n(2^{-j}y + N)\frac{(2^{-j}y)^n}{n!} \right\} \psi(y)dy
$$

$$
= \psi^{(n)}(N) \int_{-\infty}^{\infty} 2^{-jn}\frac{y^n}{n!}\psi(y)dy + \int_{-\infty}^{\infty} r_n(2^{-j}y + N)\frac{2^{-jn}y^n}{n!}\psi(y)dy.
$$

We now multiply both sides by $2^{jn}n!$ and then take the limit as $j \to \infty$ to obtain

$$
0 = \psi^{(n)}(N) \int_{-\infty}^{\infty} y^n \psi(y)dy + \int_{-\infty}^{\infty} \lim_{j \to \infty} r_n(2^{-j}y + N)y^n\psi(y)dy
$$

$$
= \psi^{(n)}(N) \int_{-\infty}^{\infty} y^n \psi(y)dy + 0,
$$

which gives us our conclusion.   □

This enables us to derive a number of properties of the scaling function and the reproducing kernel.

THEOREM 3.2

Let $\phi \in S_r$ be a real orthonormal scaling function with multiresolution analysis $\{V_m\}$; let $q_m(x, t)$ be the reproducing kernel of V_m. Then if

$\hat{\phi}(0) \geq 0$,

(i) $\hat{\phi}(2\pi k) = \delta_{ok}$, $\quad k \in \mathbb{Z}$,
(ii) $\sum_n \phi(x - n) = 1$, $\quad x \in \mathbb{R}$,
(iii) $\int_{-\infty}^{\infty} q(x, y)dy = 1$, $\quad x \in \mathbb{R}$,
(iv) $\int_{-\infty}^{\infty} q_m(x, y)y^k dy = x^k$, $\quad x \in \mathbb{R}, 0 \leq k \leq r, m \in \mathbb{Z}$.

PROOF Let f be the function such that $\hat{f}(w) = \chi_{[0,1]}(w)$. Then the projection of f onto V_m is given by

$$f_m(x) = \int_{-\infty}^{\infty} q_m(x, t)f(t)dt = \frac{1}{2\pi} \int_{-\infty}^{\infty} \hat{q}_m(x, w)\hat{f}(w)dw$$

$$= \frac{1}{2\pi} \int_0^1 \sum_n \hat{\phi}(w2^{-m})e^{-in2^{-m}w} \phi(2^m x - n)dw .$$

By Parseval's equality applied to $\{2^{m/2}\phi(2^m x - n)\}$ we have

$$\|f_m\|^2 = \frac{1}{(2\pi)^2} \sum_n \left| \int_0^1 \hat{\phi}(w2^{-m})e^{-iwn2^{-m}} dw \right|^2 2^{-m}.$$

Since $\left\{ e^{in2^{-m}w} / \sqrt{2^{m+1}\pi} \right\}$ is orthonormal on $[-2^m\pi, 2^m\pi]$ and each integral in the sum is a Fourier coefficient of $\hat{f}(w)\hat{\phi}(w2^{-m})$ we again use Parseval's equality but now with respect to this trigonometric system to obtain

$$\|f_m\|^2 = \frac{1}{2\pi} \int_0^1 |\hat{\phi}(w2^{-m})|^2 dw .$$

We now use Lebesgue's dominated convergence theorem and take the limit as $m \to \infty$. This gives us

$$\|f\|^2 = \frac{1}{2\pi} \int_0^1 |\hat{f}(w)|^2 dw = \frac{1}{2\pi} \int_0^1 |\hat{\phi}(0)|^2 dw .$$

Thus $\hat{\phi}(0) = 1$, since we have required $\hat{\phi}(0) \geq 0$. The orthonormality of $\{\phi(t - n)\}$, as we have seen, translates into

$$\sum_k |\hat{\phi}(w + 2\pi k)|^2 \equiv 1,$$

from which it follows that $\hat{\phi}(2\pi k) = 0$, $k \neq 0$, thereby proving (i).

The second result is obtained by taking the Fourier series expansion
of the periodic function

$$\sum_n \phi(x - n) = \sum_k c_k e^{i2\pi k x},$$

where

$$c_k = \int_0^1 \left(\sum_n \phi(x - n) \right) e^{-2\pi i k x} dx$$

$$= \int_{-\infty}^{\infty} \phi(x) e^{-2\pi i k x} dx = \hat{\phi}(2\pi k) = \delta_{0k}.$$

Hence we get (ii),

$$\sum_n \phi(x - n) = c_0 = 1,$$

and (iii) is a corollary to (i) and (ii).

In order to prove (iv) we use Theorem 3.1. We first observe that

$$\int_{-\infty}^{\infty} \psi(y - k) y^n dy = 0, \qquad 0 \le n \le r$$

and use the fact that the reproducing kernel of V_1 satisfies

$$q_1(x, y) = q(x, y) + r(x, y),$$

where $r(x, y)$ is the reproducing kernel of W_0. Then it follows that

$$\int q_1(x, y) y^n dy = \int q(x, y) y^n dy + \int r(x, y) y^n dy,$$

and since

$$\int r(x, y) y^n dy = \int \sum_k \psi(x - k) \psi(y - k) y^n dy = 0,$$

that

$$\int q(x, y) y^n dy = \int q_1(x, y) y^n dy = \cdots = \int q_m(x, y) y^n dy.$$

We now write this last integral as

$$\int q_m(x, y) y^n (\theta(y) + (1 - \theta(y))) dy$$

where $\theta(y) \in S_r$, $0 \le \theta(y) \le 1$, and $\theta(y) = 1$, $|y| \le 2$. Then

$$\int q_m(x,y) y^n \theta(y) dy \to x^n \theta(y) = x^n, \quad |x| \le 1$$

since $q_m(x,y)$ is a delta sequence. On the other hand, we have

$$\int q_m(x,y) y^n (1 - \theta(y)) dy = \left\{ \int_{-\infty}^{-2} + \int_2^\infty \right\} q_m(x,y) y^n (1 - \theta(y)) dy,$$

and

$$\int_2^\infty |q_m(x,y) y^n (1 - \theta(y))| dy$$
$$\le \int_2^\infty \frac{C_{n+2}}{(1+2^m|x-y|)^{n+2}} y^n dy$$
$$\le \frac{C_{n+2} 2^n}{(1+2^m(2-1))^n} \int_2^\infty \frac{1}{(1+2^m(y-1))^2} dy$$

which converges to 0 as $m \to \infty$ uniformly for $|x| \le 1$. The same is true for the other integral.

The condition $|x| \le 1$ can be removed by a change of scale to deduce that

$$\int q_m(x,y) y^n dy \to x^n, \quad \text{as } m \to \infty, \ x \in \mathbb{R}, \ n \le r.$$

Since the integral does not change with m, it follows that (iv) must hold. $\qquad\square$

3.4.1 More on example 3

The wavelet constructed in example 3 of section 1 is only the simplest example of Daubechies' wavelets. For smoother scaling functions we must use, rather than the coefficients of the dilation equation, the properties of $m_0(w)$. These are given in (3.12) and (3.13). We are interested in finding scaling functions with compact support and a high degree of smoothness. This leads to mother wavelets with a large number of vanishing moments. The compact support gives an $m_0(w)$ as a trigonometric polynomial. If ψ has N–1 vanishing moments, then

$$\widehat{\psi}^{(k)}(0) = 0, k = 0, 1, ..., N - 1$$

and hence by differentiating the right hand side of (3.16) we get

$$\frac{d^k}{dw^k}[\hat{\phi}(w)\hat{m}_0\,(w+\pi)]|_{w=0} = 0, \quad k = 0, 1, ..., N-1.$$

Since $\hat{\phi}(0) = 1$, it follows that $m_0^{(k)}(\pi) = 0, k = 0, 1, ..., N-1$, and thus m_0 must have the form

$$m_0(w) = (\frac{1+e^{-iw}}{2})^N \mathcal{L}(w)$$

where $\mathcal{L}$ is again a trigonometric polynomial. The orthogonality condition in terms of m_0, (3.13), is then used to derive properties of $\mathcal{L}(w)$

$$|m_0\,(w)|^2 + |m_0\,(w+\pi)|^2 \tag{3.21}$$

$$= \left|(\frac{1+e^{-iw}}{2})^N \mathcal{L}(w)\right|^2 + \left|(\frac{1-e^{-iw}}{2})^N \mathcal{L}(w+\pi)\right|^2$$

$$= \cos^{2N}(w/2)\,|\mathcal{L}(w)|^2 + \sin^{2N}(w/2)\,|\mathcal{L}(w+\pi)|^2 = 1.$$

In order to find an appropriate trigonometric polynomial, we first convert $\mathcal{L}$ to an ordinary polynomial so that we can use some number theoretic properties of polynomials. We do this by substituting $y = \sin^2 w/2$, which works since $|\mathcal{L}(w)|^2$ can be written as a polynomial in powers of the cosine and hence of y. Then (3.21) becomes

$$(1-y)^N P(y) + y^N P(1-y) = 1 \tag{3.22}$$

where $P(y) = |\mathcal{L}(w)|^2$.

Now we can use some results from number theory.

Lemma 3.1

Let $N > 1$; then there is a unique polynomial of degree $< N$ satisfying (3.22) given by

$$P(y) = \sum_{k=0}^{N-1} \binom{N+k-1}{k} y^k. \tag{3.23}$$

PROOF Since $(1-y)^N$ and y^N are relatively prime polynomials, it follows from Euclid's algorithm [Gr, p.181] that there exist polynomials $q_1(y)$ and $q_2(y)$ such that

$$(1 - y)^N q_1(y) + y^N q_2(y) = 1. \tag{3.24}$$

There are many such polynomials, but only one pair of degree $<N$. Because of the symmetry of our polynomials y^N and $(1 - y)^N$, we get a similar result when we replace y by $(1 - y)$ and because of their uniqueness, we find that $q_2(y) = q_1(1-y)$. Thus we know that there is a unique polynomial of degree $<N$ satisfying (3.22). In order to find it, we differentiate (3.22) $k < N$ times and evaluate the derivative at $y = 0$. This gives us

$$\sum_{j=0}^{k} \binom{k}{j} \frac{N!}{(N-j)!} (-1)^j P^{(k-j)}(0) = \delta_{0k}$$

which can be solved recursively to get the values

$$P^{(k)}(0) = \frac{(N+k-1)!}{(N-1)!}, k = 0, 1, ..., N - 1.$$

Now (3.23) follows by using Maclaurin series with these values.

This leaves us with an expression for $|m_0(w)|^2$:

$$|m_0(w)|^2 = \cos^{2N}(w/2)P(\sin^2 w/2),$$

from which we need to extract a square root. To do so we introduce another change of variable $z = e^{-iw}, \sin^2(w/2) = \frac{2-z-z^{-1}}{4}$ and denote by Q the polynomial of degree $2N - 2$

$$Q(z) := z^{N-1}P(\frac{2 - z - z^{-1}}{4}).$$

It inherits the symmetry of P and hence has the property that if z_0 is a zero of Q, then z_0^{-1} is also a zero. Hence it may be factored into

$$Q(z) = C^2 \prod_{j=1}^{N-1} (z - z_j)(z - z_j^{-1}) = C^2 Q_1(z) Q_2(z),$$

where the zeros $z_1, z_2, ..., z_{N-1}$ lie outside the unit disk and are the zeros of Q_1, and C^2 is the leading coefficient. Note that Q has no zeros on the boundary of the unit disk because of the fact that $P(\sin^2 w/2)$ is positive for all w. We now change the variable back to w again to get

$$\overline{Q_2(e^{-iw})} = \prod_{j=1}^{N-1}(e^{iw} - \overline{z_j^{-1}}) = \prod_{j=1}^{N-1}(e^{iw}\overline{z_j^{-1}})(\overline{z_j} - e^{-iw})$$

and hence

$$|\overline{Q_2(e^{-iw})}| = \prod_{j=1}^{N-1}|z_j^{-1}| \prod_{j=1}^{N-1}|\overline{z_j} - e^{-iw}| = K^2|Q_1(e^{-iw})|$$

since the complex zeros occur in complex conjugate pairs. Here $K^2 = \prod_{j=1}^{N-1}|z_j^{-1}|$. Thus we have our square root since

$$P(\sin^2 w/2) = |e^{-i(1-N)w}Q(e^{-iw})| = C^2 K^2|Q_1(e^{-iw})|^2.$$

We use it to find a choice of m_0 as

$$m_0(w) = (\frac{1 + e^{-iw}}{2})^N|C|KQ_1(e^{-iw}). \tag{3.25}$$

Now we can find the dilation equations for various values of N.

For $N = 2$, we find that $P(y) = 1 + 2y$ from the lemma. Then successively we find

$$P(\sin^2 w/2) = 1 + 2((1 - \cos w)/2) = \frac{4 - e^{iw} - e^{-iw}}{2} = \frac{4 - z^{-1} - z}{2},$$

$$Q(z) = (4z - 1 - z^2)/2 = -\frac{1}{2}(z - 2 - \sqrt{3})(z - 2 + \sqrt{3}),$$

$$Q_1(z) = (z - 2 - \sqrt{3}), C^2 = -1/2, K^2 = 1/(2 + \sqrt{3}).$$

Hence we have

$$m_0(w) = (\frac{1 + e^{-iw}}{2})^2\frac{1}{\sqrt{2}}\frac{1}{\sqrt{2 + \sqrt{3}}}(e^{-iw} - 2 - \sqrt{3})$$

$$= \frac{1}{8}(1 + \sqrt{3} + (3 + \sqrt{3})e^{-iw} + (3 - \sqrt{3})e^{-2iw} + (1 - \sqrt{3})e^{-3iw}).$$

The calculation of C and K was not really necessary since we already know that $m_0(0) = 1$, and this can be used instead to normalize the expression.

For N=3, the polynomial is $P(y) = 1 + 3y + 6y^2$, and

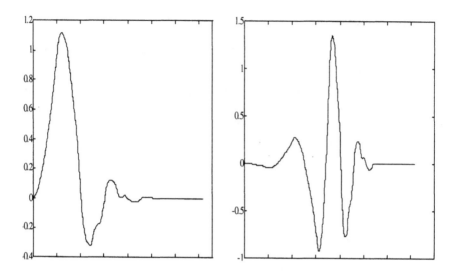

FIGURE 3.11
Daubechies scaling function and mother wavelet ($N = 4$).

$$P(\sin^2 w/2) = 1 + 3((1 - \cos w)/2) + 6((1 - \cos w)/2)^2$$
$$= 4 - \frac{9}{4}(e^{iw} + e^{-iw}) + \frac{3}{8}(e^{2iw} + 2 + e^{-2iw}),$$
$$Q(z) = \frac{3}{8} - \frac{9}{4}z + \frac{19}{4}z^2 - \frac{9}{4}z^3 + \frac{3}{8}z^4.$$

Already in this case we have a quartic polynomial whose zeros are complex and must be found numerically. The ones outside the unit disk are $z_{1,2} = 2.71275 \pm 1.44389i$ approximately. Then

$$Q_1(z) = z^2 - 5.424597z + 9.443814 = z^2 - a_1 z + a_2,$$
$$m_0(w) = \left(\frac{1 + e^{-iw}}{2}\right)^3 \left(\frac{e^{-2iw} - a_1 e^{-iw} + a_2}{1 - a_1 + a_2}\right),$$

which can be expanded to get the Fourier series coefficients of m_0 which are also the coefficients of the dilation equation. Once we have the dilation equation we can proceed recursively to find the scaling function ϕ as described previously.

3.5 Mallat's decomposition and reconstruction algorithm

The orthogonality of translates of the the scaling function together with the dilation equations enable us to construct a simple algorithm [Ma] relating the coefficients at different scales. For the space V_1, we have two distinct orthonormal bases:

$$\left\{ \sqrt{2}\phi(2x - n) \right\}_{n=-\infty}^{\infty}$$

and

$$\{\phi(x - n), \psi(x - k)\}_{n,k=-\infty}^{\infty}.$$

Hence each $f \in V_1$ has an expansion

$$f(x) = \sum_n a_n^1 \sqrt{2}\phi(2x - n)$$

$$= \sum_n a_n^0 \phi(x - n) + \sum_n b_n^0 \psi(x - n).$$

By (3.3) and (3.18) we have

$$\phi(x - n) = \sqrt{2} \sum_k c_k \phi(2x - 2n - k),$$

$$\psi(x - n) = \sqrt{2} \sum_k (-1)^{k-1} c_{1-k} \phi(2x - 2n - k).$$

We substitute this into the expansion to obtain

$$f(x) = \sum_n a_n^0 \sqrt{2} \sum_k c_k \phi(2x - 2n - k)$$

$$+ \sum_n b_n^0 \sqrt{2} \sum_k (-1)^{k-1} c_{1-k} \phi(2x - 2n - k)$$

$$= \sum_n a_n^0 \sum_j c_{j-2n} \phi(2x - j)\sqrt{2}$$

$$+ \sum_k b_n^0 \sum_j (-1)^{j-1} c_{1-j+2n} \phi(2x - j)\sqrt{2}.$$

Hence by equating the coefficients, we have the algorithm

$$a_n^1 = \sum_k c_{n-2k} a_k^0 + \sum_k (-1)^{n-1} c_{1-n+2k} b_k^0, \qquad (3.26)$$

which is the reconstruction part. The decomposition is even easier: we need merely use the formula for a_n^0 and b_n^0 to find

$$
\begin{aligned}
(a)\ a_n^0 &= \int_{-\infty}^{\infty} f(x)\phi(x-n)dx = \int_{-\infty}^{\infty} f(x) \sum_k c_k \sqrt{2}\phi(2x-2n-k)dx \\
&= \sum_k c_k a_{2n+k}^1 = \sum_k a_k^1 c_{k-2n}, \\
(b)\ b_n^0 &= \sum_k a_k^1 (-1)^{k-1} c_{1-k+2n}.
\end{aligned}
$$
$$(3.27)$$

This works at each scale to give us the tree algorithm for decomposition (3.27),

and for reconstruction (3.26)

Thus we need calculate the coefficients from the function $f(t)$ only once at the finest scale of interest. Then we work down to successively coarser scales by using this decomposition algorithm, with the error at each successive scale corresponding to the wavelet coefficients.

3.6 Filters

A continuous linear system is a device which transforms a signal $x(t) \in L^2(\mathbb{R})$ linearly. It has an associated "impulse response" $h(t)$ such that the output of the system is given by the convolution,

$$y = (h * x). \qquad (3.28)$$

This convolution exists if, say, x and $h \in L^1(\mathbb{R})$. However if x is somewhat restricted, more general impulse responses are possible ([Pa, p.402]).

A linear system is a <u>filter</u> if it passes signals of certain frequencies and attenuates others. This is best seen by taking the Fourier transform of (3.28) to get

$$\hat{y}(w) = H(w)\hat{x}(w), \qquad (3.29)$$

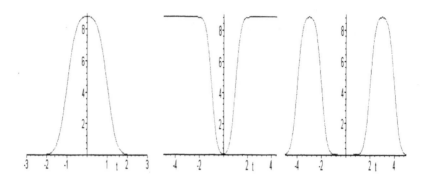

FIGURE 3.12
The system functions of some continuous filters: low-pass, high-pass and band-pass.

where $H(w)$ is the system transfer function. If $H(w) = 0$, $|w| \geq w_0$, then H is a <u>low-pass filter</u> while on the other hand, if $H(w) = 0$ for $|w| \leq w_1$, then H is a <u>high-pass filter</u>. A <u>band-pass</u> filter passes a band $w_0 \leq |w| \leq w_1$. This describes perfect filters; practical filters have small values for the frequencies they want to remove, but are not exactly zero.

Similar concepts apply to discrete signals $x_n \in \ell^2$. Then we have the discrete impulse response $\{h_n\}$ and the output is given by

$$ y_n = \sum_k h_{n-k} \chi_k. $$

By taking Fourier series we find

$$ Y(w) = \sum_n y_n e^{iwn} = H(w)X(w), $$

where each of these functions now is 2π periodic. The same concepts apply except everything is restricted to the interval $[-\pi, \pi]$. Thus, a low-pass filter would correspond to, say, $H(w) = 0$ for $\frac{\pi}{2} \leq |w| \leq \pi$.

A number of operations associated with wavelets correspond to filters. The first is the projection onto the subspace V_m given by

$$ f_m(t) := P_m f(t) = \int_{-\infty}^{\infty} q_m(t, y) f(y) dy. $$

For $m = 0$ this is given by

$$f_0(t) = \frac{1}{2\pi} \int \sum_n \phi(t-n)\overline{\hat{\phi}(w)e^{-iwn}}\, \hat{f}(w)dw$$

$$= \sum_n \phi(t-n)\frac{1}{2\pi} \int \overline{\hat{\phi}(w)}\, \hat{f}(w)e^{iwn}dw\,,$$

and hence

$$\hat{f}_0(\xi) = \sum_n \hat{\phi}(\xi)e^{-i\xi n} \tag{3.30}$$

$$\cdot \frac{1}{2\pi} \int_0^{2\pi} \sum_k \overline{\hat{\phi}(w+2\pi k)}\, \hat{f}(w+2\pi k)e^{iwn}dw$$

$$= \hat{\phi}(\xi) \sum_k \overline{\hat{\phi}(\xi+2\pi k)}\, \hat{f}(\xi+2\pi k),$$

since the integral in (3.30) is just a Fourier coefficient of

$$\sum_k \overline{\hat{\phi}(\xi+2\pi k)}\hat{f}(\xi+2\pi k).$$

In the case of Meyer's wavelets $\hat{\phi}(\xi) = 0$ for $|\xi| > \pi + \epsilon$ and

$$\hat{f}_0(\xi) = |\hat{\phi}(\xi)|^2 f(\xi) + \hat{\phi}(\xi)\overline{\hat{\phi}(\xi+2\pi)}\, f(\xi+2\pi)$$

$$+\hat{\phi}(\xi)\overline{\hat{\phi}(\xi-2\pi)}\, f(\xi-2\pi).$$

The last two terms have support on intervals of length $\leq 2\epsilon$ and can therefore be made arbitrarily small. Thus, the projection of f onto V_0 is, except for this small term, a low-pass filter with system function $H(w) = |\hat{\phi}(w)|^2$.

The projection onto the subspaces W_m can, by similar arguments, be shown to be approximately a band-pass filter. For wavelets other than Meyer's, these conclusions are still approximately valid but with a poorer approximation.

The coefficient sequences are also related to discrete filters. The decomposition algorithm gives the coefficient at the coarser scale in terms of the finer scale. We have at $m = 0$, by (3.27)

$$a_n^0 = \sum_k c_{k-2n}a_k^1, \quad b_n^0 = \sum_k (-1)^{k-1}c_{1-k+2n}a_k^1.$$

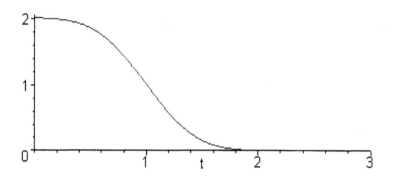

FIGURE 3.13
The system function of discrete lowpass (halfband) filter.

These operations can be decomposed into a filter

$$e_n^0 = \sum_k c_{k-n} a_k^1$$

followed by decimation (take only even terms)

$$a_n^0 = e_{2n}^0,$$

and similarly for b_n. The filter has impulse response $\{c_{-n}\}$ since e_n^0 is the convolution of c_{-n} and a_k^1, and hence the system function is

$$H(w) = \sum_n c_{-n} e^{iwn}.$$

But we recall that

$$\hat{\phi}(w) = \sum_k c_k e^{-iwk/2} \frac{1}{\sqrt{2}} \hat{\phi}\left(\frac{w}{2}\right) = m_0\left(\frac{w}{2}\right) \hat{\phi}\left(\frac{w}{2}\right),$$

i.e., the system function is exactly $H(w) = m_0(w)\sqrt{2}$.
  In the Meyer case again, we have seen that

$$m_0\left(\frac{w}{2}\right) = \sum_k \hat{\phi}(w + 4\pi k),$$

and $H(w)$ therefore has support on $|w| < \frac{\pi + \epsilon}{2}$ in the interval $[-\pi, \pi]$. Therefore c_k is the impulse response of a discrete low-pass filter. For

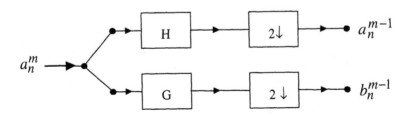

FIGURE 3.14
The decomposition algorithm.

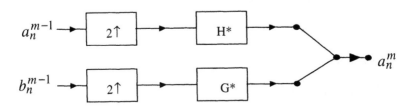

FIGURE 3.15
The reconstruction algorithm.

the wavelet coefficient, on the other hand, we get the system function for the corresponding filter

$$G(w) = \sum_k c_{k+1} e^{iw(k+1)} e^{i\pi k} = \sum_n c_{-n} e^{-i(w+\pi)n} e^{-i\pi} = e^{-i\pi} H(-w - \pi).$$

Thus the support of $G(w)$ in $[-\pi, \pi]$ is contained in the set where $|w| \geq \frac{\pi}{2} - \frac{\epsilon}{2}$, and $d_k = (-1)^{k-1} c_{1-k}$ is the impulse response of a high-pass filter. The decomposition algorithm is given symbolically by Figure 3.14, where the symbol $2 \uparrow$ denotes the removal of every other term in the sequence.

The reconstruction algorithm is treated similarly,

$$a_n^1 = \sum_k c_{n-2k} a_k^0 + \sum_k d_{n-2k} b_k^0.$$

This may be made into a pair of filters by interlacing 0's in the sequence $\{a_k^0\}$ and $\{b_k^0\}$. Then a_n^1 becomes

$$a_n^1 = \sum_j c_{n-j} a_j^{00} + \sum_j d_{n-j} b_j^{00},$$

where $\{a_j^{00}\}$ and $\{b_j^{00}\}$ are the sequences with the interlaced zeros. The first term corresponds to a low-pass and the second to a high-pass filter. It may be symbolized by Figure 3.15, where the $2 \downarrow$ denotes the operation of interlacing zeros. Note that the filters H and H^* and G and G^* have conjugate system functions, respectively.

3.7   Problems

1. Use the iterative procedure given in (3.11) to graph the Daubechies scaling functions (Example 3) for $\nu = -1/\sqrt{3}$. (You will need a computer for this; any program that handles iteration will work.)

2. Repeat Problem 1 for $\nu = -\frac{1}{2}$ and compare.

3. Find a graph of the mother wavelet by using (3.18) in Problems 1 and 2.

4. Find an expression for the Fourier transform of the scaling function for the Battle-Lemarie wavelet based on B-splines of order 2 (Example 4) by repeating the orthogonalization procedure used in Example 2.

5. Use Gram-Schmidt orthogonalization on translates of the hat function (Example 2) $\{\theta(t - n)\}$. Is this an orthogonal basis of V_0? Is this a scaling function basis?

6. Let P be the uniform probability measure on $[-\epsilon, \epsilon]$, $0 < \epsilon < \frac{\pi}{3}$. Find the corresponding Meyer wavelet (Example 7) scaling function. Graph $\hat{\phi}(w)$, $m_0(w)$, and $|\hat{\psi}(w)|$.

7. Prove that a Meyer wavelet whose Fourier transform is a C^∞ function has vanishing moments of all orders, i.e.,

$$\int_{-\infty}^{\infty} t^m \psi(t)dt = 0, m = 0, 1, 2, \cdots.$$

8. Show $\hat{\phi}(2\pi k) = \delta_{0k}$ for Problem 6 and for Problem 4.

9. Let the expansion coefficients in V_m, a_n^m for the scaling function of Problem 2 be given by $a_n^m = 2^{-n}$. Find a_n^{m-1} and b_n^{m-1} by using (3.27).

10. Check Problem 9 by using the reconstruction algorithm (3.26).

11. Find the Fourier transform of a projection of f onto V_0 in the case of Problem 6. Find an expression for the difference between this and the ideal low-pass filter.

12. Repeat Problem 11 for W_0 and a band-pass filter.

Chapter 4

Convergence and Summability of Fourier Series

In this chapter we extend and refine the introduction to trigonometric Fourier series begun in Chapter 1. Stronger pointwise convergence results are presented here, as well as rates of convergence. In addition, several new topics are taken up. These include Gibbs phenomenon which involves overshoot at discontinuities and two types of summability, Cesàro and Abel, used to overcome it. We then consider periodic tempered distributions and use them for a more general theory of Fourier series in which the distinction between formal trigonometric series and Fourier series is eliminated.

4.1 Pointwise convergence

There are many tests for the pointwise convergence of Fourier series [Z]. The one given in Chapter 1 is quite weak and excludes many functions for which we have such convergence. We present others involving Lipschitz continuity which can be used to obtain local resuts as well.

Let $f(x)$ be a 2π periodic function whose restriction to $(-\pi, \pi)$ is in $L^1(-\pi, \pi)$; f is said to satisfy a <u>Lipschitz</u> condition of order $\alpha > 0$ at x_0 if there is a constant C such that

$$|f(x) - f(x_0)| \le C|x - x_0|^\alpha \tag{4.1}$$

in some neighborhood of x_0. If the condition holds for all x_0 with the same C then f satisfies a uniform Lipschitz condition. We restrict α to be ≤ 1 since the condition implies that $f'(x) = 0$ for $\alpha > 1$. We also

have one sided versions of (4.1). If $f(x_0+)$ (resp.$f(x_0-)$) exists, and
(4.1) holds for $x > x_0$, (resp. $x < x_0$), then we say that f satisfies a
right (resp. left) hand Lipschitz condition.

We first prove the Riemann-Lebesgue lemma which we shall need for
our convergence result.

Proposition 4.1
Let $f \in L^1(-\pi, \pi)$; then the Fourier coefficients satisfy

$$|a_n| + |b_n| \to 0 \qquad \text{as } n \to \infty.$$

We already know this is true for $f \in L^2(-\pi, \pi)$ by Bessel's inequality
(1.3). Since each $f \in L^1(-\pi, \pi)$ may be approximated by some $g \in$
$L^2(-\pi, \pi)$ in the sense of $L^1(-\pi, \pi)$ the result follows.

Proposition 4.2
Let f satisfy a left and right hand Lipschitz condition at x_0; then

$$S_n(x_0) \to \frac{f(x_0+) + f(x_0-)}{2} = \overline{f}(x_0)$$

*as $n \to \infty$. In particular if f satisfies (4.1) in a neighborhood of x_0,
then $S_n(x_0) \to f(x_0)$ as $n \to \infty$.*

PROOF   We first prove the result for the particular case and then
indicate how it can be modified for the general case. The proof starts
with formula (1.9), for $x_0 \neq \pm\pi$,

$$S_n(x_0) - \overline{f}(x_0) \qquad (4.2)$$

$$= \int_{-\pi}^{\pi} \left\{ \frac{f(x_0 - u) - \overline{f}(x_0)}{u} \right\} u D_n(u) du$$

$$= \int_{-\pi}^{0} \left\{ \frac{f(x_0 - u) - f(x_0+)}{u} \right\} u D_n(u) du$$

$$+ \int_{0}^{\pi} \left\{ \frac{f(x_0 - u) - f(x_0-)}{u} \right\} u D_n(u) du$$

$$= \int_{-\pi}^{-\delta} + \int_{-\delta}^{0} + \int_{0}^{\delta} + \int_{\delta}^{\pi} = I_1 + I_2 + I_3 + I_4.$$

The first integral can be written as

$$I_1 = \int_{-\pi}^{-\delta} \left\{ \frac{f(x_0 - u) - f(x_0+)}{u} \right\} \frac{u}{2\pi \sin u/2} \sin(n + \frac{1}{2}) u du$$

$$= \frac{1}{\pi} \int_{-\pi}^{-\delta} \left\{ \frac{f(x_0 - u) - f(x_0+)}{u} \right\} \frac{u/2}{\sin u/2} (\sin nu \cos u/2$$

$$+ \cos nu \sin u/2) du$$

$$= \frac{1}{\pi} \int_{-\pi}^{\pi} g(x_0, u) \sin nu \, du + \frac{1}{\pi} \int_{-\pi}^{\pi} h(x_0, u) \cos nu \, du \, ,$$

where g and h are both L^1 functions. Hence by proposition 4.1, $I_1 \to 0$ as $n \to \infty$. A similar argument holds for I_4, while I_2 and I_3 satisfy

$$|I_2| + |I_3| \le \frac{1}{2} \int_{-\delta}^{\delta} C|u|^{\alpha-1} du = \frac{C}{\alpha} \delta^{\alpha}.$$

Thus, given $\epsilon > 0$, we may choose δ such that $\frac{C}{\alpha}\delta^{\alpha} < \frac{\epsilon}{2}$ and then choose N such that $|I_1 + I_4| < \frac{\epsilon}{2}$ for $n \ge N$. This makes

$$|S_n(x_0) - \overline{f}(x_0)| < \epsilon, \qquad n \ge N.$$

For $x_0 = \pm\pi$ we translate everything by an amount π and show that $S_n(x - \pi) \to \overline{f}(x - \pi)$ at $x = 0$.

The modification needed to prove the result for the two sided hypotheses is clear. In the integrals I_1 and I_4 we replace $f(x_0+)$ and $f(x_0-)$ by $f(x_0)$ to get the result we want. □

In order to get uniform convergence we need a uniform version of the Riemann-Lebesgue lemma.

Lemma 4.1
Let f be periodic and in $L^1[-\pi, \pi]$ and let $h \in C^1[\alpha, \beta]$ where $[\alpha, \beta] \subseteq [-\pi, \pi]$; then

$$\int_{\alpha}^{\beta} f(x - u)h(u) \sin \lambda u \, du \to 0 \qquad \text{as } \lambda \to \infty$$

uniformly in x.

PROOF We first approximate f by a periodic function $g \in C^1$ in the sense of $L^1(-\pi, \pi)$, i.e., choose g such that

$$\int_{-\pi}^{\pi} |f(x) - g(x)| dx < \epsilon.$$

Now, we use integration by parts on

$$I_1 = \int_\alpha^\beta g(x-u)h(u) \sin d\mu \, du$$

$$= -g(x-u)h(u) \frac{\cos \lambda u}{\lambda} \Big|_\alpha^\beta + \int_\alpha^\beta [g(x-u)h(u)]' \frac{\cos \lambda u}{\lambda} du.$$

Since both $g(x-u)$ and its derivative are uniformly bounded in x, this integral converges to 0 uniformly as $\lambda \to \infty$. Then we see that

$$\left| \int_\alpha^\beta f(x-u)h(u) \sin \lambda u \, du \right|$$

$$\leq \left| \int_\alpha^\beta (f(x-u) - g(x-u))h(u) \sin \lambda u \, du \right| + |I_1|$$

$$\leq \max_{\alpha \leq u \leq \beta} |h(u)| \int_\alpha^\beta |f(x-u) - g(x-u)| du + |I_1|$$

$$\leq \max_{\alpha \leq u \leq \beta} |h(u)| \epsilon + |I_1|.$$

The first term can be made arbitrarily small while the second converges to 0 as $\lambda \to \infty$.   □

As a corollary we obtain the following.

THEOREM 4.1
Let f satisfy a uniform Lipschitz condition of order $\alpha > 0$ in (a, b). Then $S_n \to f$ uniformly in any interior subinterval $[c, d] \subset (a, b)$.

PROOF   Choose $\delta < \min(c - a, b - d)$. Then the integrals I_2 and I_3 in (4.2) satisfy $|I_2| + |I_3| \leq \frac{C}{\alpha} \delta^\alpha$ as before. For I_1 we use the lemma applied to $\alpha = -\pi$, $\beta = -\delta$ and take $\lambda = \left(n + \frac{1}{2}\right)$ to deduce that

$$\frac{1}{\pi} \int_{-\pi}^{-\delta} f(x_0 - u) \frac{1}{2 \sin u/2} \sin(n + \frac{1}{2})u \, du \to 0$$

uniformly as $n \to \infty$. Since the other term in I_1 clearly converges to 0 uniformly, it follows that I_1 does as well and similarly for I_4.   □

Corollary 4.1
(Localization principle) Let f_1 and f_2 be equal in an interval (a, b); then $S_n(f_1)$ and $S_n(f_2)$ are uniformly equiconvergent in interior subintervals of (a, b).

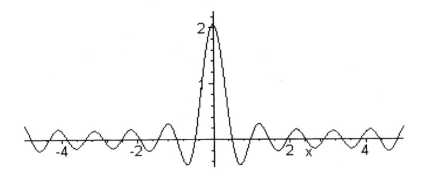

FIGURE 4.1
The Dirichlet kernel of Fourier series (n=6).

The proof is immediate since $f_1 - f_2$ satisfies the hypothesis of Theorem 4.1. This says that the local convergence of the Fourier series depends only on the local behavior of the function.

We cannot extend the above results from Lipschitz to ordinary continuity since there are many examples of continuous functions whose Fourier series diverge [Z, p. 298]. In order to treat these, various types of summability have been introduced, some of which we will cover in the next section. Many of the results in convergence and summability arise because of the properties of certain delta sequences, i.e., sequences that converge to δ in S'. For ordinary convergence the associated delta sequence is the Dirichlet kernel $D_n(t)$ restricted to $(-\pi, \pi)$. It is a delta sequence since by Proposition 4.2,

$$\int_{-\pi}^{\pi} D_n(t)\phi(t)dt = S_n(0) \to \phi(0) = (\delta, \phi) \qquad (4.3)$$

for any test function $\phi \in S$. That is, $\chi D_n \to \delta$ as $n \to \infty$ where χ is the characteristic function of $[-\pi, \pi]$. If we omit the χ, then D_n, being periodic, must converge to a periodic distribution, which we denote as $\delta^*(u)$. Clearly, $D_n \to \delta^*$ in S'.

The rate of convergence of the Fourier series can be studied by finding the rate of convergence of $\{D_n\}$ in the Sobolev space of periodic functions, $H_{2\pi}^{\alpha}$. (Compare Section 2.5.) It consists of all periodic $f \in S'$ such that

$$\sum_{n=-\infty}^{\infty} |c_n|^2 (n^2 + 1)^{\alpha} < \infty \qquad (4.4)$$

where c_n are the complex Fourier coefficients of f. For $\alpha = 0$ this is just $L^2(-\pi, \pi)$ while for $\alpha = 1, 2, \cdots$, it consists of f whose αth derivative is in $L^2(-\pi, \pi)$. For negative α, this involves the Fourier series of a periodic distribution, which we take up later in this chapter. Here we merely note that δ^*, the periodic extension of δ, has Fourier series

$$\delta^*(x) \sim \frac{1}{2\pi} \sum_{n=-\infty}^{\infty} e^{inx} \qquad (4.5)$$

and therefore $\delta^* \in H_{2\pi}^{-\alpha}$, $\alpha > \frac{1}{2}$. The norm in $H_{2\pi}^{\alpha}$ is given by the square root of (4.4) and the inner product by

$$\langle f, g \rangle_\alpha := \sum_n c_n \overline{d_n} (n^2 + 1)^\alpha.$$

It is clear that $D_n(t) \to \delta^*(t)$ in the sense of $H_{2\pi}^{-\alpha}$, $\alpha > \frac{1}{2}$. We can find the rate at which it converges, and can use this to find the rate of convergence of Fourier series. It is, since

$$D_n(x) - \delta^*(x) \sim \frac{1}{2\pi} \sum_{|k|>n} e^{ikx},$$

given by

$$\|D_n - \delta^*\|_{-\alpha}^2 = \left(\frac{1}{2\pi}\right)^2 \sum_{|k|>n} (k^2 + 1)^{-\alpha}$$

$$\leq 2\left(\frac{1}{2\pi}\right)^2 \int_{n+1}^{\infty} t^{-2\alpha}\, dt$$

$$= \frac{1}{2\pi^2} \frac{(n+1)^{-2\alpha+1}}{2\alpha - 1},$$

which we restate as

Proposition 4.3

The Dirichlet kernel $\{D_n\}$ converges to δ^* in $H_{2\pi}^{-\alpha}$ at a rate $O\left(n^{-\alpha+\frac{1}{2}}\right)$ for $\alpha > \frac{1}{2}$.

Corollary 4.2

Let $f \in H_{2\pi}^{\beta}$, $\beta > \frac{1}{2}$; then the Fourier series of f converges to f uniformly at a rate $O\left(n^{-\beta+\frac{1}{2}}\right)$ for $\beta > \frac{1}{2}$.

The proof involves merely using the Sobolev version of Schwarz's inequality, viz. if $f \in H_{2\pi}^{-\alpha}$ and $g \in H_{2\pi}^{\alpha}$

$$|(f,g)| \leq \|f\|_{2\pi}^{-\alpha} \|g\|_{2\pi}^{\alpha}$$

where (f,g) denotes the value of the functional f on the test function g. (See Chapter 2 where the test functions however were not periodic.) Since the partial sums of the Fourier series of f are given by

$$f_n(x) = (D_n * f)(x),$$

it follows that

$$|f_n(x) - f(x)| \leq \|D_n - \delta^*\|_{-\beta} \|f\|_{\beta}. \tag{4.6}$$

4.2 Summability

As we have observed, not all Fourier series of continuous functions converge pointwise. However, by replacing the partial sums of the series by other sums, we can get uniform convergence for all continuous functions.

Let $f(x)$ be a continuous periodic function whose Fourier series has partial sums $S_n(x)$; then the C-1 (Cesàro) means of the Fourier series are

$$\sigma_n(x) := \frac{1}{n} \sum_{k=0}^{n-1} S_k(x).$$

Just as with the partial sums, $\sigma_n(x)$ may be expressed as an integral involving $f(x)$

$$\sigma_n(x) = \frac{1}{n} \sum_{k=0}^{n-1} (D_k * f)(x) \tag{4.7}$$

$$= \int_{-\pi}^{\pi} \frac{1}{n} \sum_{k=0}^{n-1} D_k(x-t) f(t) dt = (F_n * f)(x),$$

where the kernel $F_n(x) := \frac{1}{n} \sum_{k=0}^{n-1} D_k(x)$ may be evaluated by the same trick used before

$$\sin^2 \frac{1}{2}x F_n(x) = \frac{1}{n}\sum_{k=0}^{n-1}\frac{\sin\frac{1}{2}x\sin\left(k+\frac{1}{2}\right)x}{2\pi} \qquad (4.8)$$

$$= \frac{1}{2\pi n}\sum_{k=0}^{n-1}\frac{\cos kx - \cos(k+1)x}{2}$$

$$= \frac{1}{2\pi n}\left(\frac{1}{2}-\frac{\cos nx}{2}\right)=\frac{1}{2\pi n}\sin^2\frac{nx}{2},$$

which gives us

$$F_n(x) = \frac{\sin^2\frac{nx}{2}}{2\pi n\sin^2\frac{1}{2}x}.$$

This is usually denoted the <u>Fejer kernel</u> and is easily shown to be a delta sequence as in the Dirichlet kernel D_n. However, it belongs to a class of such sequences, the <u>quasi-positive delta sequences</u> to which D_n does not belong. We shall discuss these in more detail in Chapter 8. The properties of $\chi_\pi F_n$, where χ_π denotes the characteristic function of $[-\pi, \pi]$ include the following:

(i) $\int_{-\pi}^{\pi}F_n(x)dx = 1,$ $n \in \mathbb{N}$
(ii) $F_n(x) \geq 0,$ $x \in \mathbb{R},\ n \in \mathbb{N}$ (4.9)
(iii) for each $\gamma > 0$, $\sup_{\pi > |x| > \gamma}F_n(x)\chi_\pi(x) \longrightarrow 0$, as $n \to \infty$.

This is actually a positive delta sequence, which is stronger than quasi-positive (see Chapter 8), which, however, is all that is needed for uniform convergence. The proof that the C–1 means converge uniformly is considerably simpler. (See, e.g., [Z, p.87].)

Another type of summability, <u>Abel summability</u>, leads to a positive delta family which depends on a parameter $0 < r < 1$. The Abel means σ_r are given by

$$\sigma_r(x) := \sum_{n=-\infty}^{\infty} c_n r^{|n|}e^{inx}\qquad 0 < r < 1. \qquad (4.10)$$

This series clearly converges uniformly in x for each such r. The kernel, the so-called Poisson kernel, is

$$P_r(x) := \frac{1}{2\pi}\sum_{n=-\infty}^{\infty} r^{|n|}e^{inx} = \frac{1-r^2}{2\pi(1-2r\cos x + r^2)}. \qquad (4.11)$$

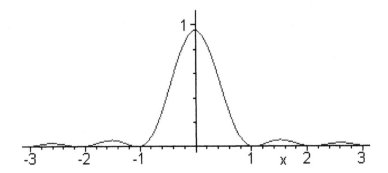

FIGURE 4.2
The Fejer kernel of Fourier series (n=6).

It satisfies the same conditions as (4.9) except that $n \to \infty$ is replaced by $r \to 1$. Since $\sigma_r = P_r * f$, it again follows that the $\sigma_r \to f$ uniformly for each continuous f.

4.3 Gibbs phenomenon

There seems to be some misunderstanding as to what constitutes Gibbs phenomenon. It deals not with the failure of the Fourier series to converge at a point of jump discontinuity but rather with the "overshoot" of the partial sums in the limit.

In order to demonstrate Gibbs phenomenon, we consider the behavior of a simple function with a jump discontinuity at 0, $f(x) = \frac{\pi}{2}\operatorname{sgn} x - \frac{x}{2}$, $|x| < \pi$. The Fourier series of $f(x)$ is

$$f(x) = \sum_{n=1}^{\infty} \frac{\sin nx}{n} = \sum_{n=1}^{\infty} \int_0^x \cos nt \, dt \qquad (4.12)$$

$$= \lim_{n \to \infty} \int_0^x \left(\frac{1}{2} + \sum_{k=1}^{n} \cos kt \right) dt - \frac{x}{2}$$

$$= \lim_{n \to \infty} \left(\pi \int_0^x D_n(t) dt \right) - \frac{x}{2}.$$

Since f satisfies a left and right Lipschitz condition at 0, and the average of the left and right hand values of f at 0 is 0, the series converges to 0 at 0 by Proposition 4.2. We wish, however, to investigate the partial sums of the series as $x \to 0^+$. The integral in (4.12) may be written as

$$\pi \int_0^x D_n(t)dt = \int_0^x \left(\frac{\sin nt \cos \frac{1}{2}t}{2 \sin \frac{1}{2}t} + \cos nt \right) dt \qquad (4.13)$$

$$= \int_0^x \frac{\sin nt}{t} dt + \int_0^x \sin nt \left(\frac{\cos t/2}{2 \sin t/2} - \frac{1}{t} \right) dt$$

$$+ \int_0^x \cos nt \, dt$$

for $x > 0$. The second and third integral converge uniformly to 0 by the Riemann-Lebesgue lemma (or by direct calculation); that leaves only the first integral, which may be written as

$$I(nx) = \int_0^{nx} \frac{\sin t}{t} dt \to \int_0^\infty \frac{\sin t}{t} dt = \frac{\pi}{2}, \quad \text{for fixed } x > 0.$$

But if we take the sequence $x_n = \frac{\pi}{n}$, the integral becomes

$$I(nx_n) = \int_0^\pi \frac{\sin t}{t} dt > \frac{\pi}{2}.$$

In fact the ratio between the two integrals is almost 1.18 [Z, p.61]. Hence the partial sums of the original function at x_n are given by

$$S_n(x_n) = I(nx_n) - \frac{x_n}{2} + \epsilon_n,$$

which converges to a value $> f(0^+)$ as $n \to \infty$. Such an overshoot appears whenever the function has a jump discontinuity [Z, p 61].

In many applications Gibbs phenomenon represents an undesirable effect. It is fortunate that we already have tools which can be used to avoid it. In fact, it does not occur if we replace the partial sums of the Fourier series by either the Cesàro or the Abel means discussed in the last section. We have the following theorem [Z, p. 87].

THEOREM 4.2
Suppose that K_n is a positive delta sequence satisfying (4.9) and $m \leq f(x) \leq M$ for $x \in (a, b)$. Then for any $\varepsilon > 0$ and $0 < \delta < (b-a)/2$ there is an N such that $\sigma_n = K_n * f$ satisfies

$$m - \varepsilon \leq \sigma_n(x) \leq M + \varepsilon$$

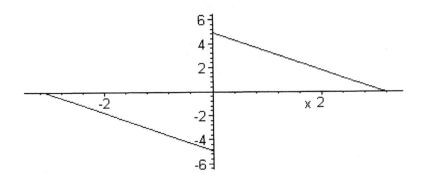

FIGURE 4.3
The saw tooth function.

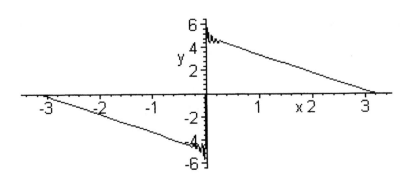

FIGURE 4.4
Gibbs phenomenon for Fourier series; approximation to the
saw tooth function using Dirichlet kernel.

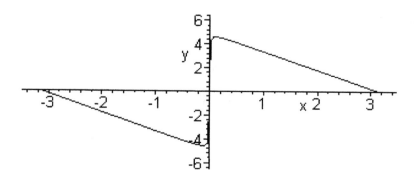

FIGURE 4.5
The approximation to the saw tooth function using Fejer kernel.

whenever $x \in (a + \delta, b - \delta)$ *and* $n > N.$

Thus if f has a jump discontinuity in the middle of the interval (a,b), there cannot be more than an ε overshoot. But since ε is arbtrarily small, there cannot be a Gibbs phenomenon for these means σ_n.

4.4   Periodic distributions

In Chapter 2, we mentioned periodic distributions and their Fourier series briefly. Here we enlarge on these results following an approach given in [K1]. The coefficients, given by (2.6), are independent of t since their derivatives are zero. Since each $f \in \mathcal{S}'$ may be expressed as

$$f = D^p F,$$

where F is a continuous function of polynomial growth, (2.6) may be written as

$$c_n = \frac{1}{2\pi} \left[T_{2\pi}(D^p F e^{-int})^{(-1)} - (D^p F e^{-int})^{(-1)} \right]. \qquad (4.14)$$

Since f is periodic it follows that $T_{2\pi} F - F$ is a polynomial R of degree less than p. We use this and integration by parts to get a bound on c_n.

Indeed

$$\left(D^p F e^{-int}\right)^{(-1)} = \sum_{j=0}^{p-1} D^j F(in)^{p-1-j} e^{-int} + (in)^p \left(F(t)e^{-int}\right)^{(-1)}$$

and hence c_n is given by

$$c_n = \frac{1}{2\pi} \left[\sum_{j=0}^{p-1} D^j R(in)^{p-1-j} e^{-int} + (in)^p \int_t^{t+2\pi} F(x)e^{-inx} dx \right]$$

from which it follows that

$$|c_n| \le \frac{1}{2\pi} \sum_{j=0}^{p} C_j(t)|n|^j, \quad |n| \ge 1,$$

where $C_j(t)$ is bounded on compact sets. Hence since c_n is constant, $c_n = 0(|n|^p)$.

The converse is also true; each series $\sum c_n e^{int}$ where $c_n = 0(|n|^p)$ converges in $\mathcal{S}'$ to a periodic distribution. This is fairly easy to show since

$$\sum_n c_n(e^{int}, \phi) = \sum_n c_n \hat{\phi}(-n) \tag{4.15}$$

for $\phi \in \mathcal{S}$. But $\hat{\phi} \in \mathcal{S}$ also and hence is rapidly decreasing so that the series (4.15) converges. Since convergence in $\mathcal{S}'$ is weak convergence, the limit must be an element $f \in \mathcal{S}'$.

We have proved parts of

THEOREM 4.3
Let $\{c_n\}_{n=-\infty}^{\infty}$ be a sequence of complex numbers; then the following are equivalent:

    (i) there is a $p \in \mathbb{Z}$ such that $c_n = 0(|n|^p)$,

    (ii) $\sum_n c_n e^{int}$ converges in the sense of $\mathcal{S}'$,

    (iii) the c_n are the Fourier coefficients of some $f \in \mathcal{S}'$.

PROOF   We have shown that $(i) \Rightarrow (ii)$ and $(iii) \Rightarrow (i)$. Thus it remains to show $(ii) \Rightarrow (iii)$. We first observe that the operations of translation and multiplication by e^{int} are continuous in $\mathcal{S}'$. Antidifferentiation is also continuous in the sense that for some constants a_n,

$f_n \to f$ in $\mathcal{S}'$ implies that $f_n^{(-1)} + a_n \to f^{(-1)}$ in $\mathcal{S}'$. Thus, the Fourier coefficients of

$$f(t) = \sum_n c_n e^{int} = \lim_{N \to \infty} \sum_{|n| \leq N} c_n e^{int}$$

are given by

$$c_n' = \frac{1}{2\pi} \int_t^{t+2\pi} f(t) e^{-int} dt,$$

which is to be interpreted as (2.6). Because of the continuity of the various operations

$$c_n' = \lim_{N \to \infty} \frac{1}{2\pi} \int_t^{t+2\pi} \sum_{|k| \leq N} c_k e^{ikt} e^{-inx} dx$$

$$= \lim_{N \to \infty} \begin{cases} c_n\,, n \leq N \\ 0, n > N \end{cases} = c_n. \qquad \square$$

There is still one property to check in order to eliminate the distinction between a trigonometric and a Fourier series. If (iii) is satisfied, then by (ii), its Fourier series must converge to some $g \in \mathcal{S}'$, which may not be the same as the f we started with.

Proposition 4.4
Let f be a periodic distribution whose Fourier coefficients are all zero; then $f = 0$.

PROOF   Since $c_0 = 0$, f has a periodic antiderivative. Moreover, its p-th order antiderivative F, the continuous function in (4.14), may be taken to be periodic. If not, we could add a polynomial P of degree $\leq p - 1$ and observe that

$$T_{2\pi}(F + P) - (F + P) = T_{2\pi}F - F + T_{2\pi}P - P = R + T_{2\pi}P - P.$$

Now R is a polynomial of degree $\leq p - 2$ since $T_{2\pi}F^{(p-1)} - F^{(p-1)}$ is the periodic antiderivative of f. Hence P may be chosen such that the polynomial

$$T_{2\pi}P - P$$

of degree $\leq p - 2$ is equal to $-R$.

If F is periodic, then the coefficients of f are given by

$$c_n = \frac{(in)^p}{2\pi} \int_t^{t+2\pi} F(x) e^{-inx} dx.$$

Thus, the Fourier coefficients of F are all zero except possibly for the 0-th coefficient. This implies that F is a constant and if $p > 0$, $f = D^p F = 0$. If $p = 0$, f is an ordinary continuous function, and the result is also true. $\square$

Since the periodic distribution $f \in \mathcal{S}'$, it must have a Fourier transform, and the transform of the series must converge to it:

$$\hat{f}(w) = \sum_n c_n \widehat{(e^{int})}(w) = 2\pi \sum_n c_n \delta(w - n). \tag{4.16}$$

Thus the support of $\hat{f}$ must be on the set of integers. In particular for $f = \delta^*$, we have

$$\delta^*(t) = \frac{1}{2\pi} \sum_n e^{int},$$

and hence

$$\hat{\delta}^*(w) = \frac{2\pi}{2\pi} \sum_n \delta(w - n) \tag{4.17}$$

$$= \sum_n 2\pi\, \delta(2\pi w - 2\pi n) = 2\pi\, \delta^*(2\pi w).$$

If we now take the value of both sides on a test function $\phi \in \mathcal{S}$ we obtain

$$(\hat{\delta}^*, \phi) = \sum_n (\delta(\cdot - n), \phi) = \sum_n \phi(n)$$

$$= (\delta^*, \hat{\phi}) = \sum_k \hat{\phi}(2\pi k).$$

This result is one form of the Poisson summation formula.

4.5 Problems

1. For which α does $f(x) = |x|^{1/2}$ satisfy a uniform Lipschitz condition of order α on $(-\pi, \pi)$?

2. Let $a_n \to 0$ and $a_1 \geq a_2 \geq \ldots \geq a_n \geq \ldots \geq 0$. Show the series $\sum_{n=1}^{\infty} a_n \sin nx$ converges for all x. (Hint: Use the summation by parts formula.

$$\sum_{n=1}^{N} a_n u_n = \sum_{n=1}^{N-1} (a_n - a_{n+1})U_n + a_N U_n,$$

where $U_n = \sum_{n=1}^{n} u_n$, and observe that for $u_n = \sin nx$, U_n is bounded.)

3. Let $\{a_n\}$ be as in Problem 2. Show that $\sum_{n=1}^{\infty} a_n \cos nx$ converges for $x \in (0, 2\pi)$.

4. Take $a_n = n^{-1/2}$ in Problem 2. Why doesn't the result of Problem 2 contradict Bessel's inequality?

5. Find the Fourier series of $f(x) = \chi_{[-1,1]}(x)$ on $(-\pi, \pi)$. Show that none of the even terms in the series are zero at $x = \frac{\pi}{2}$. Does this contradict the localization principle?

6. Find the C–1 means of the Fourier series of $\delta^*(4.5)$ and show they converge to 0 for $x \neq 0$, $x \in (-\pi, \pi)$.

7. Find the Abel means of the Fourier series of $\delta^{*\prime}$ and show they converge to zero for $x \neq 0$, $x \in (-\pi, \pi)$ as $r \to 1$.

8. Let $f(x)$ be the jump function whose Fourier series is given by (4.12). Let $\sigma_n(x)$ denote the $C - 1$ means of $f(x)$. Show that for this function,

$$\sigma_n(x) = \pi \int_0^x F_n(t)dt - \frac{x}{2}, \quad 0 < x < \pi.$$

Use this to show that $|\sigma_n(x)| \leq \frac{\pi}{2}$ and hence Gibbs phenomenon fails to hold.

9. Let $f(x)$ be as in Problem 8. Show that $\delta^*(x) - \frac{1}{2} = Df(x)$. Use this to find a continuous second order antiderivative of $\delta^*(x)$.

10. Let $f(x)$ be a function on $\mathbb{R}$ such that $f(x) = O((1 + |x|)^{-1-\epsilon})$ and $\hat{f}(w) = O((1 + |w|)^{-1-\epsilon})$ for some $\epsilon > 0$. Show that $\hat{f}^*(w) = \sum_{k=-\infty}^{\infty} \hat{f}(w+2\pi k) \in L^1(-\pi, \pi)$ and find an expression for its Fourier series.

11. Use the result in Problem 10 to prove a general form of the Poisson summation formula

$$\sum_k \hat{f}(w + 2k\pi) = \sum_n f(-n)e^{iwn}.$$

12. Repeat Problem 11 but with $f(x)$ instead of $\widehat{f}(w)$ to show that

$$\sum_n f(x - n) = \sum_k \widehat{f}(2k\pi)e^{2\pi ikx}.$$

Chapter 5

Wavelets and Tempered Distributions

Just as with other orthogonal systems, it is possible to extend the expansions in orthogonal wavelets from $L^2(\mathbb{R})$ to a certain class of tempered distributions. Since we are supposing in general that the scaling function ϕ is r-regular, i.e., $\phi \in \mathcal{S}_r$, we might expect to obtain expansions for $f \in \mathcal{S}'_r$ as indeed we do. Since the mother wavelet $\psi \in \mathcal{S}_r$ if ϕ is, the expansion coefficients with respect to both $\{\phi(t-n)\}$ and $\{\psi(t-n)\}$ are well defined. In fact, since $f \in \mathcal{S}'_r$ is characterized by

$$f = D^r\{(t^2+1)^k \mu\},$$

where μ is a bounded measure on $\mathbb{R}$, r, $k \in \mathbb{Z}$, the coefficients may be found to satisfy

$$
\begin{aligned}
a_n &= (f, \phi(\cdot - n)) \\
&= (D^r\{(t^2+1)^k \mu\}, \phi(\cdot - n)) \\
&= \int_{-\infty}^{\infty} (t^2+1)^k (-1)^r \phi^{(r)}(t-n) d\mu.
\end{aligned}
\tag{5.1}
$$

Since $(t^2+1)^k \leq (|t|+1)^{2k}$ and since $|t|+1 \leq (|n|+1)(|t-n|+1)$ it follows that

$$|a_n| \leq \int_{-\infty}^{\infty} (|n|+1)^{2k} (|t-n|+1)^{2k} |\phi^{(r)}(t-n)| d|\mu| = O(|n|^{2k}).$$

Similarly $b_n = (f, \psi(t-n))$ satisfies the same sort of growth condition.

The coefficients of $\theta \in \mathcal{S}_r$ can similarly be calculated as

$$\int_{-\infty}^{\infty} \theta(t)\phi(t-n)dt = O(|n|^{-p}), \qquad \text{for all } p \in \mathbb{Z}.$$

Thus the expansion of f converges in the sense of $\mathcal{S}'_r$ to some $f_0 \in \mathcal{S}'_r$, i.e.,

$$\sum_n a_n \int \phi(t-n)\theta(t)dt = (f_0, \theta).$$

But the convergence of the expansion of f is much stronger than this since $\sum a_n \phi(t-n)$ converges uniformly on bounded sets as do its first r derivatives. This follows from the inequality

$$\sum_n |a_n \phi^{(j)}(t-n)| \leq C \sum_n \frac{(|n|+1)^k}{(|t-n|+1)^{k+2}} \tag{5.2}$$

$$\leq C \sum_n \sum_m \binom{k}{m} \frac{(|n-t|+1)^{k-m}|t|^m}{(|n-t|+1)^{k+2}}$$

$$\leq C \sum_n \frac{1}{(|n-t|+1)^2}(1+|t|)^k$$

$$\leq C'(1+|t|)^k.$$

In fact we have shown that the limiting function and its r derivatives are continuous functions of polynomial growth on $\mathbb{R}$. These results enable us to imitate the multiresolution analysis of $L^2(\mathbb{R})$ in $\mathcal{S}'_r$.

We can also turn things around and try to expand functions in wavelets composed of distributions. Rather than requiring $\phi \in \mathcal{S}_r$, we suppose merely that $\phi \in S'$. Then, of course, we cannot orthogonalize ϕ in the sense of $L^2(\mathbb{R})$ but must use another Hilbert space. An appropriate one that works in certain cases is the Sobolev space H^{-s}. One scaling distribution we consider will be composed of linear combinations of delta functions. In this case, we again obtain a version of a multiresolution analysis. In the more general case, where we consider distribution solutions of dilation equations, we are unable to obtain an orthogonal multiresolution and so only have partial results.

5.1   Multiresolution analysis of tempered distributions

In this section we develop the program described at the start above, namely we assume that $\phi \in \mathcal{S}_r$, and extend the multiresolution analysis from $L^2(\mathbb{R})$ to S_r'. Most of the results are straightforward extensions, but there are a few differences.

Definition 5.1 *Let $\{a_n\}$ be a sequence of complex numbers such that $a_n = O(|n|^k)$ for some $k \in \mathbb{Z}$; then $T_0 := \{f | f(t) = \sum_n a_n \phi(t - n)\}$ and $U_0 := \{g | g(t) = \sum_n a_n \psi(t - n)\}$.*

Hence both T_0 and U_0 are composed of functions in C^r of polynomial growth and are therefore in S_r'. We denote by T_m and U_m their corresponding dilation spaces ($f \in T_0 \Leftrightarrow f(2^m t) \in T_m$). Then we may expect that a multiresolution analysis of S_r' exists, namely,

$$\cdots \subset T_{-1} \subset T_0 \subset T_1 \subset \cdots \subset T_m \subset \cdots S_r' \tag{5.3}$$

and

$$\overline{\bigcup_m T_m} = S_r' \,,$$

where the closure is in the topology of S_r', but we must check a few things. Since $\phi(t) \in V_1 \subset T_1$ where V_1 is part of the $L^2(\mathbb{R})$ multiresolution analysis, it follows that all partial sums $\sum_{|n| \leq N} a_n \phi(t - n) \in T_1$.

However, the corresponding series may not be in T_1 unless the dilation equation coefficients c_k in (3.3) are sufficiently rapidly decreasing. We therefore assume $c_k = O(|k|^{-m})$ for all $m \in \mathbb{Z}$, i.e., $\{c_k\}$ is rapidly decreasing. Then the series becomes

$$\sum_n a_n \phi(t - n) = \sum_n a_n \sum_k \sqrt{2} c_k \phi(2t - 2n - k) \tag{5.4}$$

$$= \sum_n a_n \sum_j c_{j-2n} \sqrt{2} \phi(2t - j)$$

$$= \sum_j \left(\sum_n a_n c_{j-2n} \right) \sqrt{2} \phi(2t - j)$$

$$= \sum_j a_j^1 \sqrt{2} \phi(2t - j).$$

Clearly if a_n is of polynomial growth ($= O(|n|^p)$), so is a_n^1.

It can also be shown that T_0 is closed in the sense of S_r'. This follows from the fact that $f_m \to 0$ in S_r' is equivalent to

$$f_m = D^r \left[(1 + t^2)^p \mu_m \right] \quad \text{and} \quad \int d|\mu_m| \to 0 \quad [\text{Ze}] \,,$$

where $\{\mu_m\}$ is a sequence of bounded measures on $\mathbb{R}$. Hence by (5.1) we deduce that $a_{mn} = (f_m, \phi(\cdot - n)) \to 0$ and $|a_{mn}| \leq C(n^2 + 1)^p$. From this it follows that if $f_m \to f$ in S_r', the coefficients of f are of polynomial growth and hence its series $\sum_n a_n \phi(t - n) \in T_0$.

Just as in the L^2 case, we can express each $f_1 \in T_1$ uniquely as $f_1 = f_0 + g_0$ where $f_0 \in T_0$ and $g_0 \in U_0$, in spite of the fact that we do not have orthogonality. Indeed since

$$a_n^0 = (f_1, \phi(t - n)) = \sum_k c_k a_{2n+k}^1 = \sum_j a_j^1 c_{j-2n}, \qquad (5.5a)$$

and

$$b_n^0 = (f_1, \psi(t - n)) = \sum_k (-1)^k c_{1-k} a_{2n+k}^1 \qquad (5.5b)$$

$$= \sum_j a_j^1 c_{2n+1-j}(-1)^j,$$

we have the same decomposition algorithm as in the L^2 case (3.20). Similarly the reconstruction algorithm is given by (3.19) using the same argument.

Thus we use (5.5a) and (5.5b) to define f_0 and g_0 respectively and (3.6) to deduce that $f_1 = f_0 + g_0$. We summarize in the following.

THEOREM 5.1
Let the scaling function $\phi \in S_r$ satisfy a dilation equation (3.3) with $c_k = O(|k|^{-p})$ for all $p \in \mathbb{N}$; let ϕ have an associated multiresolution analysis in $L^2(\mathbb{R})$; let $\psi \in S_r$ be the mother wavelet given by (3.16). Then there exists a multiresolution analysis (5.3) of closed dilation subspaces $\{T_m\}$ whose union is dense in S_r'; the closed subspaces U_m (of Definition 5.1) are complementary subspaces of T_m in T_{m+1} and

$$T_m = U_0 \oplus U_1 \oplus \cdots \oplus U_m \oplus T_0$$

where $\oplus$ denotes the nonorthogonal direct sum.

The only thing we have not shown is that $\bigcup_m T_m$ is dense in S_r'. This follows from the fact that $L^2(\mathbb{R})$ is dense in S_r' and $\bigcup_m V_m \subset \bigcup_m T_m$.

The one property of the usual multiresolution analysis that is lacking is the intersection property $\bigcap_m V_m = \{0\}$. By the moment property of the reproducing kernel $q_m(x, t)$,

$$\int_{-\infty}^{\infty} q_m(x, t) t^k dt = x^k, \qquad 0 \leq k \leq r,$$

it follows that any polynomial of degree $\leq r$ belongs to each T_m and hence to $\bigcap_m T_m$. Thus, the pure wavelet series of $f \in \mathcal{S}'_r$,

$$\sum_{m,n} b_{mn} \psi_{mn}(t),$$

does not necessarily converge to f. However, the mixed expansion,

$$f = \sum_n a_n \phi(\cdot - n) + \sum_{m=0}^{\infty} \sum_n b_{mn} \psi_{mn},$$

does converge to f in the sense of $\mathcal{S}'_r$.

This theorem lets us consider, in particular, the wavelet expansions of polynomials since they all belong to $\mathcal{S}'_r$. These play a central role in many wavelet calculations such as those involving the moment properties of both scaling functions and mother wavelets. Yet they do not properly belong to the usual multiresolution subspaces, and have to be given special considerations in the standard theory.

5.2 Wavelets based on distributions

The two methods we have for constructing scaling functions and wavelets, the method based on dilation equations, and the one based on orthogonalization, can be extended to distributions as well. The latter, however, requires a Hilbert space structure, for which we take the Sobolev spaces H^{-s}, $s > \frac{1}{2}$. The dilation equation approach does not require such a structure, at least in the first steps, and we treat it first.

5.2.1 Distribution solutions of dilation equations

The solution to the dilation equation (3.3)

$$\phi(t) = \sqrt{2} \sum_k c_k \phi(2t - k)$$

may be considered a fixed point of the map

$$(T\phi)(t) = \sqrt{2} \sum_k c_k \phi(2t - k).$$

This in turn may be expressed as

$$T\phi = D_2(h * \phi)$$

where h is the distribution

$$h(t) = \sqrt{2} \sum_k c_k \delta(t - k), \tag{5.6}$$

and D_2 is the dilation operator $D_2\phi = \phi(2\cdot)$.

We assume, as we have in the last section, that $\{c_k\}$ is rapidly decreasing, real, and that

$$\frac{1}{\sqrt{2}} \sum_k c_k = 1. \tag{5.7}$$

Then T is well defined on S' and maps S' into S'. This can be seen by taking the Fourier transform of $h * \phi$, which is

$$(\widehat{h} \cdot \hat{\phi})(w) = \sqrt{2} \sum_k c_k e^{-ikw} \hat{\phi}(w) = 2m_0(w)\hat{\phi}(w). \tag{5.8}$$

Since $c_k = O(|k|^{-p})$ for all $p \in \mathbb{N}$, $m_0(w) \in C^\infty$ and is periodic and hence is a multiplier in S'. The series in (5.8) converges in S' and hence its inverse Fourier transform does too.

THEOREM 5.2
Let T be the operator on S' given by $T\phi = D_2(h * \phi)$ where h is as in (5.6) and $\{c_k\}$ is rapidly decreasing and satisfies (5.7). Then there exists a unique $g \in S'$ such that $Tg = g$ where $\hat{g}$ is continuous and of polynomial growth, and $\hat{g}(0) = 1$.

The proof is similar to that in [V] and in [C-H] under different hypotheses. The operator T under the Fourier transform maps into

$$(\widehat{T\phi})(w) = m_0\left(\frac{w}{2}\right)\hat{\phi}\left(\frac{w}{2}\right).$$

If we iterate $\hat{T}$ k times, we obtain

$$\hat{T}^k\hat{\phi}(w) = \prod_{j=1}^{k} m_0(2^{-j}w)\hat{\phi}(2^{-k}w) = G_k(w)\hat{\phi}(2^{-k}w).$$

Since $m_0 \in C^\infty$ and $m_0(0) = 1$ by (5.7), it follows that for $K > 0$, there is a constant C_1 such that

$$|m_0(w) - 1| \le C_1 |w|, \qquad -K \le w \le K$$

and hence

$$\left| \sum_{j=m+1}^{n} m_0(2^{-j}w) - 1 \right| \le C_1 K 2^{-m} \sum_{j=1}^{n-m} 2^{-j} < C_1 K 2^{-m}$$

for $0 < m < n$, $-K \le w \le K$. From this we deduce that $\{G_k(w)\}$ is uniformly Cauchy on $[-K, K]$ and hence converges uniformly to a continuous function $G(w)$, and

$$\hat{T}G(w) = G(w), \qquad w \in \mathbb{R}.$$

In order to find the growth of G, we use the procedure in [V]; we choose $C_2 \ge 1$ such that

$$|G_k(w)| \le C_2 \quad \text{for} \quad |w| \le 1, \quad k \in \mathbb{N},$$

and let $C_3 = \frac{1}{\sqrt{2}} \sum |c_n| \ge 1$. Then $|m_0(w)| \le C_3$, and it follows that for $2^{p-1} \le |w| < 2^p$, $k > p > 0$,

$$|G_k(w)| = \left| \prod_{j=1}^{p} m_0(2^{-j}w) \right| |G_{k-p}(2^{-p}w)| \qquad (5.9)$$
$$\le C_3^p C_2$$

and similarly for $k \le p$

$$|G_k(w)| \le C_3^k \le C_3^p C_2.$$

The inequality $2^{p-1} \le |w|$ is now converted to one involving C_3,

$$C_3^{p-1} = 2^{(p-1)M} \le |w|^M$$

where $M = \log_2 C_3 \ge 0$. Thus (5.9) becomes

$$|G_k(w)| \le C_3 C_2 |w|^M, \qquad 1 \le |w|.$$

By taking the limit we conclude that $G(w)$ is of polynomial growth. Hence $G \in \mathcal{S}'$ and its inverse Fourier transform $g \in \mathcal{S}'$ and is a fixed point of T.

Since T is linear, any multiple of g is also a fixed point, and we can take g such that $\widehat{g}(0) = 1$. Then

$$\widehat{g}(w) = \widehat{T}^k \widehat{g}(w) = G_k(w)\widehat{g}(2^{-k}w) \to G(w)\widehat{g}(0) = G(w)$$

uniformly for $-K \le w \le K$. This gives us the desired uniqueness.   □

Corollary 5.1

Let $g(t)$ be the distribution solution to the dilation equation satisfying $\widehat{g}(0) = 1$. Then $g \in H^{-s}$ for $s > \log_2 \Sigma |c_n|$.

The proof follows from the fact that

$$|\widehat{g}(w)| \le C(1 + |w|)^M$$

and hence $g \in H^{-s}$ for $s > M + \frac{1}{2}$.

Although the solution $g \in \mathcal{S}'$, it may not have a "nice" structure. If the dilation equation contains only a finite number of nonzero terms, then g has compact support. If it has only one term, say $c_0 = \sqrt{2}$, then the operator is $(T\phi)(t) = 2\phi(2t)$, and the fixed point may be found explicitly. It is $\phi(t) = \delta(t)$ since $2\delta(2t) = \delta(t)$ and $\widehat{\delta}(0) = 1$.

If exactly two terms are nonzero, say c_0 and c_1, then there is never a distribution solution of the form

$$\phi(t) = \sum_{k=0}^{n} a_k \delta(t - t_k).$$

We first observe that the support of ϕ must be in $[0,1]$, and hence we may take $0 \le t_0 < t_1 < t_2 < \cdots < t_n \le 1$. From the dilation equation we have

$$\sum_{k=0}^{n} a_k \delta(t - t_k) \tag{5.10}$$

$$= \sqrt{2} \sum_{k=0}^{n} \left\{ c_0 \frac{a_k}{2} \delta\left(t - \frac{t_k}{2}\right) + c_1 \frac{a_k}{2} \delta\left(t - \frac{t_k + 1}{2}\right) \right\}.$$

If $t_0 > 0$ then $\sqrt{2} \frac{c_0}{2} a_0 = 0$ and $a_0 = 0$. When $\frac{t_k}{2} < t_k < \frac{t_k + 1}{2}$, the same argument can be repeated to conclude that $a_1 = a_2 = \cdots = a_n = 0$ since the coefficients of $\delta(t - t_k)$ on both sides must be equal. If $t_0 = 0$ then $a_0 = \sqrt{2} c_0 \frac{a_0}{2}$ and $a_0 = 0$ since $c_0 \ne \sqrt{2}$ ($c_0 + c_1 = \sqrt{2}$). We can also extend the argument to the case of two other non-zero coefficients and to more than two non-zero c_k's. In fact we have used only the first sum on the right hand side of (5.10).

Proposition 5.1

The distribution solution to the dilation equation (3.3) has no solution of the form (5.10) if $c_n = 0$ for $|n| \geq N$, but at least two c_n's are not zero.

5.2.2 A partial distributional multiresolution analysis

In Section 5.1, we introduced a multiresolution analysis of S'_r with a ladder of subspaces $\{T_m\}$ each of which was composed of continuous functions of polynomial growth. We now extend this to the distribution solutions of the dilation equation. We define T_0 in a similar way:

$$T_0 = \left\{ f \in S' \mid f = \sum_n a_n \phi(\cdot - n), \quad a_n \in \ell^2 \right\},$$

where ϕ is the solution given in Theorem 5.2. The series $\sum_n a_n \phi(\cdot - n)$ converges in S' since its Fourier transform is given by

$$\sum_n a_n e^{-iwn} \hat{\phi}(w) = \alpha(w) \hat{\phi}(w),$$

where $\alpha \in L^2(0, 2\pi)$ and is 2π periodic, and hence converges in the sense of S'. Its dilation space $T_1 = \{f \in S' \mid f(2^{-1} \cdot) \in T_0\}$ contains T_0 since $f \in T_0$ is given by

$$f = \sum_n a_n \phi(\cdot - n) = \sum_n \left(\sum_k a_k c_{n-2k} \right) \sqrt{2} \phi(2 \cdot - n)$$

by the now familiar calculation. Since $\{c_k\}$ is rapidly decreasing, the inner series converges to a sequence in ℓ^2. Similarly the other dilation spaces T_m are defined and give us the usual ladder of subspaces

$$\cdots \subset T_{-1} \subset T_0 \subset T_1 \subset \cdots \subset S'.$$

However it is not clear that $\bigcup_m T_m$ is dense in S' nor can we define a mother wavelet based on the orthogonal complement since we have no Hilbert space structure. In the next section we consider one case in which such a structure is possible.

5.3   Distributions with point support

In this section we restrict ourselves to scaling functionals which have point support and belong to the Sobolev space H^{-1}. We introduce a multiresolution analysis of H^{-1} composed of dilation subspaces $\{V_m\}$ where

$$V_0 = \{f \in H^{-1} \mid \operatorname{supp} f \subseteq \mathbb{Z}\}.$$

(We revert to the V_m notation since we now have a Hilbert space.) Then clearly

$$\cdots \subset V_{-1} \subset V_0 \subset V_1 \subset \cdots \subset V_m \subset \cdots H^{-1}.$$

It can also be shown that

$$f(t) = \sum_k a_k \delta(t - k), \qquad \{a_k\} \in \ell^2$$

belongs to V_0, and in fact [W1],[W2] characterizes V_0. This also shows that

$$\bigcup_m V_m = H^{-1}$$

since partial sums of dilations of such $f(t)$ are dense in H^{-1}. This comes from the fact that their Fourier transforms are trigonometric polynomials dense in $L^2(\mathbb{R}, (w^2 + 1)^{-1})$. However

$$\bigcap_{m=-\infty}^{\infty} V_m = \{C\delta\} \neq \{0\}$$

since $\delta \in V_m$ for all m.

If we try to define the spaces W_m as in the usual approach in Chapter 3, we are confronted with the fact that H^{-s} is not homogeneous for $s \neq 0$; i.e., dilations are not isometries on H^{-s}. However they are homeomorphisms that map H^{-s} into itself with a modified norm. Hence we introduce the inner product on H^{-1} given by

$$\langle f, g \rangle_{-1,\alpha} := \frac{1}{2\pi} \int_{-\infty}^{\infty} \hat{f}(w)\overline{\hat{g}(w)} \left(w^2 + \alpha^2\right)^{-1} dw, \qquad \alpha > 0.$$

We note that the V_m are the same whatever α is, but the orthogonal complement of V_0 in V_1 depends on α. We denote by $W_{0,\alpha}$ this orthogonal complement with respect to this inner product with $W_{m,\alpha}$ the corresponding dilation spaces.

We now need to construct an orthonormal basis of V_0 composed of translation of some ϕ. A modification of Meyer's procedure does the trick. Let $\phi \in V_0$; then we find

$$\delta_{0k} = \langle \phi, \phi(\cdot - k) \rangle_{-1,\alpha} = \int_{-\infty}^{\infty} |\hat{\phi}(w)|^2 e^{-iwk} \left(w^2 + \alpha^2 \right)^{-1} dw$$

$$= \frac{1}{2\pi} \int_0^{2\pi} \left\{ \sum_n |\hat{\phi}(w + 2\pi n)|^2 \left[(w + 2\pi n)^2 + \alpha^2 \right]^{-1} \right\} e^{-iwk} dw,$$

and δ_{0k} is the Fourier coefficient of the function inside the integral in brackets, which belongs to $L^2(0, 2\pi)$. Hence it must be equal to its Fourier series (in the L^2 sense),

$$\sum |\hat{\phi}(w + 2\pi n)|^2 \left[\sum (w + 2\pi n)^2 + \alpha^2 \right]^{-1} = \sum_k \delta_{0k} e^{iwk} = 1. \quad (5.11)$$

But $|\hat{\phi}|^2$ is periodic (2π) and hence we may solve (5.11) for $|\hat{\phi}|^2$, which is

$$|\hat{\phi}(w)|^2 = \left[\sum_n ((w + 2\pi n)^2 + \alpha^2)^{-1} \right]^{-1}. \quad (5.12)$$

The positive square root of (5.12) is one solution; the general solution is obtained by multiplying it by a unimodular periodic function. We denote the positive solution by $\hat{\phi}(w, \alpha)$,

$$\hat{\phi}(w, \alpha) := \left[\sum_n ((w + 2\pi n)^2 + \alpha^2)^{-1} \right]^{-1/2} \quad (5.13)$$

and shall concentrate on it. We have

Proposition 5.2
Let $\phi(t, \alpha)$ be given by (5.13); then $\{\phi(t - n, \alpha)\}$ is an orthonormal basis of V_0 with the inner product $\langle \, , \, \rangle_{-1,\alpha}$.

The fact that it is complete in V_0 follows easily from the definition.

The next step is to find an orthonormal basis of $W_{0,\alpha}$, i.e., the wavelets themselves. We look for an element $\psi \in W_{0,\alpha}$ whose translates are orthonormal and complete in $W_{0,\alpha}$. This may be translated into conditions similar to (5.11);

(i) $\sum_n \hat{\psi}(w + 2\pi n)((w + 2\pi n)^2 + \alpha^{-2})^{-1} = 0$
(ii) $\sum_n |\hat{\psi}(w + 2\pi n)|^2((w + 2\pi n)^2 + \alpha^2)^{-1} = 1.$ (5.14)

The first condition is a restatement of the fact that $\psi(t)$ is orthogonal to $\delta(t-k)$ for all $k \in \mathbb{Z}$.

Now however $\hat{\psi}(w)$ is periodic (π) since $\psi \in V_1$; thus both series in (5.14) may be simplified by separating the even and odd terms, which gives us

$$\begin{aligned}(i)\ &\hat{\psi}(w)a(w,\alpha) + \hat{\psi}(w+\pi)a(w+2\pi,\alpha) = 0,\\ (ii)\ &|\hat{\psi}(w)|^2 a(w,\alpha) + |\hat{\psi}(w+\pi)|^2 a(w+2\pi,\alpha) = 1\end{aligned} \tag{5.15}$$

where

$$a(w,\alpha) := \sum_n ((w+4\pi n)^2 + \alpha^2)^{-1}. \tag{5.16}$$

These two equations are easily solved for $|\hat{\psi}|^2$,

$$|\hat{\psi}(w)|^2 = \frac{a(w+2\pi,\alpha)}{a(w,\alpha)(a(w+2\pi,\alpha)+a(w,\alpha))},$$

but the positive square root does not satisfy (5.15)(i), and must be multiplied by $e^{-iw/2}$ in order to do so. We define

$$\hat{\psi}(w,\alpha) := e^{-iw/2}\left[\frac{a(w+2\pi,\alpha)}{a(w,\alpha)(a(w+2\pi,\alpha)+a(w,\alpha))}\right]^{1/2}. \tag{5.17}$$

Proposition 5.3
Let $\psi(t,\alpha)$ be given by (5.17); then $\{\psi(t-n,\alpha)\}$ is an orthonormal basis of $W_{0,\alpha}$.

Again the completeness is straightforward. To find an orthonormal basis of the dilation space V_1, we choose α' such that the map $D_2 f = \sqrt{2}f(2\cdot)$ is a multiple of an isometry from V_0 to V_1. The proper choice is $\alpha' = 2\alpha$, since

$$\|f\|_{-1,\alpha} = \frac{1}{2}\|D_2 f\|_{-1,2\alpha},$$

and hence $\{2^{3/2}\phi(2t-n,\alpha'1/2)\}$ is an orthonormal basis of V_1.

We now start with $\alpha' = 1$ in each V_m and find the corresponding α in V_0 so that we can get an orthonormal basis of V_m. This gives us

$$\phi_{m,n}(t) = 2^{3m/2}\phi(2^m t - n, 2^{-m}), \qquad n = 0,\pm1,\cdots. \tag{5.18}$$

For the orthogonal complement we begin with $W_0 = W_{0,1},\cdots, W_m = W_{m,1},\cdots$ and thereby obtain similarly

$$\psi_{m,n}(t) = 2^{3m/2}\psi(2^m t - n, 2^{-m}), \qquad n = 0,\pm1,\cdots. \tag{5.19}$$

Corollary 5.2

Let $\phi_{m,n}(t)$ and $\psi_{m,n}(t)$ be given by (5.18) and (5.19) respectively; then $\{\phi_{0,n}(t)\}_{n=-\infty}^{\infty}$ and $\{\psi_{m,n}(t)\}_{m=0,\ n=-\infty}^{\infty}$ form an orthonormal basis of H^{-1} with the usual inner product.

Hence we have the usual multiresolution analysis

$$H^{-1} = V_0 \oplus \bigoplus_{m=0}^{\infty} W_m = \{\delta\} \oplus \bigoplus_{m=-\infty}^{\infty} W_m \qquad (5.20)$$

where $\{\delta\}$ is the space spanned by the δ function.

Our theory is almost complete. We still need to get simpler expressions for $\psi(t, \alpha)$ by summing the series for $a(w, \alpha)$ of (5.16). We begin with the fact that

$$\sum_{n=-\infty}^{\infty} \frac{1}{x + 2\pi n} = \frac{1}{2} \frac{\cos \frac{x}{2}}{\sin \frac{x}{2}}$$

from the Fourier series of e^{-ixt} on $(0,1)$ evaluated at $t = 0$. Then

$$
\begin{aligned}
a(w, \alpha) &= \frac{i}{4\alpha} \sum_n \frac{1}{\frac{w+\alpha i}{2} + 2\pi n} - \sum \frac{1}{\frac{w-\alpha i}{2} + 2\pi n} \\
&= \frac{i}{8\alpha} \left\{ \frac{\cos(w + i\alpha)/4}{\sin(w + i\alpha)/4} - \frac{\cos(w - \alpha i)/4}{\sin(w - \alpha i)/4} \right\} \\
&= \frac{i}{8\alpha} \frac{\sin(-i\alpha/2)}{\sin^2 w/4 \cos^2 \alpha i/4 - \cos^2 w/4 \sin^2 \alpha i/4} \\
&= \frac{1}{4\alpha} \frac{\sinh \alpha/2}{\cos \alpha/2 - \cos w/2}.
\end{aligned}
$$

This is substituted into (5.17) to obtain

$$\hat{\psi}(w, \alpha) = e^{-iw/2}(\cosh \alpha/2 - \cos w/2)\sqrt{\frac{4\alpha}{\sinh \alpha}} \qquad (5.21)$$

after a little manipulation. The inverse Fourier transform is easily found to be, for $\alpha = 2^{-m}$,

$$\psi(t, 2^{-m}) = C \left\{ \cosh(2^{-m-1})\delta(t - \frac{1}{2}) - \frac{1}{2}\delta(t) - \frac{1}{2}\delta(t - 1) \right\} \qquad (5.22)$$

where C is a positive constant.

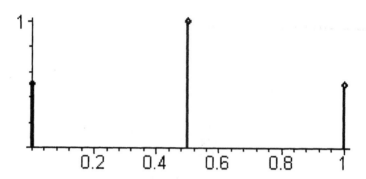

FIGURE 5.1
A mother wavelet with point support. The vertical bars represent delta functions.

We can also find $\phi(t, \alpha)$ to be given by

$$\hat{\phi}(w, \alpha) = \left[\frac{2\alpha(\cosh\alpha - \cos w)}{\sinh\alpha}\right]^{1/2}.$$

However this leads to an infinite Fourier series whose sum is not as simple [W1]. This really doesn't matter that much since the important result is (5.22), which gives an orthonormal basis of H^{-1} each of whose elements has support on three points (or fewer in the case of $\delta(t)$).

5.4 Approximation with impulse trains

In this section, rather than trying to find an orthogonal basis of the space V_0 of the last section (composed of sequences of delta functions on the integers, an impulse train), we consider direct approximations of functions by such sequences. Our impulse trains will be of a special kind, those whose coefficients are the sampled values of the functions, i.e., we want to determine in what sense a continuous function $f(t)$ can be approximated by

$$f^*(t) = \sum_n f(n)\delta(t - n) \tag{5.23}$$

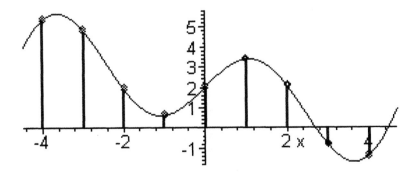

FIGURE 5.2
A continuous function and its impulse train.

and its dilations. As we shall see in Chapter 9, these are related to the
Shannon sampling theorem. The process of going from $f(t)$ to $f^*(t)$
corresponds to converting an analog signal to a digital signal. These
impulse trains in most cases belong to the subspace V_0 of Section 5.3.

We first address the question of equality of $f(t)$ and $f^*(t)$. Clearly
they are not equal pointwise, but may be equal in some other sense. In
fact if $f \in L^1(\mathbb{R})$ and $\{f(n)\} \in l^1$, then they may be equal in some
average sense:

$$\int f(t)dt = \sum_n f(n) \int \delta(t-n)dt = \sum_n f(n) \qquad (5.24)$$

obtained by integrating over the real line. But it is easy to find coun-
terexamples to see that this is not always true. (Take a continuous
function which is non-negative but is zero on the integers.) On the
other hand we have

Proposition 5.4
  *Let $f \in L^1(\mathbb{R})$ and be continuous, let $\{f(n)\} \in l^1$; then (5.24) holds if
and only if the Fourier transform of f satisfies $\widehat{f}(2\pi k) = 0, k \neq 0, k \in \mathbb{Z}$.*

The proof involves using the filtering property of δ, $f(t)\delta(t-a) = f(a)\delta(t-a)$, and the Fourier series of the periodic extension of δ,

$$\delta^*(t) := \sum_n \delta(t-n) = \sum_k e^{2\pi kit}$$

Then the integral of the difference

$$\int [f(t) - \sum_n f(n)\delta(t - n)]dt$$

$$= \int f(t)(1 - \sum_n \delta(t - n))dt$$

$$= \int f(t)(1 - \sum_k e^{2\pi kit})dt$$

$$= \widehat{f}(0) - \sum_k \widehat{f}(-2\pi k)$$

$$= \widehat{f}(0) - \widehat{f}(0) = 0,$$

which completes the proof.

Such a function generates a *partition of unity*, i.e., $\sum_n f(t - n) = 1$. (See [W23] for more details.)

However we are interested in the degree of approximation rather than these averages. We might expect that if we sample more frequently we would get a better approximation. That is, we wish to determine the sense in which

$$\varepsilon_m(t) = f(t) - \sum_n f(n2^{-m})\delta(2^m t - n) \qquad (5.25)$$

converges to 0 as $m \to \infty$. Clearly the usual pointwise and mean square types of convergence cannot hold because of the presence of the delta function. But there are other ways to define convergence for which it does make sense.

We consider two types of convergence for which it does make sense, the first of which is convergence in the sense of S'. Clearly if f and $\widehat{f}$ are both in $L^1(\mathbb{R})$, then $\varepsilon_m \in S'$ since $f(n2^{-m})$ must be a bounded sequence in n. Then we have, for $\theta \in S$,

$$(\varepsilon_m, \theta) = \int f(t)\theta(t)dt - \sum_n f(n2^{-m})\theta(t - n2^{-m})2^{-m} \qquad (5.26)$$

which must be shown to be convergent to 0 as $m \to \infty$. This can be shown by first truncating both the integral and sum to finite values and then using the properties of the Riemann sum of the integral.

The second type of convergence is in the sense of the Sobolev space $H^p, p \in \mathbb{R}$. We have the following result whose proof may be found in [W23].

Proposition 5.5

Let $f \in H^3$, and let $f(t) = O(|t|^{-1-\alpha})$ for some $\alpha > 0$; then the series in (5.25) converges in H^{-1} and the error satisfies

$$\|\varepsilon_m\|_{-1} = O(2^{-m}).$$

Other properties of these series may also be found in [W23] as well.

5.5   Problems

1. Find the scaling function expansion $\sum_n a_n \phi(t - n)$ of $\delta(t - a)$ for any 1-regular ϕ.

2. Repeat Problem 1 for $t^j \delta^{(k)}(t - a)$ $k \leq r$ for any r-regular ϕ.

3. Use the result in Problem 2 for $k = 0$, to show that

$$(t^j - a^j) q_m(a, t) \to 0 \quad \text{as} \quad m \to \infty$$

weakly (i.e., in the sense of $\mathcal{S}'$).

4. For 1-regular ϕ, the function $t \in T_0$. (Why?) Show that its expansion is

$$t = \sum (n + \mu) \phi(t - n),$$

where $\mu = \int\limits_{-\infty}^{\infty} t\phi(t) dt$.

5. Conjecture and prove a similar result to Problem 4 for t^k, $k \leq r$ when ϕ is r-regular.

6. Find the approximation to $\delta(t - a)$ in terms of its expansion in $T_0 \oplus U_0 \oplus U_1$.

7. Let the coefficients in the dilation equation be $c_0 = c_1 = c_2 = c_3 = \frac{1}{2\sqrt{2}}$. Find a closed form expression for $m_0(w)$. Then use the argument of Theorem 5.2 to show that $\hat{\phi}(w)$ is bounded.

8. It can be shown [C-H] that $\phi(t) \in L^2(\mathbb{R})$ for ϕ of Problem 7. Show that $\{\phi(t-n)\}$ is not orthogonal however.

9. Find the expansion coefficients of $f(x) = e^{-|x|}$ with respect to the wavelet basis $\{\psi_{m,n}\}$ of Corollary 5.7 in H^{-1}. (Caution: this uses the Sobolev inner product; you will need the Fourier transform $\hat{f}(w) = 2/(1+w^2)$.)

10. Find the impulse train series (5.25) of $f(x) = e^{-|x|}$ for any m and show the series converges in H^{-1}.

Chapter 6

Orthogonal Polynomials

Just as with trigonometric series, there is a large body of literature on orthogonal polynomials. The most complete source is Szegö's book [Sz], which in particular gives a detailed exposition of Jacobi polynomials (see also [Ca], [As], and [Jc]). In this chapter we present elements of the theory concentrating on those parts that are useful in applications or which are related to topics covered in other chapters. We first present some of the general theory including interpolating polynomials and then take up the classical Jacobi, Laguerre, and Hermite systems.

6.1 General theory

We begin with a nonnegative (Borel) measure μ on the interval (a, b) such that $\mu(a, b) < \infty$. The space in which we now work is $L^2(\mu; (a, b))$ consisting of all functions such that

$$\int_a^b |f|^2 d\mu < \infty.$$

The notation $\langle f, g \rangle_\mu$ will now mean the inner product in this space with $\|f\|_\mu$ denoting the norm. In most cases $d\mu$ will be given by a weight function

$$d\mu(x) = w(x)dx,$$

although in some discrete cases we assume it to be

$$d\mu(x) = \sum a_i \delta(x - x_i)dx.$$

In order that polynomials belong to the space, we shall also require that the moments $\int_a^b x^n d\mu$, $n = 0, 1, \cdots$, exist. This is no problem if (a, b) is a finite interval but it may be if it is infinite.

There generally exist many orthogonal polynomial systems associated with μ, but we shall usually consider systems where $P_n(x)$ is of degree n. This may be ensured by orthogonalizing the sequence

$$1, x, x^2, \cdots, x^n, \cdots.$$

That is, we let $P_0(x) = 1$ and then subtract from x its projection on the space spanned by $P_0(x)$. This gives us $P_1(x)$. We then subtract from x^2 its projection on the space spanned by $P_0(x)$ and $P_1(x)$ to get $P_2(x)$, etc. This will give us a sequence $\{P_n\}$ of polynomials that are of exact degree n and orthogonal with respect to $d\mu$. They may be made into an orthonormal system by normalization

$$p_n(x) = P_n(x)/\|P_n\|_\mu.$$

If the interval (a, b) is finite, and the support of μ is an infinite set, the orthonormal system is always complete. This follows from the Weierstrass approximation theorem [Sz, p.5], which says that a continuous function on $[a, b]$ can be uniformly approximated by polynomials. Since each $f \in L^2(\mu; (a, b))$ can be approximated by continuous functions in the sense of the norm $\| \ \|_\mu$, it follows that f can be approximated by polynomials in the sense of $\| \ \|_\mu$. Furthermore, since the partial sums of the orthogonal polynomial expansion gives the best approximation (Chapter 1) to f, it follows that

$$\left\| f - \sum_{n=0}^{N} \langle f, p_n \rangle_\mu p_n \right\|_\mu \to 0$$

as $N \to \infty$ and hence $\{p_n\}$ is complete.

If the support of μ is a finite set, as happens in interpolation problems, then $1, x, \cdots, x^n, \cdots$ spans a finite dimensional space, and the sequence of orthogonal polynomials terminates after a finite number.

Some examples of orthogonal polynomials (which are usually not normalized) are:

1 The Legendre polynomials $\{P_n(x)\}_{n=0}^\infty$, $a = -1$, $b = 1$ and $d\mu(x) = dx$.

2 The Hermite polynomials $\{H_n(x)\}_{n=0}^\infty$, $a = -\infty$, $b = \infty$ and $d\mu(x) = e^{-x^2} dx$.

3 The Krawtchouk polynomials $\{K_n(x)\}_{n=0}^N$, $a = 0$, $b = N$ and $d\mu(x) = \sum_{k=0}^N \binom{N}{k} q^k (1-q)^{N-k} \delta(x-k)$, $0 < q < 1$. Here the closed interval $[0, N]$ is needed since the endpoints have positive mass.

A surprising property of orthogonal polynomials is that they always satisfy a three term <u>recurrence relation</u>

$$x p_n(x) = a_n p_{n+1}(x) + b_n p_n(x) + c_n p_{n-1}(x), \quad n = 1, 2, \cdots. \qquad (6.1)$$

To prove this, we first choose a_n such that $x p_n - a_n p_{n+1}$ is a polynomial of degree n, which we may then write as a linear combination of $p_0, p_1, \cdots, p_n$. But $\int_a^b p_k(x) p_n(x) d\mu = 0$ for $k < n-1$ since p_n is orthogonal to all polynomials of lower degree (in particular to $x p_k(x)$), and the linear combination consists of the terms $b_n p_n(x) + c_n p_{n-1}(x)$.

This is used to find the kernel $K_n(x, y)$ of convergence, the Christoffel-Darboux kernel, which is analogous to the Dirichlet kernel of Fourier series. The partial sums of the expansion of $f \in L^2(\mu; (a, b))$ are

$$S_n(x) = \sum_{k=0}^n \int_a^b f(y) p_k(y) d\mu(y) p_k(x) \qquad (6.2)$$

$$= \int_a^b f(y) \sum_{k=0}^n p_k(y) p_k(x) d\mu(y)$$

$$= \int_0^b f(y) K_n(x, y) d\mu(y)$$

and the kernel is

$$(x - y) K_n(x, y) \qquad (6.3)$$

$$= \sum_{k=0}^n x p_k(x) p_k(y) - \sum_{k=0}^n y p_k(y) p_k(x)$$

$$= \sum_{k=0}^n (a_k p_{k+1}(x) + b_k p_k(x) + c_k p_{k-1}(x)) p_k(y)$$

$$- \sum_{k=0}^n (a_k p_{k+1}(y) + b_k p_k(y) + c_k p_{k-1}(y)) p_k(x)$$

$$= \sum_{k=0}^n a_k (p_{k+1}(x) p_k(y) - p_{k+1}(y) p_k(x))$$

$$+ \sum_{k=0}^{n} c_k (p_{k-1}(x)p_k(y) - p_{k-1}(y)p_k(x))$$

where we have taken $c_0 = 0$. We now use the fact that

$$\int_a^b x p_n p_{n+1} d\mu = \int (a_n p_{n+1}^2 + b_n p_n p_{n+1} + c_n p_{n-1} p_{n+1}) d\mu = a_n$$

and similarly

$$c_n = \int_a^b x p_n p_{n-1} d\mu = a_{n-1}.$$

Thus the sums in (6.3) cancel each other except for one term, and we get

$$K_n(x,y) = \frac{a_n(p_{n+1}(x)p_n(y) - p_n(x)p_{n+1}(y))}{x - y}, \qquad (6.4)$$

the <u>Christoffel-Darboux</u> formula.

Just as with trigonometric Fourier series, we can prove pointwise convergence by using $K_n(x,y)$. For example, if $f(x) \in L^2(\mu; (a,b))$ and satisfies a Lipschitz condition of order α at $x_0 \in (a,b)$, $\frac{1}{2} < \alpha \le 1$, $b - a < \infty$, then in many cases

$$f_n(x_0) := \int_a^b K_n(x_0, y) f(y) d\mu(y) \to f(x_0).$$

The argument is the same as in Chapter 4 for Fourier series and needs only a suitable bound on $p_n(x)$.

This kernel can also be used to determine the behavior of the real zeros of $\{p_n(x)\}$ when the support of μ is an infinite set [Sz]. By taking $y = x$ in (6.4) we obtain

$$K_n(x,x) = a_n(p_{n+1}'(x)p_n(x) - p_n'(x)p_{n+1}(x)) \qquad (6.5)$$

$$= \sum_{k=0}^{n} p_k^2(x) > 0.$$

Since by the construction the leading coefficients of $p_n(x)$ are positive and since

$$x p_n(x) - a_n p_{n+1}(x)$$

is a polynomial of degree $\le n$, it follows that a_n must be a quotient of two leading coefficients and hence must be > 0. Hence we have

$$p_{n+1}'(x)p_n(x) - p_n'(x)p_{n+1}(x) > 0,$$

from which we deduce that if $p_n(x_0) = 0$, then

$$p_{n+1}(x_0)p'_n(x_0) < 0 \quad \text{and} \quad p_{n-1}(x_0)p'_n(x_0) > 0. \qquad (6.6)$$

Also from the recurrence formula (6.1) we have

$$a_n p_{n+1}(x_0) + a_{n-1} p_{n-1}(x_0) = 0$$

and hence

$$a_n p_{n+1}(x_0) p_{n-1}(x_0) = -a_{n-1} p_{n-1}^2(x_0) < 0. \qquad (6.7)$$

From (6.6), it follows that the zeros are simple, and p_n and p_{n+1} cannot have common zeros. Let x_0 and x_1 be two successive zeros of $p_n(x)$; then $p'_n(x_0)$ and $p'_n(x_1)$ have opposite signs and by (6.6) so do $p_{n+1}(x_0)$ and $p_{n+1}(x_1)$.

Hence, there is an odd number of zeros of p_{n+1} between x_0 and x_1. By a similar argument involving the other inequality in (6.6), there is an odd number of zeros of $p_n(x)$ between two successive zeros of $p_{n+1}(x)$. Thus, there is exactly one zero between x_0 and x_1. For the largest zero of $p_n(x)$, say x_m, $p'_n(x_m) > 0$ and $p_{n+1}(x_m) < 0$ by (6.6). Since the leading coefficient of p_{n+1} is positive, $p_{n+1}(x) > 0$ for x sufficiently large and hence p_{n+1} must have a zero larger than x_m; similarly it has a zero smaller than the smallest zero of p_m. This gives p_{n+1} one more zero than p_n. Since p_1 has one zero (otherwise $\langle p_1, p_0 \rangle_\mu > 0$), p_2 must have two zeros, p_3, three, ...

Proposition 6.1
The zeros of $p_n(x)$ are real, simple, lie in (a,b) and are interlaced with those of $p_{n+1}(x)$.

We have not shown that they lie in (a, b). If only $x_0, x_1, \ldots, x_m$, $m + 1 < n$ lie in (a, b) then

$$\int_a^b p_n(x) \prod_{j=0}^m (x - x_j) d\mu(x) \neq 0$$

since the integrand does not change sign in (a, b). But $\prod_{j=0}^m (x - x_j)$ is a polynomial of degree $m < n$ and therefore orthogonal to p_n which gives a contradiction.

If the measure μ is discrete and finite, the orthogonal polynomials can be taken to be interpolation polynomials. That is, if

$$d\mu(x) = \sum_{i=0}^{n} a_i \delta(x - x_i) dx, \quad a < x_0 < \cdots < x_n < b,$$

then the polynomials $\ell_i(x)$ of degree n, $i = 0, 1, \cdots, n$ where $\ell_i(x_j) = \delta_{ij}$ will be orthogonal with respect to $d\mu$. Indeed, we see that

$$\int_a^b \ell_i(x)\ell_j(x)d\mu(x) = \sum_k a_k \ell_i(x_k)\ell_j(x_k) = a_i\delta_{ij}, \quad i,j = 0,1,\cdots,n.$$

It is easy to define these polynomials by

$$\ell_i(x) := \prod_{j=0}^{n} (x - x_j) / \prod_{i \neq j} (x_i - x_j).$$

These $\{\ell_i\}$ constitute a complete orthogonal set in $L^2(d\mu, (a,b))$; they are usually referred to as <u>Lagrange interpolation polynomials</u> at the nodes $x_0, x_1, \cdots, x_n$. Any polynomial of degree n has a representation

$$P(x) = \sum_{i=0}^{n} P(x_i)\ell_i(x).$$

This is a simple example of a sampling theorem that will be taken up in Chapter 9.

6.2   Classical orthogonal polynomials

These classical orthogonal polynomials are usually taken to be on the interval $(-1, 1)$, $(0, \infty)$, or $(-\infty, \infty)$. In the first case, the measure is

$$d\mu = w(x)dx = (1 - x)^\alpha (1 + x)^\beta dx$$

which gives for general $\alpha, \beta > -1$, the <u>Jacobi</u> polynomials, for $\alpha = \beta = 0$, the <u>Legendre</u> polynomials, for $\alpha = \beta = \lambda - \frac{1}{2}$, the <u>Gegenbauer</u> polynomials, for $\alpha = \beta = \frac{1}{2}$, the <u>Tchebychev</u> polynomials. We shall discuss only the Legendre case in detail. On the infinite intervals, we have the Laguerre and Hermite polynomials. More details may be found in [Sz].

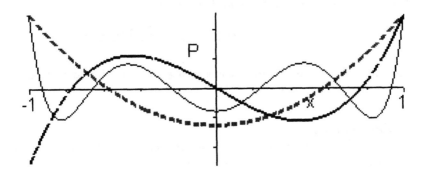

FIGURE 6.1
Some Legendre polynomials (n=2, 3, and 6).

6.2.1 Legendre polynomials

The Legendre polynomials $\{P_n\}$ are usually normalized so that $P_n(1) = 1$. They are obtained as before by orthogonalizing $1, x, x^2, \cdots$ with respect to dx. It is easy to find that

$$P_0(x) = 1, \ \ P_1(x) = x, \ \ P_2(x) = \frac{1}{2}(3x^2 - 1),$$

but then it becomes a little harder. The others are given most easily by the recurrence formula (6.2), which in this case becomes

$$xP_n(x) = \frac{n+1}{2n+1}P_{n+1}(x) + \frac{n}{2n+1}P_{n-1}(x). \tag{6.8}$$

This does not have the exact form of (6.1) since the $P_n(x)$ do not have norm 1; in fact $\|P_n\|^2 = 1/\left(n + \frac{1}{2}\right)$.

Another approach may be based on the differential equation [Jc, p.49]

$$((1 - x^2)y')' + n(n+1)y = 0 \tag{6.9}$$

which is satisfied for $y = P_n(x)$. Both the recurrence formula and the differential equation may be based on the generating function [Jc, p.45]

$$H(x,r) = (1 - 2xr + r^2)^{-1/2}.$$

The power series expansion of $H(x,r)$ is for $|r| < 1$ given by

$$H(x,r) = \sum_{n=0}^{\infty} P_n(x)r^n. \tag{6.10}$$

The Christoffel-Darboux formula (6.4) becomes

$$K_n(x,y) = \frac{n+1}{2} \frac{P_{n+1}(y)P_n(x) - P_n(y)P_{n+1}(x)}{y-x}.$$

This may be used to obtain a convergence theorem similar to that for trigonometric Fourier series.

Proposition 6.2

Let $f \in L^2(-1,1)$ *and satisfy a Lipschitz condition of order* $\alpha > \frac{1}{2}$ *at* $x \in (-1,1)$, *then the Legendre expansion*

$$\sum_{n=0}^{\infty} a_n P_n(x), \quad a_n = \left(n + \frac{1}{2}\right) \int_{-1}^{1} f(x)P_n(x)dx,$$

converges to $f(x)$.

PROOF   The proof uses the fact that

$$S_n(x) - f(x) = \int_{-1}^{1} (f(t) - f(x))K_n(x,t)dt$$

and is similar to the one for Fourier series. The necessary bounds on the kernel are obtained from the integral formula for the P_n [Jc, p.61], which is proved by showing that it satisfies the same recurrence formula as the $P_n(x)$,

$$P_n(\cos\theta) = \frac{1}{\pi} \int_0^{\pi} [\cos\theta + i\sin\theta \sin\varphi]^n d\varphi. \qquad (6.11)$$

It is clear from this integral formula that

$$|P_n(\cos\theta)| \leq 1,$$

but it can also be used to prove a stronger inequality

$$|P_n(x)| \leq C_\epsilon n^{-1/2}, \quad n = 1,2,\cdots, |x|^2 < 1 - \epsilon.$$

This follows from

$$|P_n(\cos\theta)| \leq \frac{1}{\pi} \int_0^{\pi} [\cos^2\theta + \sin^2\theta\cos^2\varphi]^{n/2} d\varphi$$

$$= \frac{2}{\pi} \int_0^{\pi/2} [1 - \sin^2\theta\sin^2\varphi]^{n/2} d\varphi$$

$$\le \frac{2}{\pi} \int_0^{\pi/2} \left[1 - \sin^2 \theta \left(\frac{2\varphi}{\pi} \right)^2 \right]^{n/2} d\varphi$$

$$\le \frac{2}{\pi} \int_0^{\pi/2} \left[e^{-\sin^2 \theta (2\varphi/\pi)^2} \right]^{n/2} d\varphi$$

$$= \frac{2}{\pi} \frac{\pi}{\sin \theta \, 2\sqrt{\frac{n}{2}}} \int_0^{\sin \theta \sqrt{n/2}} e^{-t^2} dt$$

$$\le \frac{\sqrt{2}}{\sqrt{n} \sin \theta} \int_0^{\infty} e^{-t^2} dt, \qquad 0 < \theta < \pi.$$

If $|x|^2 = \cos^2 \theta < 1 - \epsilon$, then $\sin^2 \theta > \epsilon$, and the inequality follows. We then return to the difference, and express it as

$$|S_n(x) - f(x)| \le \left| \int_{-1}^{x-\delta} \right| + \left| \int_{x-\delta}^{x+\delta} \right| + \left| \int_{x+\delta}^{1} (f(t) - f(x)) K_n(x,t) dt \right|$$

where δ is sufficiently small that $(x - \delta, x + \delta) \subset (-1, 1)$. The middle integral is dominated by

$$\frac{n+1}{2} \int_{x-\delta}^{x+\delta} |x - t|^\alpha \left| \frac{P_n(x) P_{n+1}(t) - P_n(t) P_{n+1}(x)}{x - t} \right| dt$$

$$\le C \int_{x-\delta}^{x+\delta} |x - t|^{\alpha-1} dt = \frac{C_2 \delta^\alpha}{\alpha},$$

since both x and t are in an interval $(-1 + \epsilon, 1 - \epsilon)$. The other integrals are made up of terms of the form

$$(n+1) \int_{-1}^{1} g(x,t) P_n(x) P_{n-1}(t) dt \tag{6.12}$$

where $g(x, \cdot) \in L^2(-1, 1)$. Since $\left\{ \sqrt{n + \frac{3}{2}} P_{n+1}(t) \right\}$ is orthonormal, it follows by Bessel's inequality that

$$\sqrt{n+1} \int_{-1}^{1} g(x,t) P_{n+1}(t) dt \to 0 \text{ as } n \to \infty.$$

Also since $\sqrt{n+1} P_n(x)$ is bounded in n, the integral in (6.12) is the product of a bounded function and one that converges to 0 and hence also converges to 0 as $n \to \infty$. Hence for any $\epsilon > 0$, we may choose δ so that $\frac{2C}{\alpha}\delta^{\alpha} < \frac{\epsilon}{2}$ and then choose n_0 so large that the remaining integrals are less than $\frac{\epsilon}{2}$ for $n \geq n_0$, thereby concluding that

$$|S_n(x) - f(x)| < \epsilon \qquad \text{for } n \geq n_0.$$

$\square$

This same formula (6.11) can be used to find the derivatives of P_n to deduce that

$$\left| \frac{d^k}{d\alpha^k} P_n(\cos\alpha) \right| \leq C \left(n + \frac{1}{2} \right)^k, \qquad n = 0, 1, 2, \cdots \qquad (6.13)$$

for some constant C. This gives results similar to Proposition 6.2 for distribution expansions.

Proposition 6.3

Let $f \in \mathcal{S}'$ and have support in the interior of $(-1, 1)$, let $a_n = (f, P_n\theta)$, where $\theta \in \mathcal{S}$, $\theta(x) = 1$ for $x \in supp(f)$, and $supp(\theta) \subset (-1, 1)$. Then
(i) $a_n = O(n^p)$ for some $p \in \mathbb{Z}$
(ii) $\sum_n a_n P_n\theta$ converges to f in $\mathcal{S}'$.

PROOF

If f has support on $[\alpha, \beta]$, then there exists a function $G \in L^2(-1, 1)$ with the same support and $f = D^p G + \sum_{i=0}^{p-1} C_i \delta^{(i)}$ (see Chapter 2). We then use integration by parts (i.e., the definition of distribution derivative) on the coefficients

$$a_n = (D^p G, P_n\theta) + \sum_{i=0}^{p-1} C_i P_n^{(i)}(0)$$

$$= (-1)^p (G, P_n^{(p)}\theta) + \sum_{i=0}^{p-1} C_i P_n^{(i)}(0)$$

which by (6.13) satisfy the conditions needed for (i).
For (ii) we have to show that

$$\sum_{n=0}^{\infty} a_n (P_n\theta, \varphi)$$

converges for each $\varphi \in \mathcal{S}$. Since

$$n(n+1)(P_n, \theta\varphi) = -(((1 - x^2)P_n')', \theta\varphi) = -(P_n, ((1 - x^2)(\theta\varphi)')')$$

and may be iterated any number of times, $(P_n, \theta\varphi) = O(n^{-q})$ for any $q \in \mathbb{N}$. Thus the series converges. That it converges to f involves the same sort of considerations and Parseval's equality,

$$(f, \theta\varphi) = (-1)^p(G, (\theta\varphi_1^{(p)}) + \sum C_i(\theta\varphi)^{(i)}(0).$$

$\square$

6.2.2   Jacobi polynomials

The Jacobi polynomials $\{P_n^{(\alpha,\beta)}\}$ are orthogonal with respect to the weight function $w(x) = (1-x)^\alpha(1+x)^\beta$, $-1 < \alpha, \beta$. As with the Legendre polynomials, they may be obtained by orthogonalizing $\{1, x, x^2, \cdots\}$. This gives us $P_0^{(\alpha,\beta)}(x) = 1$ while

$$P_1^{(\alpha,\beta)}(x) = \frac{\alpha + \beta + 2}{2}x + \frac{\alpha - \beta}{2}$$

with the usual normalization given by $P_n^{(\alpha,\beta)}(1) = \binom{n+\alpha}{n}$.

The remaining $P_n^{(\alpha,\beta)}$ may be obtained from the recurrence formula

$$xP_n^{(\alpha,\beta)} = a_nP_{n+1}^{(\alpha,\beta)} + b_nP_n^{(\alpha,\beta)} + c_nP_{n-1}^{(\alpha,\beta)},\qquad(6.14)$$

where

$$a_n = 2(n + 1)(\lambda + n + 1)/(\lambda + 2n + 1)(\lambda + 2n + 2),$$
$$b_n = (\beta^2 - \alpha^2)/(n + 2n)(\lambda + 2n + 2),$$
$$c_n = 2(\alpha + n)(\beta + n)/(\lambda + 2n)(\lambda + 2n + 1),$$

and $\lambda = \alpha + \beta$. The differential equation is

$$((1 - x^2)y')' + [\beta(1 - x) - \alpha(1 + x)]y' + n(n + \alpha + \beta + 1)y = 0,$$

and the normalization constant $\delta_n^{(\alpha,\beta)} = \left\|P_n^{(\alpha,\beta)}\right\|_w^2$ is

$$\delta_n^{(\alpha,\beta)} = \frac{2^{\lambda+1}}{\lambda + 2n + 1}\frac{\Gamma(\alpha + n + 1)\Gamma(\beta + n + 1)}{n!\Gamma(\lambda + n + 1)}.$$

Other formulae for the Legendre polynomials have analogues for the Jacobi polynomials. Convergence of the expansion of functions and distributions is also proved in the same way. An exhaustive treatment of Jacobi polynomials may be found in [Sz].

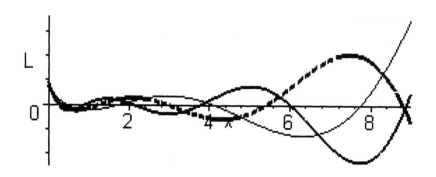

FIGURE 6.2
Some Laguerre polynomials (alpha=1/2, n= 5, 7, and 10).

6.2.3 Laguerre polynomials

The Laguerre polynomials $\left\{L_n^{(\alpha)}(x)\right\}$ are orthogonal on $(0, \infty)$ with respect to the weight function $w(x) = x^\alpha e^{-x}$, $\alpha > -1$. Their first two elements are
$$L_0^{(\alpha)}(x) = 1, \quad L_1^{(\alpha)}(x) = -x + \alpha + 1,$$
while the recurrence formula is
$$xL_n^{(\alpha)} = -(n+1)L_{n+1}^{(\alpha)} + (2n+1+\alpha)L_n^{(\alpha)} - (n+\alpha)L_{n-1}^{(\alpha)}. \tag{6.15}$$

They satisfy the differential equation
$$(xy')' + (\alpha - x)y' + ny = 0,$$
and their normalization factor $\delta_n^\alpha = \int_0^\infty |L_n^\alpha(x)|^2\, w(x)dx$ is
$$\delta_n^\alpha = \Gamma(n+\alpha+1)/\Gamma(n+1).$$

Since the interval is infinite, they are not automatically a complete system. However, in this case the completeness may be proved using the Laplace transform:
$$F(s) = \int_0^\infty e^{-st} f(t)dt. \tag{6.16}$$

If $f(x)(x^2+1)^{-p} \in L^2(0, \infty)$ for some $p \in \mathbb{Z}$, then $F(s)$ is a holomorphic on the half-plane Re $s > 0$. We have to show that if
$$\int_0^\infty e^{-x/2} x^n x^{\alpha/2} g(x)dx = 0, \quad n = 0, 1, 2, \cdots, \tag{6.17}$$

for $g \in L^2(0, \infty)$, then $g = 0$ a.e. This would imply that $\{x^n x^{\alpha/2} e^{-x/2}\}$ and hence $\{L_n^{(\alpha)}(x) x^{\alpha/2} e^{-x/2}\}$ are complete in $L^2(0, \infty)$.

We now take $f(t) = t^{\alpha/2} g(t)$ in the Laplace transform (6.16) and use the fact that

$$F'(s) = - \int_0^\infty t e^{-st} f(t) dt$$

to express (6.17) as

$$(-1)^n F^{(n)}\left(\frac{1}{2}\right) = 0, \qquad n = 0, 1, \cdots.$$

This implies that $F(s) \equiv 0$ for Re $s > 0$ and hence that $g(t) = 0$, a.e.

Again local and global convergence theorems can be proved by the same tricks. If anything the formulae and bounds are simpler than in the Jacobi case and may again be found in [Sz].

6.2.4   Hermite polynomials

The Hermite polynomials $\{H_n(x)\}$ are orthogonal on $(-\infty, \infty)$ with respect to the weight function $w(x) = e^{-x^2}$. They may be obtained by orthogonalizing $1, 2x, (2x)^2, \cdots$ with respect to this weight, or as before, by using a recurrence formula together with

$$H_0(x) = 1, \quad H_1(x) = 2x.$$

It is

$$x\, H_n(x) = \frac{1}{2} H_{n+1}(x) + n H_{n-1}(x), \quad n = 1, 2, \cdots.$$

An alternate approach uses the differential equation

$$y'' - 2xy' + 2ny = 0$$

whose only polynomial solution is a multiple of $H_n(x)$. In the usual normalization, $H_n(x)$ has leading coefficient 2^n; this makes its norm equal to

$$\int_{-\infty}^\infty H_n^2(x) e^{-x^2} dx = \pi^{1/2} 2^n n!, \qquad n = 0, 1, 2, \cdots.$$

Rather than use these polynomials for our analysis we combine them with the weight to get the <u>Hermite functions</u>.

$$h_n(x) = \frac{H_n(x) e^{-x^2/2}}{\pi^{1/4} 2^{n/2} (n!)^{1/2}}, \quad n = 0, 1, 2, \cdots, x \in \mathbb{R}.$$

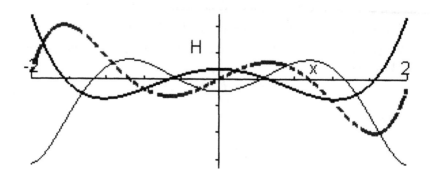

FIGURE 6.3
Some Hermite polynomials (modified by constant multiples,
n= 4, 5, and 7).

These form a complete orthonormal sequence in $L^2(\mathbb{R})$ and are widely known as forming a solution to the linear oscillator equation,

$$(x^2 - D^2)y = (2n+1)y, \qquad n = 0, 1, \cdots .$$

However, they also have a number of other uses in mathematics that concern us here. As we observed in Chapter 2, they belong to the space S and can be used to characterize the space of tempered distributions S'.

The recurrence formula for the polynomials can be restated in terms of the $\{h_n(x)\}$ as

$$x\, h_n(x) = \sqrt{\frac{n}{2}}\, h_{n-1}(x) + \sqrt{\frac{n+1}{2}}\, h_{n+1}(x). \qquad (6.18a)$$

In addition the derivatives satisfy

$$h'_n(x) = \sqrt{\frac{n}{2}}\, h_{n-1}(x) - \sqrt{\frac{n+1}{2}}\, h_{n+1}(x). \qquad (6.18b)$$

The $h_n(x)$ are uniformly bounded for real x. Indeed, we have

$$|h_n(x)| \le Cn^{-1/12}, \qquad n = 1, 2, \cdots$$

for a constant C ([Sz, p.242]). With this bound and the recurrence formulae we can get a pointwise convergence theorem

Proposition 6.4

Let $f(x) \in L^2(\mathbb{R})$, be absolutely continuous, and satisfy $(x + D)f(x) \in L^2(\mathbb{R})$; then the Hermite series $\sum a_n h_n(x)$ of $f(x)$ converges uniformly on $\mathbb{R}$ to $f(x)$.

PROOF By taking the difference between (6.18a) and (6.18b) we obtain

$$(x - D)h_n(x) = \sqrt{2n + 2}\, h_{n+1}(x) \qquad (6.19)$$

and, hence, the coefficients of $(x + D)f$ are $a_{n+1}\sqrt{2n + 2}$.

Then the Hermite series of $f(x)$ can be bounded by

$$\sum_{n=1}^{\infty} |a_n h_n(x)| \leq \sum_n \left| a_n \sqrt{2n} \frac{h_n(x)}{\sqrt{2n}} \right|$$

$$\leq \left\{ \sum_n \left| a_n \sqrt{2n} \right|^2 \right\}^{1/2} \left\{ \sum_n \frac{h_n^2(x)}{2n} \right\}^{1/2}$$

$$\leq \|(x + D)f\| \left\{ \sum_n \frac{C}{(2n)^{7/6}} \right\}^{1/2},$$

and hence the series converges absolutely and uniformly on $\mathbb{R}$. The limit function, say $g(x)$, must be bounded since the partial sums are bounded.

The Hermite coefficients of g must be given by

$$\langle g, h_m \rangle = \sum_n a_n \langle h_n, h_m \rangle = a_m,$$

i.e., must be the same as those of f. We deduce that $f(x) = g(x)$ by the following uniqueness theorem.

Lemma 6.1

Let $F(x)$ be a continuous function of polynomial growth whose Hermite coefficients $\langle F, h_n \rangle = 0$, $n = 0, 1, \cdots$; then $F \equiv 0$.

PROOF Since F is of polynomial growth $\left| F(x)e^{-x^2/2} \right| \leq Ce^{-\epsilon|x|}$ for some $\epsilon > 0$ and constant C. Thus, the Fourier transform of $F(x)e^{-x^2/2}$ is analytic in a strip $-\delta < \operatorname{Im} w < \delta$. By the hypothesis we have

$$\int_{-\infty}^{\infty} F(x)e^{-x^2/2} H_n(x)dx = 0, \qquad n = 0, 1, 2, \cdots$$

and, hence, since x^n is a linear combination of $H_0(x), \cdots, H_n(x)$,

$$\int_{-\infty}^{\infty} F(x)e^{-x^2/2}x^n dx = 0, \qquad n = 0, 1, 2, \cdots.$$

Therefore the Fourier transform of $F(x)e^{-x^2/2}$ will have all of its derivatives zero at 0, and hence it must be identically zero. So must $F(x)e^{-x^2/2}$ by the Fourier integral theorem.   □

   This uniqueness theorem may be extended to all of S' by an argument of J. Korevaar [K1].

Corollary 6.1
   Let $f \in S'$ have Hermite coefficients $(f, h_n) = 0$, $n = 0, 1, \cdots$, then $f \equiv 0$.

PROOF   Each $f \in S'$ may be expressed as $f = D^p F$ where F is continuous of polynomial growth. We proceed by induction on p. The result is true for $p = 0$ by the Lemma and we suppose it true for $p - 1$. Then $f = Dg = D(D^{p-1}F)$ satisfies $(f, h_n) = (Dg, h_n) = -(g, h'_n) = 0$. But by the recurrence formula

$$(g, h'_n) = \sqrt{\frac{n}{2}}\, (g, h_{n-1}) - \sqrt{\frac{n+1}{2}}\, (g, h_{n+1}) = 0$$

or

$$(g, h_{n+1}) = \sqrt{\frac{n}{n+1}}\, (g, h_{n-1}), \qquad n = 0, 1, \cdots.$$

Thus g has the same Hermite coefficients as a constant function since $-(C, h_{n+1})\sqrt{\frac{n+1}{2}} + (C, h_{n-1})\sqrt{\frac{n}{2}} = 0$. By the induction hypothesis, since $g - C$ has zero coefficients, $g - C = 0$. Thus $Dg = 0$ and we are finished.   □

THEOREM 6.1
Let $f \in S'$; then
   (i) $(f, h_n) = 0(n^p)$ *for some $p \in \mathbb{Z}$,*
   (ii) $\sum_n (f, h_n)h_n = f$ *in the sense of S,*
   (iii) *there is $G \in L^2(\mathbb{R})$ such that $f = (x^2 - D^2)^r G$ for some $r \in \mathbb{Z}$.*

PROOF   For (i), we use the fact $|(f, h_n)| \le C \sup_t (1 + |t|)^r |h_n^{(r)}(t)|$ for some r (Chapter 2). But the recurrence formulae may each be applied

r times to get first

$$\left| h_n^{(r)}(t) \right| \le \sum_{k=-r}^{r} c_{n,k} \left| h_{n+k}(t) \right| \quad \text{where } c_{n,k} = O(n^{r/2})$$

from (6.18b) and then

$$(1+|t|)^r \left| h_n^{(r)}(t) \right| \le \sum_{k=-2r}^{2r} d_{nk} \left| h_{n+k}(t) \right|, \quad d_{nk} = O(n^r)$$

from (6.18a). Since $|h_n(t)|$ is uniformly bounded in t and n, the conclusion follows.

The series in (ii) clearly converges in S' since for $\varphi \in S$,

$$(h_n, \varphi) = (2n+1)^{-q}((x^2 - D^2)^q h_n, \varphi) = (2n+1)^{-q}(h_n, (x^2 - D^2)^q \varphi)$$

and $(x^2 - D^2)^q \varphi \in L^2(\mathbb{R})$ for any $q = 0, 1, 2, \cdots$. By taking $q = p+2$ we get the convergence of

$$\sum_n (f, h_n)(h_n, \varphi).$$

If this series in (ii) converges to, say g, then g and f both have the same Hermite coefficients and hence are the same.

The third conclusion follows from (i) by forming the series

$$\sum_n \frac{(f, h_n)}{(2n+1)^{p+1}} h_n = G.$$

Since the coefficients are in ℓ^2, $G \in L^2(\mathbb{R})$. Then by applying the operator $(x^2 - D^2)^{p+1}$ to both sides we obtain

$$\sum (f, h_n) h_n = (x^2 - D^2)^{p+1} G.$$

Since this series also converges to f and $(x^2 - D^2)^{p+1} G \in S'$, they must be equal. □

It should be noted that condition (i) and (iii) are also sufficient for $f \in S'$.

Another important property of Hermite functions is their relation to Fourier transforms. We first observe the well known fact that the Fourier transform of $e^{-x^2/2}$ is a multiple of the same function; in fact

$$\int_{-\infty}^{\infty} e^{-x^2/2} e^{-iwx} dx = \sqrt{2\pi}\, e^{-w^2/2}.$$

Thus, $\widehat{h}_0(w) = \sqrt{2\pi}\, h_0(w)$, and for other h_n we use the fact that

$$\int_{-\infty}^{\infty} e^{-iwx}(x - D)\varphi(x)dx = (-i)(w - D)\int_{-\infty}^{\infty} e^{-iwx}\varphi(x)dx$$

for $\varphi \in S$. Hence by taking $\varphi = h_n$, we get by (6.19)

$$\int_{-\infty}^{\infty} e^{-iwx}\sqrt{2n+2}\, h_{n+1}(x)dx = (-i)(w - D)\widehat{h}_n(w), \quad n = 0, 1, 2, \cdots,$$

or

$$\widehat{h}_{n+1}(w) = (-i)\frac{1}{\sqrt{2n+2}}(w - D)\widehat{h}_n(w).$$

Thus, for $n = 0$, we have $\widehat{h}_1(w) = (-i)\frac{\sqrt{2\pi}}{\sqrt{2}}(w-D)h_0(w) = -i\sqrt{2\pi}\, h_1(w)$ and by induction

$$\widehat{h}_n(w) = (-i)^n\sqrt{2\pi}\, h_n(w).$$

Thus the Hermite functions are eigenfunctions of the Fourier transform. This is the starting point for a simple theory of Fourier transforms.

Proposition 6.5
   Let $f \in L^2(\mathbb{R})$ (resp. $\in S'$) with Hermite series

$$f \sim \sum_n a_n h_n;$$

then the Fourier transform $\hat{f}$ is given by

$$\hat{f}(w) \sim \sqrt{2\pi}\sum_n (-i)^n a_n h_n(w)$$

with convergence in the sense of $L^2(\mathbb{R})$ (resp. S').

   The proof is obvious. The reader may wish to explore the implications of this result.

6.3   Problems

1. Find a complete set of orthogonal polynomials on $(-\infty, \infty)$ with respect to $d\mu = \frac{1}{4}\delta(x - 1) + \frac{1}{4}\delta(x + 1) + \frac{1}{2}\delta(x)$.

2. Use the recurrence formula (6.8) for the Legendre polynomials to find $P_3(x)$ and $P_4(x)$, and show their orthogonality by direct calculation.

3. Show directly that the zeros of $P_3(x)$ and $P_4(x)$ are interlaced.

4. Find the expansion of the hat function $f(x) = 1 - |x|$ on $(-1, 1)$ with respect to Legendre polynomials.

5. The Legendre polynomials satisfy Rodrigues' formula

$$P_n(x) = \frac{1}{2^n n!} \frac{d^n (x^2 - 1)^n}{dx^n}.$$

Use this to prove the orthogonality.

6. Use Rodrigues' formula to prove

$$P'_{n+1}(x) - P'_{n-1}(x) = (2n + 1)P_n(x).$$

7. Find the first five Tchebychev polynomials (Jacobi polynomials with $\alpha = \beta = \frac{1}{2}$) by using the recurrence formula (6.14).

8. Expand x^4 in terms of Tchebychev polynomials.

9. Find the first five Laguerre polynomials for $\alpha = 0$, and find the expansion of x^4 in terms of Laguerre polynomials.

10. Show that $L_n^{(0)}(x)$ satisfies the Rodrigues' formula

$$L_n^{(0)}(x) = (-1)^n \frac{d^n (x^n e^{-x})}{dx^n} e^x.$$

11. Prove the recurrence relation (6.15) for $\alpha = 0$ from the formula in Problem 10.

12. Expand $\delta(x)$ in terms of Hermite functions $\{h_n\}$. Use (6.18a) to establish a formula for the coefficients.

13. Expand $f(x) = 1$ in terms of Hermite functions. Use (6.18b) to establish a formula for the coefficients.

14. Show that $\hat{\delta}(w) = 1$ by combining Problems 12 and 13 and using Proposition 6.8.

15. Find the Hermite expansion of x and x^2 by using (6.18a) and (6.18b).

16. Prove Parseval's identity $\|f\|^2 = \frac{1}{2\pi} \|\hat{f}\|^2$ by using Proposition 6.8.

Chapter 7

Other Orthogonal Systems

Orthogonal systems other than the trigonometric, orthogonal polynomials, and wavelets also play an important role in applications. Chief among these are the systems arising from <u>boundary value problems</u>, which are primarily used to solve partial differential equations. We present here only the barest essentials of these systems. A more complete (much more) description may be found in [T] or [C-L].

Another system consists of <u>eigenfunctions of integral operators</u>; these have proved to be useful in stochastic processes [Ro] as well as differential equations. These systems are related to Sturm-Liouville systems since the latter are also eigenfunctions of the integral operator with the associated Green's function [C-L]. In one case, that of the <u>prolate spheroidal functions</u>, the system is composed of eigenfunctions of both a differential operator and an integral operator which is not associated with the Green's function [Sl], [W6].

Two systems related to the Haar system, the Rademacher and Walsh systems, are also discussed in this chapter. The latter has been used for digital transmission of information (see [Ha], [O]).

Periodic wavelets that give orthonormal bases of finite intervals are included in this chapter. These are obtained by periodizing wavelets on the real line [D]. Such periodic wavelets can also be constructed directly by using Fourier series [W-C1].

Finally, one of the predecessors to wavelets, the "localized sine (or cosine) basis" of Coifman and Meyer, is constructed [C-M], [A-W-W]. This is an orthonormal version of a windowed Fourier transform.

7.1 Self adjoint eigenvalue problems on finite intervals

One of the classical techniques used to solve partial differential equations with associated boundary conditions is separation of variables, which converts the partial differential equation into ordinary differential equations. These equations in turn also inherit certain boundary conditions and have solutions (eigenfunctions) only for certain values (eigenvalues) of a parameter.

A simple example is the wave equation

$$\frac{\partial^2 u}{\partial x^2} - \frac{\partial^2 u}{\partial t^2} = 0, \qquad u(0,t) = u(\pi,t) = 0, \qquad (7.1)$$
$$0 \le x \le \pi, \ 0 \le t.$$

The separation of variables consists of looking for solutions of the form

$$u(x,t) = f(x)g(t),$$

which must satisfy

$$f''g - fg'' = 0, \qquad f(0) = f(\pi) = 0.$$

Then (7.1) is written as

$$\frac{f''(x)}{f(x)} = \frac{g''(t)}{g(t)} = \lambda, \qquad f(0) = f(\pi) = 0,$$

where λ is a constant. The eigenvalue problem then is

$$y'' - \lambda y = 0, \qquad y(0) = y(\pi) = 0,$$

which $y = f$ must satisfy. The values of λ for which a solution exists are $\lambda_k = -k^2$, $k \in \mathbb{Z}$ and the resulting solutions are $y_k = \sin kx$. These y_k's are orthogonal on $(0, \pi)$ and are complete in $L^2(0, \pi)$. (This follows directly from considerations of Fourier series in Chapter 4.)

To solve the original equation one also has to find the $g(t)$, which in this case for $\lambda = -k^2$ is also of the form

$$g_k(t) = A_k \cos kt + B_k \sin kt.$$

Thus the general solution to the boundary value problem is

$$w(x,t) = \sum_{k=1}^{\infty} (A_k \cos kt + B_k \sin kt) \sin kx.$$

The coefficients A_k and B_k are found from initial conditions that do not concern us here.

The procedure can be imitated in a large number of other cases. We shall consider only the second order case on a finite interval (a, b). Let L be the differential operator

$$Ly = -(py')' + qy,$$

where $p \in C^1[a, b]$ and $q \in C[a, b]$. Then the problem is

$$Ly = \lambda y, \qquad \alpha_1 y(a) + \alpha_2 y'(a) = 0$$
$$\beta_1 y(b) + \beta_2 y'(b) = 0. \qquad (7.2)$$

If p, q, α_1, α_2, β_1 and β_2 are real, and $p > 0$ on $[a, b]$, then (7.2) is a self adjoint, second order, linear eigenvalue problem commonly called a Sturm-Liouville problem. By self adjoint is meant the condition

$$\langle Lf, g \rangle = \langle f, Lg \rangle$$

for f and g satisfying the boundary conditions in (7.2). Such problems have solutions $\{\lambda_k, y_k\}$ such that the eigenfunctions $\{y_k\}$ are a complete orthonormal system in $L^2(a, b)$. See [C-L, p.197].

Indeed the same result is true for higher order linear differential operators and more general boundary values, but the Sturm-Liouville (S-L) problems are the most important in applications. We have described regular S-L problems that involve a finite interval on which $p \neq 0$. There are also singular S-L problems, which sometimes (but not always) have complete systems of eigenfunctions. These occur when the coefficients are singular, or $p(x) = 0$ for some $x \in [a, b]$, or the interval is infinite. Various standard orthogonal systems such as Hermite functions, Legendre polynomials, and Bessel functions arise as examples of solutions for such singular S-L problems. They are covered in great detail elsewhere and will not be discussed further here. See [T], [Jc].

A remarkable additional property of regular S-L problems is the existence of a Green's function. This is a function $g(t, s)$, $t, s \in [a, b]$, which forms the kernel of an integral operator that inverts the differential operator L (provided 0 is not an eigenvalue). It satisfies

$$L_t g(t, s) = \delta(t - s) \qquad t, s \in (a, b)$$

as well as the boundary conditions in (7.2) and is continuous on $[a, b] \times [a, b]$. The associated integral operator is

$$(Gf)(t) := \int_a^b g(t, s) f(s) ds$$

and is an inverse to L in the sense that if for $f \in C[a, b]$, $Lw = f$ where w satisfies the boundary conditions, then $w = Gf$. (See [C-L, p.192].)

This operator G is a special case of a self adjoint compact operator, which we consider in the next section.

7.2   Hilbert-Schmidt integral operators

We here consider integral operators with kernels $q(t, s)$, $t, s \in (a, b)$, (a, b) finite, which belong to $L^2[(a, b) \times (a, b)]$ and which are symmetric, $q(t, s) = q(s, t)$. The operator Q given by

$$(Qf)(t) := \int_a^b q(t, s) f(s) ds$$

is a Hilbert-Schmidt operator and clearly maps $L^2(a, b)$ into itself. These operators have been extensively studied and have the following properties [R-N, p.227] [R, p.203]:

(i)   There is at least one nonzero eigenvalue λ_1 with an eigenfunction $\varphi_1 (Q\varphi_1 = \lambda_1 \varphi_1)$.

(ii) Each nonzero eigenvalue λ_n has finitely many independent eigenfunctions.

(iii) The set of eigenfunctions $\{\varphi_n\}$ corresponding to nonzero eigenvalues may be taken to be orthogonal.

(iv) Each $g \in L^2(a, b)$ has a representation
$g(t) = h(t) + \sum_n \langle g, \varphi_n \rangle \varphi_n(t)$, where $Qh(t) = 0$.

(v) $q(t, s) = \sum_n \lambda_n \varphi_n(t) \overline{\varphi_n(s)}$.

Here the convergence is in the sense of $L^2(a, b)$ for s fixed. Since the symmetric Green's functions of the last section are particular cases, these operators are a rich source of orthogonal systems. However, their practical utility is limited by the difficulty of calculating the $\varphi_n(t)$.

If Q is positive definite, i.e., if $\langle Qf, f \rangle > 0$ for $f \neq 0$, then (iv) reduces to

$$g(t) = \sum_n \langle g, \varphi_n \rangle \varphi_n(t),$$

i.e., $\{\varphi_n\}$ is a complete orthonormal system. In this case the eigenvalues λ_n are all positive, and the $\varphi_n(t)$ may be taken to be real valued functions (both the real and imaginary parts are eigenfunctions).

The eigenfunctions may be approximated by means of Fourier series. By means of a translation and dilation if necessary, we may assume the interval is $(0, 2\pi)$. Then we have

$$q(t, s) = \sum_m \sum_k q_{mk} e^{imt} e^{-iks}$$

convergent in $L^2(0, 2\pi)^2$. If λ and $\varphi(t)$ are an eigenvalue-eigenfunction pair, and $\varphi(t)$ has Fourier series

$$\varphi(t) = \sum_k c_k e^{ikt},$$

then

$$\int_0^{2\pi} q(t, s)\varphi(s)ds = \sum_m \left(\sum_k q_{mk} c_k \right) e^{imt}$$
$$= \lambda \sum_m c_m e^{imt}.$$

The series $\sum_k q_{mk} c_k$ converges by Cauchy's inequality to a sequence in ℓ^2. Hence by truncating both series we have the algebraic eigenvalue problem

$$\sum_{|k| \leq N} q_{mk} c_k = \lambda c_m, \qquad |m| \leq N.$$

We then solve this for λ and $\{c_k\}$ to obtain an approximation to the true λ and $\varphi(t)$. However, if the $q(t, s)$ is the Green's function of a differential operator, the properties of the latter can be used to obtain better approximations.

It would be nice if we could associate with Q a differential operator in cases when $q(t, s)$ is not a Green's function. This does occur in at least one case, but as we shall see in the next section, this is the only case. If each eigenvalue has multiplicity one, then we need only require that the differential operator L commute with Q. In this case we have

$$L\,Q\varphi_n = L\,\lambda_n\varphi_n$$

or

$$Q\,L\varphi_n = \lambda_n L\varphi_n$$

so that $L\varphi_n$ is also an eigenfunction of Q with eigenvalue λ_n. Thus $L\varphi_n$ must be some multiple of φ_n, i.e., φ_n is also an eigenfunction of L but possibly belonging to a different eigenvalue.

7.3   An anomaly: the prolate spheroidal functions

The prolate spheroidal (wave) functions $\{\varphi_n\}$ are the eigenfunctions of the integral operator

$$(S_{\tau,\sigma}\varphi)(t) := \int_{-\tau}^{\tau} \frac{\sin\sigma(t-x)}{\pi(t-x)}\varphi(x)dx, \quad \tau,\sigma > 0. \qquad (7.3)$$

The kernel is Hilbert-Schmidt on the interval $(-\tau,\tau)$, and thus the eigenfunctions may be taken to be orthonormal on $(-\tau,\tau)$. It is also positive definite so that $\{\varphi_n\}$ is complete in $L^2(-\tau,\tau)$. By (7.3), they belong to the Paley-Wiener space of σ band limited functions since the kernel is exactly the reproducing kernel of B_σ (prototype II in Chapter 1).

We may write the eigenvalue problem in (7.3) as a convolution

$$\lambda\varphi = (\chi_\tau\varphi) * \left(\frac{\sin\sigma t}{\pi t}\right)$$

and hence by taking Fourier transforms

$$\lambda\hat{\varphi} = (\widehat{\chi_\tau\varphi}) \cdot \chi_\sigma$$

since the Fourier transform of $(\sin\sigma t)/\pi t$ is the characteristic function χ_σ of $[-\sigma,\sigma]$. This last equation implies that the support of $\hat{\varphi}$ is in $[-\sigma,\sigma]$, which therefore must satisfy

$$\hat{\varphi} = \hat{\varphi}\chi_\sigma.$$

By taking the inverse Fourier transform we see that

$$\varphi = \varphi * \left(\sin \frac{\sigma t}{\pi t} \right)$$

or

$$\varphi(t) = \int_{-\infty}^{\infty} \frac{\sin \sigma(t - x)}{\pi(t - x)} \varphi(x) dx. \tag{7.4}$$

Thus we find that each φ_n is an eigenfunction of the same kernel but on the infinite interval $(-\infty, \infty)$ as well as on $(-\tau, \tau)$. They all belong to the same eigenvalue 1 however. This is no contradiction since the kernel is not Hilbert-Schmidt on $(-\infty, \infty)$. A further surprising property is that the $\{\varphi_n\}$ are orthogonal on $(-\infty, \infty)$ also. To see this we use (7.3) in

$$\int_{-\infty}^{\infty} \varphi_n \varphi_m = \int_{-\infty}^{\infty} \varphi_n(t) \frac{1}{\lambda_m} \int_{-\tau}^{\tau} \varphi_m(x) \frac{\sin \sigma(t - x)}{\pi(t - x)} dx \, dt$$

$$= \frac{1}{\lambda_m} \int_{-\tau}^{\tau} \varphi_m(x) \int_{-\infty}^{\infty} \varphi_n(t) \frac{\sin \sigma(t - x)}{\pi(t - x)} dt \, dx$$

$$= \frac{1}{\lambda_m} \int_{-\tau}^{\tau} \varphi_m(x) \varphi_n(x) dx = \frac{1}{\lambda_m} \delta_{mn}.$$

The interchange of the order of integration is justified since the integrand is absolutely integrable over the strip $[-\tau, \tau] \times (-\infty, \infty)$.

There are many more interesting properties of these $\{\varphi_n\}$ that make them useful in signal analysis. (See [Pa, p.205] or [Sl].) But perhaps the most interesting is the fact that they are also eigenfunctions of yet a third operator.

7.4 A lucky accident?

The prolate spheroidal wave functions $\{\varphi_n\}$ may also be characterized as eigenfunctions of the differential operator [Sl]

$$P_{\tau,\sigma}\varphi := (\tau^2 - t^2) \frac{d^2 \varphi}{dt^2} - 2t \frac{d\varphi}{dt} - \sigma^2 t^2 \varphi. \tag{7.5}$$

Many of the useful properties of the $\{\varphi_n\}$ derive from the fact that the two operators commute

$$P_{\tau,\sigma} S_{\tau,\sigma}\varphi = S_{\tau,\sigma} P_{\tau,\sigma}\varphi$$

and hence that the $\{\varphi_n\}$ are the eigenfunctions of either one. The first $(S_{\tau,\sigma})$ shows they are entire functions of exponential type while the second is used to determine local properties. This is the "lucky accident" referred to by Slepian. However both operators are related to a third operator consisting of multiplication by the characteristic function χ_σ of an interval $[-\sigma, \sigma]$ and to the Fourier transform

$$(\mathcal{F}\varphi)(w) = \frac{1}{\sqrt{2\pi}} \int_{-\infty}^{\infty} \varphi(t)e^{-iwt}dt = \hat{\varphi}(w).$$

Here we have added the factor $\frac{1}{\sqrt{2\pi}}$ in order to obtain symmetric formulae for both the Fourier transform and its inverse which we denote by $\check{\varphi}$. We do this only in this section. We shall show that (7.5) is the unique differential operator such that both it and its Fourier transform are second order and commute with characteristic functions.

Let $P(x, D)$ be the linear second order differential operator with polynomial coefficients given by

$$P(x, D)\varphi := \rho_0(x)D^2\varphi + \rho_1(x)D\varphi + \rho_2(x)\varphi.$$

It is easy to derive conditions under which $P(x, D)$ commutes with multiplication by the characteristic functions χ_σ (with differentiation in the sense of distributions) as given in the following lemma.

Lemma 7.1
$P(x, D)(\chi_\sigma(x)\varphi(x)) = \chi_\sigma(x)P(x, D)\varphi(x)$ *for all $\varphi \in C^2$ if and only if*

$$\rho_0(\pm\sigma) = 0, \qquad \rho_1(\pm\sigma) = \rho_0'(\pm\sigma).$$

PROOF By direct computation we have

$$P(\cdot, D)(\chi_\sigma\varphi) = \rho_0(\chi_\sigma''\varphi + 2\chi_\sigma'\varphi') + \rho_1\chi_\sigma'\varphi + \chi_\sigma P(\cdot, D)\varphi$$
$$= \rho_0\varphi(\delta_{-\sigma}' - \delta_\sigma') + (2\rho_0\varphi' + \rho_1\varphi)(\delta_{-\sigma} - \delta_\sigma)$$
$$+ \chi_\sigma P(\cdot, D)\varphi.$$

Hence the two operators commute if and only if

$$(\rho_0\varphi)(\delta_{-\sigma}' - \delta_\sigma') + (2\rho_0\varphi' + \rho_1\varphi)(\delta_{-\sigma} - \delta_\sigma) = 0 \qquad (7.6)$$

for all φ. In particular if φ has support on $(0, \infty)$ and (7.6) holds we see that

$$\rho_0\varphi\delta_\sigma' + (2\rho_0\varphi' + \rho_1\varphi)\delta_\sigma = 0$$

and since $f\delta'_\sigma = f(\sigma)\delta'_\sigma - f'(\sigma)\delta_\sigma$, $f\delta_\sigma = f(\sigma)\delta_\sigma$, it follows that

$$\rho_0(\sigma)\varphi(\sigma)\delta'_\sigma -$$
$$[\rho'_0(\sigma)\varphi(\sigma) + \rho_0(\sigma)\varphi'(\sigma) - 2\rho_0(\sigma)\varphi'(\sigma) - \rho_1(\sigma)\varphi(\sigma)]\delta_\sigma$$
$$= 0.$$

Hence if $\varphi(\sigma) \neq 0$, $\rho_0(\sigma) = 0$ and $\rho'_0(\sigma) - \rho_1(\sigma) = 0$. A similar argument works for $-\sigma$.

These conditions are also clearly sufficient for (7.6). □

Example 1

The operator $P_{\tau,\sigma}$ of (7.5) commutes with χ_τ.

Example 2

The operator of the Jacobi polynomials $\{P_n^{(\alpha,\beta)}\}$ [Chapter 6]

$$Py = (1 - x^2)D^2y + [(\beta - \alpha) - (\alpha + \beta + 2)x]Dy$$

commutes with χ_1 if and only if $\alpha = \beta = 0$, i.e., the Legendre case.

Example 3

The operator of the Laguerre polynomials $\{L_n^\alpha\}$

$$Py = xD^2y + (\alpha + 1 - x)Dy$$

commutes with the characteristic function of $[0, \infty)$ if and only if $\alpha = 0$.

If the operator $P(x, D)$ has quadratic polynomial coefficients, then it assumes a particular simple form under the hypothesis of Lemma 7.1.

Corollary 7.1

Let $\rho_0(x)$ and $\rho_1(x)$ be quadratic polynomials; then $P(x, D)$ commutes with χ_σ if and only if

$$\rho_0(x) = a(x^2 - \sigma^2), \qquad \rho_1(x) = 2ax + b(x^2 - \sigma^2)$$

for some constant a and b.

The Fourier transform maps differential operators $P(x, D)$ of the form (7.5) into operators $P(iD, ix)$. That is

$$(\widehat{P(\cdot, D)\varphi})(w) = P(iD, iw)\hat{\varphi}(w).$$

In particular if $P(x, D)$ has quadratic coefficients, then $P(iD, ix)$ is also a second order linear operator with quadratic coefficients. By Corollary 7.2 those $P(x, D)$ which commute with χ_σ have the form

$$P(x, D) = [x^2 \ x \ 1] \begin{bmatrix} a & 0 & c_1 \\ -ib & 2a & c_2 \\ -a\sigma^2 & -b\sigma^2 & c_3 \end{bmatrix} \begin{bmatrix} D^2 \\ D \\ 1 \end{bmatrix}.$$

Clearly $P(iD, ix)$ has a similar form

$$P(iD, ix) = [x^2 \ x \ 1] \begin{bmatrix} a & 0 & -a\sigma^2 \\ -ib & 2a & ib\sigma^2 \\ -c_1 & i(c_2 - 2b) & ic_3 \end{bmatrix} \begin{bmatrix} D^2 \\ D \\ 1 \end{bmatrix}.$$

This is shown by the calculation

$$
\begin{aligned}
&P(iD, ix) \\
&= a(D^2 - \sigma^2)x^2 - ib(D^2 - \sigma^2)x - 2aDx - c_1 D^2 + ic_2 D + c_3 \\
&= a(x^2 D^2 + 4xD + 2 - \sigma^2 x^2) - ib(xD^2 + 2D - \sigma^2 x) \\
&\quad -2a(xD + 1) - c_1 D^2 + ic_2 D + c_3.
\end{aligned}
$$

If in addition $P(iD, ix)$ commutes with χ_τ, then the expression is further simplified since Lemma 7.1 may be applied to both operators. This gives us

Lemma 7.2

If $P(x, D)$ commutes with χ_σ and $P(iD, ix)$ commutes with χ_τ, then

$$
\begin{aligned}
P(x, D) &= a(x^2 - \sigma^2)D^2 + 2axD + a\tau^2 x^2 + c \\
P(iD, ix) &= a(x^2 - \tau^2)D^2 + 2axD + a\sigma^2 x^2 + c
\end{aligned}
$$

and conversely.

The operator $P_{\tau,\sigma}$ (7.5) is an operator of the form considered here. Its Fourier transform operator is $P_{\sigma,\tau}(x, D) = P_{\tau,\sigma}(iD, ix)$. Hence $P_{\tau,\sigma}$ commutes with χ_τ and $P_{\sigma,\tau}$ with χ_σ.

The integral operator $S_{\tau,\sigma}$ of (7.3) may be expressed as

$$S_{\tau,\sigma}\varphi = \sqrt{2\pi}\, \hat{\chi}_\sigma * (\chi_\tau \varphi).$$

Hence we have

$$P_{\tau,\sigma}(\hat{\chi}_\sigma * (\chi_\tau\varphi)) = \hat{\chi}_\sigma * (\chi_\tau P_{\tau,\sigma}\varphi). \tag{7.7}$$

PROOF The conjugate (inverse) Fourier transform of the left side of (7.7) is

$$P_{\sigma,\tau}(\chi_\sigma(\tilde{\chi}_\tau * \tilde{\varphi})) = \chi_\sigma P_{\sigma,\tau}(\tilde{\chi}_\tau * \tilde{\varphi}).$$

By taking the Fourier transform of the right side of this expression, we obtain

$$\hat{\chi}_\sigma * P_{\tau,\sigma}(\chi_\tau\varphi) = \hat{\chi}_\sigma * (\chi_\tau P_{\tau,\sigma}\varphi)$$

which is the right side of (7.7). □

Corollary 7.2
Let $P(x, D)$ be a linear second order differential operator with quadratic coefficients such that $P(x, D)$ commutes with χ_σ, and $P(iD, ix)$ commutes with χ_τ; then

$$P(x, D) = aP_{\sigma,\tau} + b$$

for constants a and b.

This is merely a restatement of Lemma 7.2.

We can come full circle by showing that if Corollary 7.1 holds for an operator $P(x, D)$ then the hypothesis of Corollary 7.2 is satisfied.

THEOREM 7.1
Let $P = P(x, D)$ be a second order linear differential operator with quadratic coefficients; then P commutes with the integral operator $S_{\tau,\sigma}$, if and only if

$$P = aP_{\tau,\sigma} + b.$$

PROOF We must show that $P(x, D)$ commutes with χ_σ and $P(iD, ix)$ with χ_τ. The commutator of $S_{\tau,\sigma}$ and P applied to φ is

$$P(\hat{\chi}_\sigma * (\chi_\tau\varphi)) - \hat{\chi}_\sigma * (\chi_\tau P\varphi) = 0.$$

That of P and χ_τ is

$$P(\chi_\tau\varphi) - \chi_\tau P\varphi = a_{-\tau}\delta'_{-\tau} + a_\tau\delta'_\tau + b_{-\tau}\delta_{-\tau} + b_\tau\delta_\tau \tag{7.8}$$

for some constants $a_{-\tau}$, a_τ, $b_{-\tau}$ and b_τ by calculations similar to those in Lemma 7.1. Similarly the commutator of $\hat{P} = P(iD, ix)$ and χ_σ has the same form

$$\hat{P}(\chi_\sigma\psi) - \chi_\sigma\hat{P}\psi = c_{-\sigma}\delta'_{-\sigma} + c_\sigma\delta'_\sigma + d_{-\sigma}\delta_{-\sigma} + d_\sigma\delta_{-\sigma}. \tag{7.9}$$

We now express the difference of the two sides of (7.7) as

$$
\begin{aligned}
PS_{\tau,\sigma}\varphi - S_{\tau,\sigma}P\varphi &= P(\hat{\chi}_\sigma * (\chi_\tau \varphi)) - \hat{\chi}_\sigma * P(\chi_\tau \varphi) \\
&\quad + \hat{\chi}_\sigma * P(\chi_\tau \varphi) - \hat{\chi}_\sigma * (\chi_\tau P\varphi) \\
&= \mathcal{F}^{-1}\left[(\hat{P}\chi_\sigma - \chi_\sigma \hat{P})(\hat{\chi}_\tau * \hat{\varphi})\right] \\
&\quad + \hat{\chi}_\sigma * (P(\chi_\tau \varphi) - \chi_\tau P\varphi)
\end{aligned}
\tag{7.10}
$$

where $\mathcal{F}^{-1}$ denotes the inverse Fourier transform and we have used the fact that the Fourier transform of $\hat{\chi}_\sigma$ is χ_σ. Therefore (7.10) is a linear combination of the eight functions obtained by substituting (7.8) and (7.9) in (7.10).

These in turn are linear combinations of

$$
\begin{aligned}
&te^{-i\sigma t}, \ te^{i\sigma t}, \ e^{-i\sigma t}, \ e^{i\sigma t}, \ \frac{\sin\sigma(t-\tau)}{(t-\tau)}, \ \frac{\sin\sigma(t+\tau)}{(t+\tau)}, \\
&\frac{(t-\tau)\cos\sigma(t-\tau)-\sin\sigma(t-\tau)}{(t-\tau)^2}, \\
&\frac{(t+\tau)\cos\sigma(t+\tau)-\sin\sigma(t+\tau)}{(t+\tau)^2},
\end{aligned}
$$

which are linearly independent whenever $\sigma\tau \neq 0$. Since by the hypothesis, (7.10) is zero for all φ, it follows that all the coefficients must be zero. Hence (7.8) and (7.9) must be zero, and the hypothesis of Corollary 7.2 is satisfied.   □

Thus, this is indeed a "lucky accident".

7.5   Rademacher functions

The Haar functions, which constituted the first prototype example of wavelets, are orthogonal on $(-\infty, \infty)$ if one takes both m and n ranging over $\mathbb{Z}$ in $\psi_{mn}(t) = 2^{m/2}\psi(2^m t - n)$. However a subset together with $\phi(t)$ constitute an orthogonal system in $[0, 1]$ complete in $L^2(0, 1)$. This subset consists of all $\psi_{mn}(t)$ with support contained in $[0, 1]$, i.e., all $m \geq 0$ and $0 \leq n \leq 2^m - 1$. The <u>Rademacher functions</u> are an orthogonal system on $[0, 1]$ obtained by adding up all the Haar functions at the same scale, i.e.,

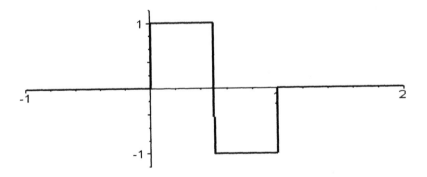

FIGURE 7.1
The Haar mother wavelet.

$$r_0(t) := \phi(t),$$
$$r_1(t) := \psi(t),$$
$$r_{m+1}(t) := \sum_{n=0}^{2^m-1} \psi(2^m t - n).$$

These are clearly orthogonal because of the orthogonality of the sub-spaces $V_0, W_0, W_1, \cdots, W_m, \cdots$ to which $r_0, r_1, \cdots, r_{m+1}, \cdots$ belong. These are also not complete since $\langle \psi_1, r_k \rangle = 0$ for all $k \neq 2$ but ψ_1 is not a multiple of r_2.

Their principal use is to study series of the form

$$\sum_{n=0}^{\infty} \pm c_n \varphi_n(x) \tag{7.11}$$

where the $+$'s and $-$'s are chosen at random and $\{\varphi_n\}$ is another or-thogonal system. This may be rewritten as

$$\sum_{n=0}^{\infty} c_n r_n(t) \varphi_n(x) \tag{7.12}$$

since, except for dyadic rational numbers, $r_n(t) = \pm 1$. In fact, if t has an infinite dyadic expansion

$$t = 0. a_1 a_2 a_3 \cdots$$

and the $+$'s correspond $a_n = 1$ and the $-$'s to $a_n = 0$, then (7.11) and (7.12) are the same. (The sign of c_0 is always taken to be $+$.)

The principal result [Al, p.53] is that (7.11) is convergent almost everywhere for almost all choices of sign when $c_n \in \ell^2$. There are a number of other results related to probability, which may be found in [Z, p.212].

7.6   Walsh function

The Rademacher functions were obtained by combining the Haar functions by simply adding them at a given scale. The Walsh functions take sums and differences of the Haar functions to obtain a complete system. We define

$$w_0(t) := \phi(t), \ w_1(t) := \psi(t), \tag{7.13}$$
$$w_2(t) := \psi(2t) + \psi(2t - 1),$$
$$w_3(t) := \psi(2t) - \psi(2t - 1),$$
$$\cdots$$
$$w_{2n}(t) := w_n(2t) + w_n(2t - 1),$$
$$w_{2n+1}(t) := w_n(2t) - w_n(2t - 1),$$
$$\cdots .$$

Thus these Walsh functions also belong to the wavelet subspaces of the Haar system:

$$w_0 \in V_0, \ w_1 \in W_0, \ w_2, w_3 \in W_1, \ w_4, w_5, w_6, w_7 \in W_2, \cdots$$

$$w_{2^m}, \ w_{2^m+1}, \cdots, w_{2^{m+1}-1} \in W_m, \cdots .$$

Notice that these defining relations (7.13) are exactly the same as those in the two dilation equations of the Haar system,

$$\phi(t) = \phi(2t) + \phi(2t - 1), \tag{7.14a}$$

$$\psi(t) = \phi(2t) - \phi(2t - 1). \tag{7.14b}$$

Since all functions defined by (7.14a) are orthogonal to all defined by (7.14b), it follows that w_{2n} and w_{2n+1} are orthogonal. Also if w_n and w_m are orthogonal so are w_{2n} and w_{2m}. Hence by an induction argument $w_{2^m}, w_{2^m+1}, \cdots, w_{2^{m+1}-1}$ are orthogonal in W_m. Since all of these

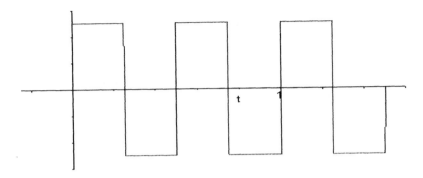

FIGURE 7.2
One of the Rademacher functions.

functions have support contained in [0,1], the $\{w_n\}$ are an orthogonal system in $L^2(0,1)$. Moreover, there are exactly 2^m Haar functions in W_m whose support lies in [0,1], and therefore the Walsh functions in W_m form a basis of this space. Since the Haar functions are complete in $L^2(0,1)$ so are the Walsh functions.

These Walsh functions play an important role in communications theory in network analysis and synthesis. (See, e.g., [Ha].) They also serve as a model for some of the wavelet packets of Coifman, Meyer, and Wickerhauser [C-M̀-W].

7.7 Periodic wavelets

Up to now all the wavelets we have considered are bases of $L^2(\mathbb{R})$. However, in many applied problems, one is more interested in orthonormal systems on a finite interval. One such system can be found by periodizing the scaling functions and wavelets [D, p.304] on the real line. But these and other periodic wavelets can also be constructed directly without reference to wavelets on the real line.

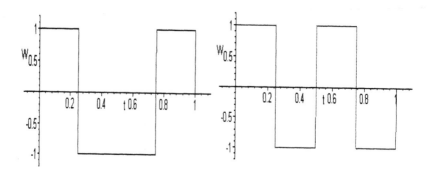

FIGURE 7.3
Two orthogonal Walsh functions.

7.7.1 Periodizing wavelets

Let $\phi \in S_r$ be a scaling function with associated MRA $\{V_m\}$ and mother
wavelet ψ. We define ϕ^*_{mn} and ψ^*_{mn} as the periodized versions of ϕ_{mn}
and ψ_{mn},

$$\phi^*_{mn}(t) := \sum_k \phi_{mn}(t - k) = 2^{m/2}\sum_k \phi(2^m t - 2^m k - n),$$

$$\psi^*_{mn}(t) := \sum_k \psi_{mn}(t - k) = 2^{m/2}\sum_k \psi(2^m t - 2^m k - n).$$

Both are clearly periodic of period 1, and

$$\phi^*_{00}(t) = \sum_k \phi(t - k) = 1.$$

For negative values of m we get nothing new since, e.g.,

$$\phi^*_{-1,n}(t) = \sum_k 2^{-1/2}\phi\left(\frac{t}{2} - n - \frac{k}{2}\right)$$

$$= 2^{-1/2}\left(\sum_j \phi\left(\frac{t}{2} - n - j\right) + \sum_j \phi\left(\frac{t}{2} - n - \frac{1}{2} - j\right)\right)$$

$$= 2^{-1/2}(1 + 1) = 2^{1/2},$$

and similarly for other $\phi^*_{m,n}(t)$ when $m < 0$ which are all in the space
spanned by $\phi^*_{00}(t)$. Thus, the $\phi^*_{mn}(t)$ need only be considered for $m \geq 0$.

We denote by V_m^* the space spanned by $\{\phi_{m,n}^*\}_{n\in\mathbb{Z}}$. This is now a finite dimensional space since $\phi_{m,n}^* = \phi_{m,n+2^m}^*$. In fact, $\phi_{m,0}^*, \phi_{m,1}^*, \cdots, \phi_{m,2^m-1}^*$ are an orthogonal basis in the sense of $L^2(0,1)$ of V_m^* since

$$\int_0^1 \phi_{m0}^*(t)\phi_{mn}^*(t)dt$$

$$= \int_0^1 \sum_k \phi(2^m(t-k)) \sum_j \phi(2^m(t-j)-n)2^m \, dt$$

$$= \sum_j 2^m \sum_k \int_k^{k+1} \phi(2^m t)\phi(2^m(t-j+k)-n)dt$$

$$= 2^m \sum_k \int_k^{k+1} \phi(2^m t) \sum_j \phi(2^m(t-j)-n)dt$$

$$= \sum_j 2^m \int_{-\infty}^{\infty} \phi(2^m t)\phi(2^m(t-j)-n)dt = 0,$$

for $0 < n < 2^m$. These V_m^*'s have the nested property

$$V_0^* \subset V_1^* \subset \cdots \subset V_m^* \subset \cdots \subset L^2(0,1)$$

and

$$\overline{\bigcup_m V_m^*} = L^2(0,1).$$

Again the V_m^*'s are RKHS with RK q_m^* given by

$$q_m^*(x,t) := \sum_{n=0}^{2^m-1} \phi_{mn}^*(x)\phi_{mn}^*(t), \quad x,t \in [0,1].$$

By using the result for $q_m(x,t)$, it can be shown that$\{q_m^*(x,t)\chi_{[0,1]}(x) \chi_{[0,1]}(t)\}$ is a quasi-positive delta sequence. Hence we have

Proposition 7.1
Let $f(t)$ be continuous and periodic, then the projection f_m^* of f onto V_m^* converges to f uniformly.

The corresponding wavelets themselves are considerably simpler in the periodic case as well. If we denote by ψ_0 the constant function

$$\psi_0(t) := \phi_{00}^*(t) = 1$$

and by ψ_n the functions

$$\psi_n(t) = \psi_{2^m + k}(t) = \psi^*_{m,k}(t), \quad 0 \le m, \ k = 0, 1, \cdots 2^m - 1,$$

then the $\{\psi_n\}$ are an orthonormal basis of $L^2(0,1)$ consisting of periodic functions. It is clear that

$$\{\psi_0, \psi_1, \cdots, \psi_{2^m - 1}\}$$

is also an orthonormal basis of V^*_m since there are 2^m of them, and they are orthonormal. The orthogonal expansion of an $f \in L^2(0,1)$ given by

$$f \sim \sum_{n=0}^{\infty} \langle f, \psi_n \rangle \psi_n$$

has as its 2^mth partial sum f^*_m. The other partial sums can be squeezed between f^*_m and f^*_{m+1} for some m, and hence for f continuous and periodic, the series converges to f uniformly. As we observed in Chapter 4, this is not true for trigonometric Fourier series.

7.7.2   Periodic wavelets from scratch

We now construct periodic wavelets by using some of the properties of the periodic wavelets constructed above. We denote by $\phi^*_m = \phi^*_{m,0}$ and observe that it must satisfy the following properties:

(1) It must have a Fourier series $\sum_k \alpha^m_k e^{2\pi i k x}$ whose cofficients satisfy

$$\sum_n |\alpha^m_n|^2 e^{-2\pi i k n 2^{-m}} = \delta_{0k}, k = 0, 1, ..., 2^m - 1,$$

by the orthogonality of translates of ϕ^*_m.

(2) The ϕ^*_m must satisfy a recurrence formula based on their definition

$$\phi^*_{m+1}(x) + \phi^*_{m+1}(x - 1/2) = \sqrt{2} \phi^*_m(2x), 0 \le x \le 1/2,$$

and hence the coefficients must satisfy

$$\alpha^{m+1}_{2k} = \frac{1}{\sqrt{2}} \alpha^m_k, k = 0, \pm 1, ...$$

(3) The dilation equation for the $\phi(x)$ translates into the Fourier coefficients as

$$\alpha_k^m = \alpha_k^{m+1}\mu_{m+1}(k), \text{ where } \mu_{m+1}(k) = \sum_n c_n e^{-2\pi i n k 2^{-m}}$$

and the c_n are the dilation equation coefficients of ϕ.

We now try to find ϕ_m^* which satisfy these three conditions and are as simple as possible. Since these are periodic we look for Fourier series composed of only a finite number of terms, i.e., a trigonometric polynomial. We can then try to define the wavelets themselves in such a way that the usual properties apply.

Example 1.

Clearly we must have $\phi_0^*(x) = 1$. Then we look for ϕ_1^* of the form $\phi_1^*(x) = \alpha_0^1 + \alpha_1^1 e^{i2\pi x} + \alpha_{-1}^1 e^{-i2\pi x}$ which is the simplest possible. By (2), we see that $\alpha_0^1 = \frac{1}{\sqrt{2}}$ and by (1) we deduce that $|\alpha_0^1|^2 - |\alpha_1^1|^2 - |\alpha_{-1}^1|^2 = 0$. If, in addition we would like ϕ_1^* to be real, then the only choice is $\alpha_1^1 = \overline{\alpha_{-1}^1} = \frac{1}{2}e^{i\theta}$. The θ is only a shift parameter, which we may take equal to 0 to get $\phi_1^*(x) = \frac{1}{\sqrt{2}} + \cos 2\pi x$. At the next scale we can use the same type of argument to obtain $\alpha_0^2 = \frac{1}{2}, \alpha_{\pm 2}^2 = \frac{1}{2\sqrt{2}}, \alpha_{2k}^2 = 0, k \neq 0, 1$ but α_{2k-1}^2 arbitrary. These latter coefficients cannot be taken to be zero since that would contradict property (1). Again we look for the simplest choice which is $\alpha_{\pm 1}^2 = \frac{1}{2}$, with the remaining odd coefficients equal to 0. This gives us

$$\phi_2^*(x) = (\frac{1}{\sqrt{2}})^2(1 + 2\cos 2\pi x + \sqrt{2}\cos 2^2 \pi x).$$

We continue in this way and find that

$$\phi_m^*(x) = (\frac{1}{\sqrt{2}})^m(1 + \sqrt{2}\cos 2^m \pi x + 2\sum_{k=1}^{2^{m-1}-1}\cos 2k\pi x) \qquad (7.15)$$

should be a plausible choice for the general form. Now we must check that everything works.

THEOREM 7.2

Let $\phi_m^*(x)$ be given by (7.15) for $m = 0, 1, ...$; then

$$\{\phi_m^*(x - k2^{-m})|k = 0, 1, ..., 2^m - 1\}$$

is periodic and orthonormal on (0,1); its linear span V_m^* consists of trigonometric polynomials of degree $\leq 2^m \pi$ and satisfies:

1) $V_0^* \subset V_1^* \subset \cdots \subset V_m^* \subset \cdots \subset L^2(0,1)$
2) $\bigcup_{m=0}^{\infty} V_m^*$ *is dense in* $L^2(0,1)$.

The associated wavelets ψ_m^* are given by

$$\psi_m^*(x) = \sum_{k=-2^m}^{2^m} \beta_k^m e^{2\pi i k x} \tag{7.16}$$

where

$$\beta_k^m = 2^{\frac{m}{2}+1}[\alpha_{k+2^m}^m + \alpha_{k-2^m}^m]\alpha_k^{m+1} e^{-2\pi i k 2^{-m-1}}. \tag{7.17}$$

Together with $\phi_0^*(x) = 1$, their translates

$$\{\psi_m^*(x - k2^{-m})|k = 0, 1, ..., 2^m - 1, m = 0, 1, ...\}$$

constitute an orthonormal basis of $L^2(0,1)$.

The proof which is quite technical, but not difficult, may be found in[W-C1]. It is observed there that this particular periodic wavelet is not obtained from the periodization of a wavelet on the real line. Many other examples are possible since the odd coefficients may be chosen with considerable freedom. A few are presented in [W-C1]; we present one here and others in Chapter 9.

Example 2.

In example 1, we considered only real scaling functions (even so, we did not consider the possibility that $\phi_1^*(t) = \frac{1}{\sqrt{2}} + \sin 2\pi t$). If we weaken the requirement that $\phi_m^*(t)$ be real, we can find even simpler coefficients by taking $\alpha_n^m = 0$ for $n < 0$. Then $\phi_1^*(t) = \frac{1}{\sqrt{2}} + \frac{1}{\sqrt{2}} e^{2\pi i t}$ satisfies the condition needed for property (1). Similarly for

$$\phi_2^*(t) = \frac{1}{2} + \frac{1}{2} e^{2\pi i t} + \frac{1}{2} e^{4\pi i t} + \frac{1}{2} e^{6\pi i t},$$

we can check that for $k = 1, 2, 3$ the coefficients of the trigonometric series satisfy

$$\sum_{n=0}^{3} |\alpha_n^2|^2 e^{-2\pi i n \frac{k}{4}} = \frac{1}{4} + \frac{1}{4} e^{-i\frac{\pi}{2}k} + \frac{1}{4} e^{-i\pi k} + \frac{1}{4} e^{-i\frac{3}{2}\pi k}$$

$$= \frac{1}{4}\left(\frac{1 - e^{-i2\pi k}}{1 - e^{-i\frac{\pi}{2}k}}\right) = 0,$$

while for $k = 0$, this sum reduces to $\sum_{n=0}^{3} |\alpha_n^2|^2 e^{-2\pi i n \frac{k}{r}} = 1$.

In general, we choose the α_n^m to be constant for $0 \le n \le 2^{m-1}$ to give us

$$\phi_m^*(t) = 2^{-\frac{m}{2}} \sum_{k=0}^{2^m-1} e^{i2\pi kt} = 2^{-\frac{m}{2}} \frac{1 - e^{i2\pi 2^m t}}{1 - e^{2\pi i t}},$$

which enables us to show that the orthogonality condition (1) is satisfied. Furthermore, we claim that

$$e^{2\pi i n t} \in V_m^*, \quad n = 0, 1, \dots, 2^m - 1. \tag{7.18}$$

This is shown by using the fact that

$$\phi_m^* \left(t - \frac{k}{2^m} \right)$$

$$= 2^{-\frac{m}{2}} \left[1 + e^{2\pi i \left(t - \frac{k}{2^m} \right)} + \dots + e^{2\pi i (2^m - 1)\left(t - \frac{k}{2^m} \right)} \right]$$

$$= 2^{-\frac{m}{2}} \left[1 + e^{2\pi i t} e^{-2\pi \frac{k}{2^m}} + \dots + e^{2\pi i (2^{m-1})t} e^{-2\pi i (2^m - 1) \frac{k}{2^m}} \right].$$

If we look for a linear combination

$$l_0 \phi_m^*(t) + l_1 \phi_m^* \left(t - \frac{1}{2^m} \right) + \dots + l_{2^m - 1} \phi_m^* \left(t - \frac{2^m - 1}{2^m} \right) = e^{2\pi i n t},$$

then the coefficient matrix A is

$$A = 2^{-\frac{m}{2}} \begin{bmatrix} 1 & 1 & 1 & \cdots & 1 \\ 1 & e^{-2\pi i \frac{1}{2^m}} & e^{-2\pi i \frac{2}{2^m}} & \cdots & e^{-2\pi i \frac{2^m-1}{2^m}} \\ \cdots & \cdots & \cdots & & \cdots \\ 1 & e^{-2\pi i \frac{2^m-1}{2^m}} & e^{-2\pi i \frac{2(2^m-1)}{2^m}} & \cdots & e^{-2\pi i \frac{(2^m-1)(2^m-1)}{2^m}} \end{bmatrix}$$

which is symmetric. The inverse of A is found to be $\bar{A}$ by observing that

$$1 + e^{2\pi \frac{l-k}{2^m}} + \dots + e^{2\pi i \frac{(2^m-1)(l-k)}{2^m}} = \frac{1 - e^{2\pi i (l-k)}}{1 - e^{2\pi i \frac{l-k}{2^m}}}.$$

Then $\psi_m^*(t - n2^{-m}) \in V_{m+1}^*$ by the claim and it is orthogonal for different values of n as well as being orthogonal to V_m^*. But $\bigcup_{m=0}^{\infty} V_m^*$ is not dense in $L^2(0,1)$ since it contains no trigonometric polynomial with negative coefficients. However it follows from the claim that $\bigcup_{m=0}^{\infty} V_m^* = H^2$, the Hardy space of all analytic functions in D (unit disk) and with boundary values in $L^2(\partial D)$.

7.8   Local sine or cosine basis

A number of procedures have been proposed for using the windowed
Fourier transform to construct an orthonormal basis of $L^2(\mathbb{R})$. This is
given by

$$\int_{-\infty}^{\infty} f(t)e^{iakt}w(t-bn)dt, \qquad n,k \in \mathbb{Z}$$

where $w(t)$ is the window. In general these functions $\{e^{iakt}w(t-bn)\}$
will not be orthogonal. However if $a=1$ and $b=2\pi$ with $w(t)$ given by
the characteristic function of $[0,2\pi)$, $\chi(t)$, it is easy to see that

$$u_{k,n}(t) \; := \; \frac{e^{ikt}}{\sqrt{2\pi}}\chi(t-2\pi n)$$

is orthonormal in $L^2(\mathbb{R})$. Indeed, the restriction $f_n(t)$ of $f(t)$ to the
interval $[2\pi n, 2\pi(n+1)]$ has a Fourier series convergent in the L^2 sense
to the periodic extension of $f_n(t)$. However, even if f is C^∞ this Fourier
series will converge very slowly unless $f(2\pi(n+1))$ and $f(2\pi n)$ match
up, and we will also get Gibbs phenomenon at these end points.

   This is only one of the shortcomings of this basis. The other is that
the Fourier transform of $u_{k,n}(t)$ given by

$$\hat{u}_{k,n}(w) = \frac{e^{-i(w-k)n}}{\sqrt{2\pi}}\left\{\frac{e^{-i(w-k)2\pi}-1}{i(k-w)}\right\}$$

decays very slowly as $|k| \to \infty$. To overcome this we must use a smooth
window $w(t)$. But then it is difficult to obtain orthonormality. A number
of procedures have been proposed. We present one, the local sine and
cosine bases of Coifman and Meyer [C-M]. A detailed exposition of the
construction may be found in [A-W-W].

   The construction begins with a bell $b(x)$ similar to that used in the
construction of the Meyer Wavelets in Chapter 3. Let $h(x)$ be a probability function (or distribution) with support on $[-1,1)$ and $h_\epsilon, \epsilon > 0$, be
its dilation $h_\epsilon(x) = \frac{1}{\epsilon}h(x/\epsilon)$. Then the (ϵ, ϵ') bell over $[\alpha, \beta]$ is defined
to be the positive square root of

$$b^2_{\alpha,\beta}(x) \; := \; \int_{-\epsilon}^{x-\alpha} h_\epsilon - \int_{-\epsilon'}^{x-\beta} h_{\epsilon'}, \qquad (7.19)$$

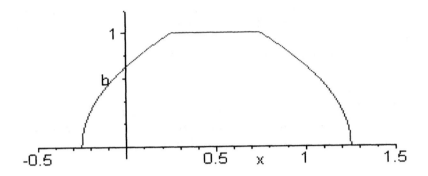

FIGURE 7.4
A bell used for a local cosine basis.

which depends on α, β, ϵ, and ϵ'. If $\epsilon = \epsilon'$ this reduces to the expression

$$b^2(x) = \int_{x-\beta}^{x-\alpha} h_\epsilon.$$

It clearly has support on $[\alpha - \epsilon, \beta + \epsilon']$ and $b(x)b(2\alpha - x)$ has support on $[\alpha - \epsilon, \beta + \epsilon]$. Moreover, $b^2(x) + b^2(2\alpha - x) = 1$ for $2\alpha - \beta + \epsilon' < x < \beta - \epsilon'$, and similar results hold near β as well.

Two adjacent intervals $[\alpha, \beta]$ and $[\beta, \gamma]$ have compatible bells if the (ϵ, ϵ') bell over $[\alpha, \beta]$ and the (ϵ', ϵ'') bell over $[\beta, \gamma]$ satisfy $\gamma - \epsilon'' \geq \beta + \epsilon'$.

We can associate with the bell over $[\alpha, \beta]$ a pair of projection operators $P^\pm = P^\pm_{[\alpha,\beta]}$ on $L^2(\mathbb{R})$ given by

$$(P^\pm f)(x) := b(x)\{b(x)f(x) \pm b(2\alpha - x)f(2\alpha - x) \qquad (7.20)$$
$$\mp b(2\beta - x)f(2\beta - x)\}.$$

The properties of $b(x)$ transform into properties of these operators. Indeed we have

(i)　　$P^{\pm 2} = P^\pm$　(so that $P^\pm$ is a projection)

(ii)　　$P^\pm_{[\alpha,\beta]} + P^\pm_{[\beta,\gamma]} = P^\pm_{[\alpha,\gamma]}$

(iii)　　$P^\pm_{[\alpha,\beta]} P^\pm_{[\beta,\gamma]} = P^\pm_{[\beta,\gamma]} P^\pm_{[\alpha,\beta]} = 0$　(orthogonality).

We now take an increasing sequence $\{a_k\}$ and a sequence $\{\epsilon_k\}$ of positive numbers such that $a_{\pm k} \to \pm\infty$ and

$$a_k + \epsilon_k < a_{k+1} - \epsilon_{k+1}, \qquad k \in \mathbb{Z}. \qquad (7.21)$$

Then for each interval $[a_k, a_{k+1}]$ we find the $(\epsilon_k, \epsilon_{k-1})$ bell $b_k(x)$ and form the projectors $P_k^{\pm}$. Then these projectors are mutually orthogonal and

$$L^2(\mathbb{R}) = \bigoplus_k P_k^+(L^2(\mathbb{R})) \quad (P_k^- \text{ respectively}).$$

See [A-W-W] for more details.

Thus if we can find an orthogonal basis of the Hilbert spaces $\mathcal{H}^{\pm} = P^{\pm}L^2(\mathbb{R})$ we can get an orthogonal basis of all of $L^2(\mathbb{R})$. For simplicity we take the interval $[\alpha, \beta]$ to be $[0, 1]$. Then

$$\left\{ \sqrt{2}\, b(x) \cos\left(k + \frac{1}{2}\right)\pi x \right\}_{k=0}^{\infty}$$

$$\left(\text{resp.} \left\{ \sqrt{2}\, b(x) \sin\left(k + \frac{1}{2}\right)\pi x \right\}_{k=0}^{\infty} \right)$$

is an orthonormal basis of $\mathcal{H}^+$ (resp. $\mathcal{H}^-$).

It is easy to show that $\{\sqrt{2} \cos\left(k + \frac{1}{2}\right)\pi x\}$ is an orthonormal basis of $L^2(0, 1)$ (Chapter 4). Hence on $[0, 1]$,

$$g(x) \; := \; b(x)f(x) + b(-x)f(-x) - b(2 - x)f(2 - x)$$

$$= \sum_{k=0}^{\infty} \sqrt{2}\, c_k \cos\left(k + \frac{1}{2}\right)\pi x$$

where $f \in L^2(\mathbb{R})$ with convergence in $L^2[0, 1]$. Hence on $[0, 1]$

$$(P^+ f)(x) = b(x)g(x) = \sum_{k=0}^{\infty} \sqrt{2}\, c_k b(x) \cos\left(k + \frac{1}{2}\right)\pi x.$$

This can be extended to $[-\epsilon, 1 + \epsilon']$ since both $g(x)$ and $\cos\left(k + \frac{1}{2}\right)\pi x$ are even about $x = 0$ and odd about $x = 1$.

It remains to be shown that $\{w_k\}$, where $w_k(x) = \sqrt{2}\, b(x) \cos\left(k + \frac{1}{2}\right)\pi x$, is orthogonal in $\mathcal{H}^+ = P^+L^2(\mathbb{R})$. By direct calculation

$$\int_{-\epsilon}^{1+\epsilon'} w_k(x)w_j(x)dx$$

$$= \int_{-\epsilon}^{\epsilon} + \int_{\epsilon}^{1-\epsilon'} + \int_{1-\epsilon'}^{1+\epsilon'}$$

$$= 2\int_0^{\epsilon} \{b^2(x) + b^2(-x)\} \cos\left(k + \frac{1}{2}\right)\pi x \cos\left(j + \frac{1}{2}\right)\pi x\, dx$$

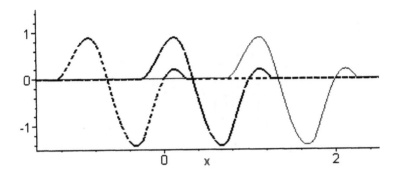

FIGURE 7.5
Three elements in the local cosine basis with bell of Figure 7.4.

$$+2\int_{\epsilon}^{1-\epsilon'}\cos\left(k+\frac{1}{2}\right)\pi x \cos\left(j+\frac{1}{2}\right)\pi x\, dx$$

$$+2\int_{1-\epsilon'}^{1}\left(b^2(x)=b^2(2-x)\right)\cos\left(k+\frac{1}{2}\right)\pi x \cos\left(j+\frac{1}{2}\right)\pi x\, dx$$

$$=2\int_{0}^{1}\cos\left(k+\frac{1}{2}\right)\pi x \cos\left(j+\frac{1}{2}\right)\pi x\, dx = \delta_{kj}$$

since $b^2(x)+b^2(-x)=1$ on $[0,\epsilon]$ and $b^2(x)=b^2(2-x)=1$ on $[1-\epsilon',1]$.

Now by shifting the interval from $[0,1]$ to $[a_k, a_{k+1}]$ we can obtain an orthonormal basis of $\mathcal{H}_k^+ = P_k^+(L^2(\mathbb{R}))$ and hence of $L^2(\mathbb{R})$, and similarly for $\mathcal{H}_k^-$.

THEOREM 7.3
Let $\{a_k\}$ be a sequence of real numbers and $\{\epsilon_k\}$ of positive numbers such that $a_{\pm k} \to \pm\infty$ and

$$a_k + \epsilon_k < a_{k+1} - \epsilon_{k+1};$$

let $b_k(x)$ be the $(\epsilon_k, \epsilon_{k+1})$ bell over $[a_k, a_{k+1}]$; then $\{u_{k,j}\}$ where

$$u_{k,j}(x) = \sqrt{\frac{2}{a_{k+1}-a_k}}\, b_k(x)\cos\frac{(2j+1)\pi(x-a_k)}{2(a_{k+1}-a_k)},$$

$$k = 0, \pm 1, \pm 2, \cdots, \; j = 0, 1, 2, \cdots$$

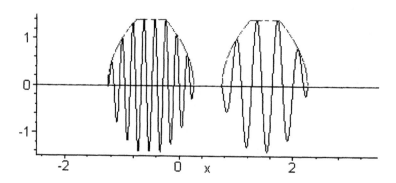

FIGURE 7.6
Two additional elements of the local cosine basis showing the bell.

and $\{v_{k,j}\}$ *where*

$$v_{k,j}(x) = \sqrt{\frac{2}{a_{k+1} - a_k}} b_k(x) \sin \frac{(2j+1)\pi(x - a_k)}{2(a_{k+1} - a_k)},$$

$$k = 0, \pm 1, \pm 2, \cdots, \quad j = 0, 1, 2, \cdots$$

are each an orthonormal basis of $L^2(\mathbb{R})$.

These are very versatile systems since the $\{a_k\}$ are quite general. If it is known that a signal changes its frequency structure at points $\{a_k\}$, then these are points to use. There is also a "merging" algorithm [A-W-W] that enables one to choose an optimal sequence $\{a_k\}$ by removing points successively.

7.9   Biorthogonal wavelets

Each of the techniques used in Chapter 3 to construct orthogonal scaling functions and consequently orthogonal wavelets may be modified to construct biorthogonal pairs of scaling functions. Such constructions allow greater flexibility but at the cost of greater complexity.

The biorthogonal pair will be denoted as $(\phi, \tilde{\phi})$ where each is a function (or distribution) on $\mathbb{R}$ and satisfies

$$\int \phi(t)\overline{\tilde{\phi}(t-n)}dt = \delta_{on}. \tag{7.22}$$

We expect that $\{\phi(t-n)\}$ (resp. $\{\tilde{\phi}(t-n)\}$) is a Riesz basis of its closed linear span V_0 (resp. $\tilde{V}_0$). This leads to a pair of multiresolution analyses,

$$\cdots \subset V_{-1} \subset V_0 \subset V_1 \subset \cdots$$
$$\cdots \subset \tilde{V}_{-1} \subset \tilde{V}_0 \subset \tilde{V}_1 \subset \cdots,$$

a pair of dilation equations,

$$\phi(t) = \sqrt{2}\sum_n h_n\phi(2t-n)$$
$$\tilde{\phi}(t) = \sqrt{2}\sum_n \tilde{h}_n\tilde{\phi}(2t-n),$$

(or in the transformed version

$$\hat{\phi}(w) = m_0(\frac{w}{2})\hat{\phi}(\frac{w}{2}) \tag{7.23}$$
$$\hat{\tilde{\phi}}(w) = \tilde{m}_0(\frac{w}{2})\hat{\tilde{\phi}}(\frac{w}{2})),$$

and a pair of mother wavelets

$$\psi(t) = \sqrt{2}\sum_n \tilde{h}_{-n-1}(-1)^n\phi(2t-n) \tag{7.24}$$
$$\tilde{\psi}(t) = \sqrt{2}\sum_n h_{-n-1}(-1)^n\tilde{\phi}(2t-n).$$

We would like to place conditions on ϕ and $\tilde{\phi}$ such that each $f \in L^2(\mathbb{R})$ has an expansion

$$f = \sum\langle f, \tilde{\psi}_{kj}\rangle\psi_{kj} = \sum\langle f, \psi_{kj}\rangle\tilde{\psi}_{kj},$$

i.e., that $\{\psi_{kj}\}$ and $\{\tilde{\psi}_{kj}\}$ be dual Riesz bases of $L^2(\mathbb{R})$.

There are many possible procedures for choosing ϕ and $\tilde{\phi}$ in order to obtain such a result. The simplest is the one used to construct the Franklin wavelets in Chapter 3. It begins with the hat function

$$\phi(t) := (1 - |t|)\chi_{[-1,1]}(t),$$

156

7. *Other Orthogonal Systems*

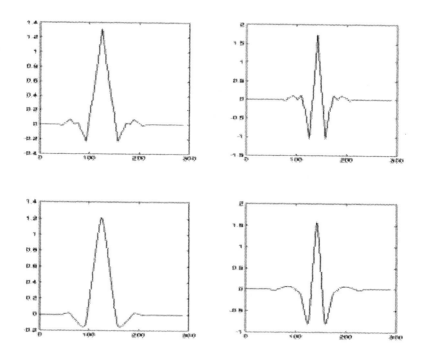

FIGURE 7.7
Two biorthogonal pairs of scaling functions with the same
MRA.

which is not orthogonal to its translates. However $\{\phi(t-n)\}$ is a Riesz basis of V_0. Its Fourier transform is found to be

$$\hat{\phi}(w) = \left[\frac{\sin w/2}{w/2}\right]^2.$$

We must choose $\tilde{\phi}$ such that (7.22) is satisfied or in terms of the Fourier transform such that

$$\sum_k \hat{\phi}(w + 2\pi k)\, \overline{\widehat{\tilde{\phi}}(w + 2\pi k)} = 1. \tag{7.25}$$

We substitute $\hat{\phi}(w)$ into (7.25) and solve for $\tilde{\phi} \in V_0$, i.e.,

$$\widehat{\tilde{\phi}}(w) = \sum_k a_k e^{-iwk}\hat{\phi}(w) = a(w)\hat{\phi}(w).$$

Then (7.25) becomes

$$\sum_k |\hat{\phi}(w + 2\pi, k)|^2 \overline{a(w)} = 1,$$

and since we know the sum of this series (section 3.1, example 2), it follows that

$$a(w) = \frac{1}{1 - \frac{2}{3}\sin^2 \frac{w}{2}}.$$

Hence $\tilde{\phi}$ is given by

$$\widehat{\tilde{\phi}}(w) = \frac{\sin^2 w/2}{(1 - \frac{2}{3}\sin^2 \frac{w}{2})(w/2)^2}.$$

Since $\widehat{\tilde{\phi}}(w)$ is not an entire function $\tilde{\phi}(t)$ cannot have compact support. However $\tilde{\phi}$ is analytic in a strip about the real axis so that the coefficients $\{a_n\}$ and hence $\tilde{\phi}(t)$ have exponential decay.

In this case the multiresolution analysis (MRA) of $\tilde{\phi}$ is identical to that of ϕ since $\tilde{\phi} \in V_0$ and $\{\tilde{\phi}(t-n)\}$, being the biorthogonal sequence, is the dual Riesz basis. The mother wavelets defined by (7.24) lead to a pair of Riesz bases of $L^2(\mathbb{R})$. This and many other biorthogonal bases associated with B-splines are found in the book by C. Chui [C].

A second approach due to Cohen, Daubechies and Feaveau [C-D-F] makes possible the construction of a $\tilde{\phi}$ with compact support (but not in V_0). We illustrate their procedure again with the hat function. The function $m_0(w)$ in (7.23) in this case is given by

$$m_0(w) = \cos^2 \frac{w}{2}.$$

We try to find an $\tilde{m}_0(w)$ such that the biorthogonality condition expressed in terms of m_0 and $\tilde{m}_0$

$$m_0(w)\overline{\tilde{m}_0(w)} + m_0(w + \pi)\overline{\tilde{m}_0(w + \pi)} = 1 \qquad (7.26)$$

is satisfied.

The simplest choice is $\widetilde{m_0}(w) = 1$. This does not lead to a biorthogonal function in $L^2(\mathbb{R})$ but rather the delta distribution. In fact this choice gives another approach to the sampling theorems in Chapter 9.

Another choice is

$$\tilde{m}_0(w) = \cos^2 \frac{w}{2} \left(1 + 2\sin^2 \frac{w}{2}\right)$$

for which (7.26) becomes

$$\cos^4 \frac{w}{2} \left(1 + 2\sin^2 \frac{w}{2}\right) + \sin^2 \frac{w}{2} \left(1 + 2\cos^2 \frac{w}{2}\right) = 1.$$

This is a special case of an expression in [C-D-F] for B-splines of order $2n - 1$ for which

$$m_0(w) = [\cos \frac{w}{2}]^{2n}.$$

It is

$$\tilde{m}_0(w) = \left[\cos \frac{w}{2}\right]^{2\tilde{n}} \sum_{m=0}^{n+\tilde{n}-1} \binom{n + \tilde{n} + m - 1}{m} \left[\sin^2 \frac{w}{2}\right]^m,$$

where $\tilde{n} \geq n$. Each $\tilde{m}_0$ is a trigonometric polynomial, and thus the dilation equation (7.23) has a finite number of terms. Thus by the argument in Chapter 5, the corresponding $\tilde{\phi}$ has compact support. It is shown in [C-D-F] that $\tilde{\phi}$ becomes progressively smoother as $\tilde{n}$ increases.

There is no *a priori* reason that the mother wavelets given in (7.24) generate a Riesz basis of $L^2(\mathbb{R})$. However in this special case they do [C-D-F]. Then one has the representation

$$f(x) = \sum <f, \psi_{kj}> \tilde{\psi}_{kj}(x),$$

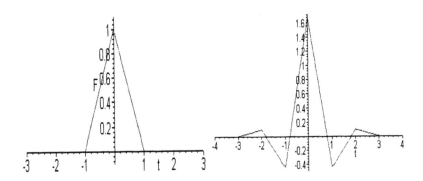

FIGURE 7.8
A biorthogonal pair of scaling functions and wavelets with compact support.

where the $\{\psi_{kj}\}$ have small support, and the $\tilde{\psi}_{kj}$ are smooth but have greater support. However, particularly in the case of splines the representation with $\tilde{\psi}$ and ψ interchanged may also be useful. For the splines of order 1 we have, for $f \in V_m$, the scaling function series

$$f(x) = \sum_n \langle f, \tilde{\phi}_{mn}\rangle \phi_{mn}(x).$$

However we also have the sampling theorem

$$f(x) = \sum_n f(2^{-m}n)\phi_{mn}(x)$$

in this case. Thus a simple quadrature formula

$$\langle f, \tilde{\phi}_{mn}\rangle = f(2^{-m}n)$$

is possible.

7.10 Problems

1. Find the eigenvalues and eigenfunctions of the S-L problems
   (a) $y'' = \lambda y$, $\quad y(0) = y'(\pi) = 0$
   (b) $y'' = \lambda y$, $\quad y(0) + y'(0) = 0$, $\quad y(\pi) = 0$.

2. Find the Green's function $g(t, s)$ for the S-L problems in Problem 1. This involves finding a general solution to $y'' = \delta(t - s)$ and then using the boundary conditions.

3. Let $q(t, s) = \begin{cases} t(1 - s), & t \leq s \\ s(1 - t), & s < t \end{cases}$ be the kernel of an integral operator on $(0, 1)$. Approximate $q(t, s)$ by the first several terms of its double Fourier series

$$\sum_{m=-\infty}^{\infty} \sum_{k=-\infty}^{\infty} q_{mk} e^{2\pi i m t} e^{-2\pi i k s}$$

and find the eigenvalues of the resulting matrix $[q_{mk}]$.

4. Observe that $q(t, s)$ in Problem 3 is the Green's function of a S-L problem. Use this fact to find the exact eigenvalues.

5. Show directly that the differential operator of the Legendre polynomials

$$Py = (1 - x^2)D^2y - 2xDy$$

commutes with multiplication by the characteristic function of $[-1, 1]$.

6. Find the periodized version of the Haar wavelets and explain why it and not the Rademacher system is complete in $L^2(0, 1)$.

7. Let $h(x)$ be the characteristic function of $[-1, 1]$, with associated $(\epsilon, \epsilon\prime)$ bell $b_{\alpha,\beta}$(7.17). Let $a_k = k$, $\epsilon_k = 1/4$. Find the associated local sine basis $v_{k,j}$.

8. Expand the function $e^{-|t|}$ in terms of the "unsatisfactory" local basis

$$w_{k,n}(t) = \frac{1}{\sqrt{2\pi}} e^{ikt} \chi(t - 2\pi n).$$

Chapter 8

Pointwise Convergence of Wavelet Expansions

Much of classical approximation theory is based on delta sequences, i.e., sequences of functions that converge to the delta distribution. For example, and the Dirichlet and Fejer kernels of Fourier series and Fourier transforms, the Christoffel-Darboux formulae for orthogonal polynomials, and the kernel function in statistics are all examples of delta sequences. The rate of convergence of certain types of these sequences in Sobolev spaces has recently been studied by Canuto and Quarteroni [C-Q] and Schomburg [Sc]. This rate is important for estimation in ill posed problems, in statistics [W-B], in numerical analysis, and for sampling and hybrid sampling series [N-W],[W-She2]. Canuto and Quarteroni studied delta sequences based on orthogonal polynomials while Schomburg considered kernel type sequences. In this chapter, we consider the approximation of the delta distribution by certain partial sums of wavelet expansions. We assume that $\phi \in \mathcal{S}_r$ and that $q_m(x,y)$ is the reproducing kernel of V_m given in Chapter 3. These reproducing kernels will constitute one of the delta sequences we shall be interested in. However, they do lead to a phenomenon which is usually undesirable, Gibbs phenomenon [Gi]. In order to remove it we introduce positive delta sequences and their associated estimators in V_m which, however, do not come from projections onto the subspace V_m.

8.1   Reproducing kernel delta sequences

We shall show the reproducing kernel delta sequences $\{q_m(x, y)\}$ to be a quasi-positive delta sequence on $(-\infty, \infty)$ and then find the rate of convergence in terms of the Sobolev norms

$$\|f\|^2_{-\alpha} = \int_{-\infty}^{\infty} |\hat{f}(w)|^2 (w^2 + 1)^{-\alpha} dw, \qquad \alpha > \frac{1}{2}.$$

Since $\hat{\delta}(w) \equiv 1$ this is finite for $f = \delta$ when $\alpha > \frac{1}{2}$ but not when $\alpha \leq \frac{1}{2}$.

These delta sequences have a number of properties not shared by those arising from other orthogonal systems. These are to some extent a consequence of the following properties of the reproducing kernel of V_0, $q(x, y)$ [M, p.33]:

(i) $q(x + 1, y + 1) = q(x, y)$,

(ii) $\left| \dfrac{\partial^{\alpha+\beta}}{\partial x^\alpha \partial y^\beta} q(x, y) \right| \leq C_k (1 + |x - y|)^{-k}, \quad 0 \leq \alpha, \beta \leq r$,

(iii) $\displaystyle\int_{-\infty}^{\infty} q(x, y) y^\alpha dy = x^\alpha, \quad 0 \leq \alpha \leq r$.

This last condition is perhaps the most surprising; it enables us to deduce that the expansion in terms of $\{\phi(x - n)\}$ of any polynomial of degree $\leq r$ converges to it, as we have seen in Chapter 3.

In particular we have

$$1 = \sum_n \phi(x - n) \int \phi(y - n) dy = \sum_n \phi(x - n) \int \phi(y) dy \qquad (8.1)$$

and hence

$$\sum_n \phi(x - n) = C, \qquad x \in \mathbb{R}$$

where C is a constant $\neq 0$. From this we get

$$C \int_0^1 dx = \int_0^1 \sum \phi(x - n) dx = \int_{-\infty}^{\infty} \phi(x) dx,$$

which when substituted into (8.1) gives $1 = C^2$. Thus we may, by changing the sign if necessary, conclude that

$$1 = \int_{-\infty}^{\infty} \phi(x) dx = \sum_n \phi(x - n). \qquad (8.2)$$

This formula will be used below to construct positive scaling functions and subsequently a positive delta sequence in V_m.

The delta sequences based on the kernel, $q_m(x,y)$, have their properties derived from the three conditions above which lead directly to the quasi-positive delta sequences discussed in the next section.

8.2   Positive and quasi-positive delta sequences

A quasi-positive delta sequence is a sequence $\{\delta_m(\cdot, y)\}$ of functions in $L^1(\mathbb{R})$ with parameter $y \in \mathbb{R}$ that satisfy:

(i)   there is a $C > 0$ such that

$$\int_{-\infty}^{\infty} |\delta_m(x,y)| dx \le C, \quad y \in \mathbb{R}, \quad m \in \mathbb{N};$$

(ii)   there is a $c > 0$ such that

$$\int_{y-c}^{y+c} \delta_m(x,y) dx \to 1$$

uniformly on compact subsets of R, as m→ ∞;

(iii)   for each $\gamma > 0$,

$$\sup_{y} \int_{|x-y| \ge \gamma} |\delta_m(x,y)| dx \to 0 \quad \text{as } m \to \infty.$$

It is easy to check and will follow from the next Lemma that $\delta_m(x,y) \to \delta(x-y)$ in the sense of $\mathcal{S}'$ for fixed y. For a positive delta sequence the first condition is replaced by

$$(\text{i}')\ \delta_m(x,y) \ge 0, \quad x, y \in \mathbb{R}.$$

Clearly a positive delta sequence is also quasi-positive since (i′) and (iii) together imply (i). An example of a positive delta sequence is the Fejer kernel,

$$F_m(x,y) = \frac{\sin^2(\frac{m+1}{2}(x-y))}{2(m+1)\pi \sin^2 \frac{x-y}{2}} \chi_{[-\pi,\pi]}(x-y),$$

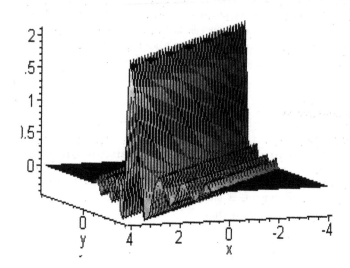

FIGURE 8.1
The delta sequences from Fourier series – the Dirichelet kernel.

while a delta sequence that is <u>not</u> quasi-positive is the Dirichlet kernel
of Fourier series,

$$D_m(x,y) = \frac{\sin((m+\frac{1}{2})(x-y))}{2\pi \sin(\frac{x-y}{2})} \chi_{[-\pi,\pi]}(x-y).$$

Quasi-positive delta sequences that are not positive also arise in Fourier
series. The kernels of (C,α) summability for $0 < \alpha < 1$ are one example
[Z, p. 94].

Lemma 8.1
   *Let $\{\delta_m(x,y)\}$ be a quasi-positive delta sequence ; let $f \in L^1(\mathbb{R})$ be
continuous on (a,b) and bounded on $\mathbb{R}$; then*

$$f_m(y) = \int_{-\infty}^{\infty} \delta_m(x,y)f(x)dx \to f(y) \qquad \text{as } m \to \infty$$

uniformly on compact subsets of (a,b).

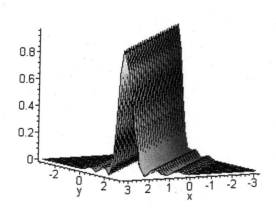

FIGURE 8.2
The delta sequences from Fourier series – the Fejer kernel.

PROOF The proof of this lemma is similar to that for positive delta sequences of the form $\delta_m(x, y) = \delta_m(x - y)$ given in [K1, p.33] or for bounded intervals in [Z, p.87]. We present it here in the interests of completeness.

Let $\gamma > 0$; then

$$
\begin{aligned}
f_m(y) &= \int_{y-\gamma}^{y+\gamma} \delta_m(x, y) f(x) dx + \int_{y+\gamma}^{\infty} + \int_{-\infty}^{y-\gamma} \\
&= f(y) \int_{y-\gamma}^{y+\gamma} \delta_m(x, y) dx \\
&\quad + \int_{y-\gamma}^{y+\gamma} \delta_m(x, y)(f(x) - f(y)) dx + \left\{ \int_{y+\gamma}^{\infty} + \int_{-\infty}^{y-\gamma} \right\} \\
&= I_1 + I_2 + I_3.
\end{aligned}
\tag{8.3}
$$

Now let $[\alpha, \beta] \subset (a, b)$ be a closed subinterval containing the compact subset, and let $y \in [\alpha, \beta]$. We choose γ in (8.4) so that $0 < \gamma < c$, $\beta + \gamma < b$, and $\alpha - \gamma > a$. Then for any $0 < \epsilon < 1$, we restrict f further so that

$$
|f(x) - f(y)| < \epsilon \quad \text{for} \quad x \in [a, b] \quad \text{and} \quad |x - y| < \gamma.
$$

From this we deduce that

$$
|I_2| \le \epsilon \int_{y-\gamma}^{y+\gamma} |\delta_m(x, y)| dx
\tag{8.4}
$$

and that

$$
|I_3| \le \sup_y \int_{\gamma \le |x-y|} |\delta_m(x, y)| dx \, \|f\|_\infty < \epsilon \|f\|_\infty, \quad m \ge M_1,
\tag{8.5}
$$

where we have chosen M_1 so large that $\int_{\gamma \le |x-y|} |\delta_m(x, y)| < \epsilon$ for $m \ge M_1$. We then choose $M_2 \ge M_1$ so large that

$$
\left| 1 - \int_{y-\gamma}^{y+\gamma} \delta_m(x, y) dx \right| < \epsilon, \quad m \ge M_2.
\tag{8.6}
$$

This is possible since by (8.3) (iii)

$$
\int_{y+\gamma}^{y+c} \delta_m(x, y) dx \to 0
$$

uniformly on $\mathbb{R}$ as $m \to \infty$. We now combine (8.4), (8.5), and (8.6) to get

$$
\begin{aligned}
|f(y) - f_m(y)| &\leq |f(y) - I_1| + |I_2| + |I_3| \\
&\leq |f(y)| \left| 1 - \int_{y-\gamma}^{y+\gamma} \delta_m(x,y)dx \right| + \epsilon \int_{-\infty}^{\infty} |\delta_m(x,y)|dx + \epsilon \|f\|_\infty \\
&\leq \sup_{y \in [\alpha,\beta]} |f(y)|\epsilon + \epsilon C + \epsilon \|f\|_\infty
\end{aligned}
$$

for $m \geq M_2$, which gives us the desired uniform convergence. □

For positive delta sequences further results are possible.

Lemma 8.2

Let $\{\delta_m(x,y)\}$ be a positive delta sequence f; let $f \in L^1(\mathbb{R})$ be bounded on $\mathbb{R}$ and satisfy $M_1 \leq f(x) \leq M_2$; then

$$
f_m(y) = \int_{-\infty}^{\infty} \delta_m(x,y)f(x)dx
$$

also satisfies $M_1 \leq f_m(x) \leq M_2$.

The proof follows directly from the definitions.

We now show that the reproducing kernels $q_m(x,y)$ constitute a quasi-positive delta sequence f [W3].

Lemma 8.3

Let $q_m(x,y)$ be the reproducing kernel of V_m, $\phi \in \mathcal{S}_r$; then $\{q_m(x,y)\}$ satisfies (8.3) (i), (ii), and (iii).

PROOF The inequality (8.3) (i) follows from the fact that

$$
\begin{aligned}
\int_{\infty}^{\infty} |q_m(x,y)|dx &= \int_{-\infty}^{\infty} 2^m |q(2^m x, 2^m y)|dx \\
&= \int_{-\infty}^{\infty} |q(x, 2^m y)|dx \leq C_2 \int_{-\infty}^{\infty} (1 + |x - 2^m y|)^{-2}dx = C.
\end{aligned}
$$

The second condition (ii) is obtained by a change of scale; let $c > 0$, then

$$
\int_{y-c}^{y+c} q_m(x,y)dx = \int_{2^m(y-c)}^{2^m(y+c)} q(x, 2^m y)dx
$$

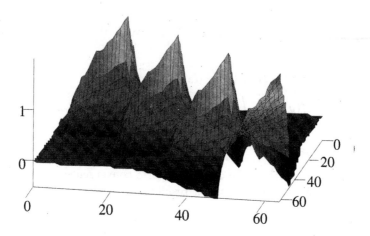

FIGURE 8.3
The quasi–positive delta sequence for the Daubechies wavelet
$_2\phi$, $m = 0$.

$$= \int_{t-2^m c}^{t+2^m c} q(x,t)dx$$

$$= 1 - \int_{t+2^m c}^{\infty} - \int_{-\infty}^{t-2^m c}$$

$$= 1 - I_1 - I_2.$$

I_1 satisfies

$$|I_1| \le C_2 \int_{t+2^m c}^{\infty} \frac{1}{1 + (t - x)^2}dx = C_2 \int_{2^m c}^{\infty} \frac{1}{1 + x^2}dx \to 0$$

as $m \to \infty$ and the same for I_2. The third condition also follows from
the same inequality for any $\gamma > 0$. $\square$

Corollary 8.1

Let $f \in L^1 \cap L^\infty(\mathbb{R})$, *continuous on* (a, b); *let* f_m *be the projection of* f
onto V_m; *then*

$$f_m \to f \qquad \text{as} \quad m \to \infty$$

uniformly on compact subsets of (a, b).

This follows from the lemmas since the projection is given by

$$f_m(x) = \int_{-\infty}^{\infty} q_m(x,y)f(y)dy.$$

8.3 Local convergence of distribution expansions

The global convergence of the expansions of distributions in (S'_r) discussed in Chapter 5 is important for theoretical purposes. However, for computational purposes, it is desirable to have some sort of local convergence. In the case of distributions we should like to have it converge to the value of the distribution at a point [Lo]. This is a weaker version of continuity of a function at a point.

Definition 8.1   Let $f \in S'$; then f is said to have a value γ of order r at x_0 if there exists a continuous function $F(x)$ of polynomial growth such that $D^r F = f$ in some neighborhood of x_0 and

$$\lim_{x \to x_o} \frac{F(x)}{(x - x_0)^r} = \frac{\gamma}{r!}.$$

EXAMPLES The δ distribution has values 0 everywhere except at 0; the function $\sin \frac{1}{x}$, considered as an element of S', has value 0 at 0.

Lemma 8.4
[W3] Let $q_m(x,t)$ be the reproducing kernel of V_m, with $\phi \in S_r$, let $0 \le \alpha \le r$; then

$$K_m(x,t) := \frac{(x-t)^\alpha}{\alpha!} \frac{\partial^\alpha}{\partial t^\alpha} q_m(x,t)$$

is a quasi-positive delta sequence on $\mathbb{R}$.

PROOF We must show that the three conditions given in (8.3) hold. To prove (i) we observe first that since $|\phi^{(\alpha)}(x)| \le C_{\alpha k}(1 + |x|)^{-k}$,

$$\int |K_0(x,t)|dt$$

$$\le \sum_{k=0}^{\alpha} \binom{\alpha}{k} \sum_n |\phi(x-n)| \, |x-n|^k \int |\phi^{(\alpha)}(t)| t^{\alpha-k} dt \le C,$$

a constant. The same holds for $K_m(x,t)$ since it is given by

$$
K_m(x,t)
$$
$$
= \frac{[2^m(x-t)]^\alpha}{\alpha!}\frac{\partial^\alpha}{\partial(2^mt)^\alpha}2^m q_\alpha(2^mx,2^mt) = 2^m K_0(2^mx,2^mt).
$$

To prove (ii) we observe first that

$$
\int_{-\infty}^{\infty} K_m(x,y)dy = \int_{-\infty}^{\infty} q_m(x,y)dy = 1
$$

by integration by parts. Thus it remains to be shown that for some $c > 0$,

$$
\left\{\int_{x+c}^{\infty} + \int_{-\infty}^{x-c}\right\} K_m(x,y)dy \to 0
$$

as $m \to 0$ uniformly for x in a bounded set. This is shown by the same sorts of tricks as in the proof of (i),

$$
\int_{x+c}^{\infty} K_m(x,y)dy
$$

$$
= \int_{2^m(x+c)}^{\infty} K_0(2^mx,z)dz
$$

$$
= \sum_n \sum_{k=0}^{\alpha}(2^mx-n)^k\binom{\alpha}{k}\phi(2^mx-n)\int_{2^m(x+c)}^{\infty}\phi^{(\alpha)}(z-n)(z-n)^{\alpha-k}dz.
$$

This last integral satisfies

$$
\left|\int_{2^m(x+c)}^{\infty}\phi^{(\alpha)}(z-n)(z-n)^{\alpha-k}dz\right|
$$

$$
\leq \int_{2^m(x+c)-n}^{\infty}\frac{C_{\alpha j}}{(1+|z|)^j}|z|^{\alpha-k}dz
$$

$$
\leq \frac{C_{\alpha j}}{(1+|2^m(x+c)-n|)^{j-\alpha+k-2}}\int_{-\infty}^{\infty}\frac{1}{(1+|z|)^2}dz,\ j \geq \alpha-k+2.
$$

Hence

$$
\left|\int_{x+c}^{\infty}K_m(x,y)dy\right| \leq C\sum_n\frac{1}{(|2^mx-n|+1)^p}\frac{1}{(1+|2^m(x+c)-n|)^p}
$$

for all $p \geq 1$. Since

$$
\frac{1}{1+|x-a|}\frac{1}{1+|x-b|} \leq \frac{1}{1+|a-b|},
$$

it follows that for $p > 2$

$$\left| \int_{x+c}^{\infty} K_m(x,y)dy \right|$$

$$\leq \frac{C_p}{1+2^m c} \sum_n \frac{1}{(1+|2^m x - n|)^{p-1}} \frac{1}{(1+|2^m(x+c)-n|)^{p-q}}$$

$$= \frac{C_p}{1+2^m c} h_p(2^m x, 2^m(x+c))$$

where

$$h_p(x,y) := \sum_n \frac{1}{(1+|x-n|)^{p-1}} \frac{1}{(1+|y-n|)^{p-1}}$$

is uniformly bounded in x and y. Hence (ii) holds.

For (iii) we use the same argument to get a bound on

$$\int_{|x-y|>\gamma} |K_m(x,y)|dy \leq \frac{C\|h_p\|_\infty 2^m}{(1+2^m\gamma)^2},$$

which clearly converges to 0 as $m \to \infty$. □

THEOREM 8.1

[W3] Let $f \in S'_{r-1}$ and have a value γ of order $\alpha \leq r$ at $x = x_0$, then the function f_m given by $f_m(x) = (f, q_m(x, \cdot))$ satisfies

$$f_m(x_0) \to \gamma \qquad \text{as } m \to \infty.$$

PROOF Each $f \in S'_{r-1}$ is a derivative of order $\beta \leq r$ of a continuous function G of polynomial growth. We may assume, by adding a polynomial of degree $< \beta$ if necessary, and by increasing α if necessary, that $G = F$ and $\alpha = \beta$. Then

$$f_m(x) = \int_{-\infty}^{\infty} (-1)^\alpha \partial_y^\alpha q_m(x,y) F(y) dy$$

$$= \int_{-\infty}^{\infty} \frac{(x-y)^\alpha}{\alpha!} \partial_y^\alpha q_m(x,y) \frac{F(y)\alpha!}{(y-x)^\alpha} dy$$

$$= \int_{x-A}^{x+A} + \int_{x+A}^{\infty} + \int_{-\infty}^{x-A}.$$

By a repetition of the argument in the proof in the Lemma 8.4 of (ii) we see that

$$\int_{x+A}^{\infty} + \int_{-\infty}^{x-A} \to 0 \qquad \text{as } m \to \infty.$$

Hence we may express f_m as

$$f_m(x) = \int_{-\infty}^{\infty} K_m(x,y) F_A(x,y)\,dy$$

where $F_A(x,y)$ is continuous for all y except for $y = x \pm A$ and has compact support. Clearly, F_A is bounded as well. Hence

$$f_m(x_0) \to F_A(x_0, x_0) \qquad \text{as } m \to \infty$$

by the property of quasi-positive delta sequences. Since $F_A(x_0, x_0) = \gamma$, the theorem is proved. $\qquad\square$

8.4   Convergence almost everywhere

If f is not continuous but is merely in $L^2(\mathbb{R})$, we cannot expect f_m to converge to f pointwise everywhere since f can be completely arbitrary on a set of measure zero. However for Fourier series we have, by Carleson's theorem [E, p.162], convergence almost everywhere to $f(x)$ for $f \in L^2(0, 2\pi)$. We would expect the same thing to hold for wavelets since their convergence properties are usually stronger than Fourier series. This was shown to be true by S. Kelly [Ke], [Ke-K-R] for $f \in L^p(\mathbb{R})$, $1 \le p \le \infty$. This is again stronger than the result for Fourier series since there exist functions in $L^1(0, 2\pi)$ whose Fourier series diverge everywhere [Z, p.310].

THEOREM 8.2
[Ke] Let $f \in L^p(\mathbb{R})$, $1 \le p \le \infty$, and let x be in the Lebesgue set of f; then

$$f_m(x) \to f(x) \qquad \text{as } m \to \infty.$$

In particular $f_m \to f$ almost everywhere.

The Lebesgue set is composed of all x such that

$$\frac{1}{r} \int_{-r}^{r} (f(x,t) - f(x))\,dt \to 0. \tag{8.7}$$

This holds almost everywhere on $\mathbb{R}$ if f is locally integrable [S-W, p.12]. The proof of the theory involves inequalities similar to those in Section 8.1 and will be omitted. A complete version may be found in [Ke-Ko-Ra].

8.5 Rate of convergence of the delta sequence

In order to find a rate of convergence of the wavelet expansion, we need to determine the rate at which $q_m(x, y)$ converges to $\delta(x - y)$. This rate will be in terms of the Sobolev norms mentioned at the start of the chapter.

The hypothesis will require some properties of the Zak transform[H-W] of the function ϕ,

$$Z\phi(t, w) := \sum_k e^{-iwk}\phi(t - k). \tag{8.8}$$

Since $\phi \in S_r$, this series will converge to a function which is periodic and C^∞ in w. By (8.2), $Z\phi(t, 0) = 1$ and hence $e^{iwt}(Z\phi)(t, w) = 1 + 0(|w|)$. Also by (8.2) $\hat{\phi}(0) = 1$, and similarly $\hat{\phi}(w) = 1 + 0(|w|)$.

In some cases, such as the Meyer wavelet or the Coiflets [D, p.258], $\hat{\phi}(w) = 1 + 0(|w|^\lambda)$ for larger values of λ. This is true also for the Zak transform. Therefore, the following definition is always satisfied for some $\lambda \geq 1$.

Definition 8.2   *Let $\phi \in S_r$ be a scaling function; then ϕ satisfies property Z_λ if*

(i) $\hat{\phi}(w)1 + 0(|w|^\lambda)$ *as $w \to 0$*
(ii) $(Z\phi)(t, w) = e^{-iwt}(1 + 0(|w|^\lambda))$ *uniformly as $w \to 0$.*

This property Z_λ is related to the derivatives of $\hat{\phi}(w)$ at $w = 0$. For $\phi \in S_r$, $\hat{\phi} \in C^\infty$ and hence if $\hat{\phi}^{(k)}(0) = 0$, $k = 1, 2, \cdots, \lambda - 1$, then

$$\hat{\phi}(w) = 1 + O(|w|^\lambda) \qquad \text{as } |w| \to 0.$$

For the Zak transform we have

$$\frac{d^k}{dw^k}e^{iwt}(Z\phi)(t, w)|_{w=0}$$
$$= \sum_n (i(t - n))^k \phi(t - n).$$

Since we have the property that

$$\int x^k q(x, t)dx = t^k, \qquad k = 0, 1, \cdots, r$$

(Chapter 3) for $\phi \in S_r$, it follows that

$$t^k = \int x^k \sum_n \phi(x-n)\phi(t-n)dx$$

$$= \sum_n \int \sum_{j=0}^{k} \binom{k}{j}(x-n)^j(n)^{k-j}\phi(x-n)dx\,\phi(t-n)$$

$$= \sum_n \sum_{j=0}^{k} \binom{k}{j}n^{k-j}\int x^j\phi(x)dx\,\phi(t-n) = \sum_n n^k\phi(t-n) \quad (8.9)$$

provided $\widehat{\phi}^{(j)}(0) = 0$, $j = 1, 2, \cdots, k$. Moreover, $\sum_n \phi(t-n) = 1$ and therefore

$$\sum_n (t^j - n^j)\phi(t-n) = 0, \qquad j = 0, 1, 2, \cdots, k.$$

From this it is a simple induction step to show that

$$\sum_n (t-n)^k \phi(t-n) = 0, \qquad k = 1, 2, \cdots \min(r, \gamma - 1). \qquad (8.10)$$

Hence the Zak transform satisfies

$$e^{iwt}(Z\phi)(t, w) = 1 + O(|w|^\gamma)$$

provided $\widehat{\phi}^{(k)}(0) = 0$, $k = 1, 2, \cdots, \min(r, \gamma - 1)$, and we can conclude that $\phi \in Z_\lambda$ if $r \geq \lambda - 1$ and $\widehat{\phi}^{(k)}(0) = 0$, $k = 1, 2, \cdots, \lambda - 1$.

Before we state the result we make two more observations. Since $\phi \in L^1(\mathbb{R})$, $|\widehat{\phi}(w)| \leq \widehat{\phi}(0) = 1$, and

$$\int_0^{2\pi} |Z\phi(t, w)|^2 dw = 2\pi \sum_k |\phi(t-k)|^2$$

$$\leq 2\pi \sum_k \frac{C_1^2}{(1+|t-k|)^2} \leq C \qquad (8.11)$$

where C is a constant.

Lemma 8.5

Let $\phi \in S_r$ be a scaling function that satisfies property Z_λ for some $\lambda > 0$; let $q_m(x, y)$ be the reproducing kernel of V_m; then

$$\|q_m(\cdot, y) - \delta(\cdot - y)\|_{-\alpha} = O(2^{-m\lambda}) \quad uniformly \ for \ y \in \mathbb{R}, \qquad (8.12)$$

where $\| \ \|_{-\alpha}$ is the Sobolev norm and $\alpha > \lambda + \frac{1}{2}$.

PROOF The proof is the same as that in [W4] with a few simplifications.

Let $\epsilon(x,y) = q(x,y) - \delta(x-y)$ and denote by ϵ_m

$$\epsilon_m(x,y) = 2^m \epsilon(2^m x, 2^m y).$$

The left side of (8.12) can be expressed as

$$\|\epsilon_m(\cdot,y)\|_{-\alpha}^2 = \int_{\mathbb{R}} (1+w^2)^{-\alpha} |\hat{\epsilon}_m(w,y)|^2 dw$$

$$= 2^m \int_{\mathbb{R}} \left(1 + 2^{2m}\xi^2\right)^{-\alpha} |\hat{\epsilon}(\xi, 2^m y)|^2 d\xi$$

$$= 2^m \left\{ \int_{|\xi| \le 1} + \int_{|\xi| > 1} \right\} = I_1 + I_2. \tag{8.13}$$

Since the Fourier transform of $q(x,y)$ is given by

$$\hat{q}(w,y) = \sum_k e^{-iwk} \hat{\phi}(w)\phi(y-k) = \hat{\phi}(w)Z\phi(y,w), \tag{8.14}$$

it follows from condition Z_λ that

$$\hat{q}(w,y) = e^{-iwy}(1 + 0(|w|^\lambda)),$$

and

$$\hat{\epsilon}(w,y) = e^{-iwy} 0(|w|^\lambda).$$

Hence the first integral in (8.13) satisfies

$$|I_1| \le 2^{m+1} \int_0^1 (1 + 2^{2m}\xi^2)^{-\alpha} C_1 |\xi|^{2\lambda} d\xi$$

$$\le 2^{-2m\lambda} C_2 \int_0^{2^m} (1+w^2)^{-\alpha} w^{2\lambda} dw = 0(2^{-2m\lambda}). \tag{8.15}$$

The second integral satisfies

$$|I_2| \le 2^m \int_{|\xi|>1} (1 + 2^{2m}\xi^2)^{-\alpha} |\hat{q}(\xi, 2^m y) - e^{-i\xi 2^m y}|^2 d\xi$$

$$\le 2^{m(1-2\alpha)+1} \int_{|\xi|>1} |\xi|^{-2\alpha} \left(|\hat{q}(\xi, 2^m y)|^2 + 1\right) d\xi.$$

Since $|\hat{\phi}(w)| \leq \|\phi\|_1$, we have by (8.14)

$$|I_2| \leq 2^{m(1-2\alpha)+1} \int_{|\xi|>1} |\xi|^{-2\alpha} \left(|\hat{\phi}(\xi)|^2 |Z\phi(2^m y, \xi)|^2 + 1 \right) d\xi$$

$$\leq 2^{m(1-2\alpha)+2} \|\phi\|_1 \sum_{k=0}^{\infty} \int_0^{2\pi} |Z\phi(2^m y, \xi + 1)|^2 |\xi + 1 + 2\pi k|^{-2\alpha} d\xi$$

$$+ \frac{2^{m(1-2\alpha)+2}}{2\alpha - 1}$$

$$\leq 2^{m(1-2\alpha)+2} \left\{ \sum_{k=0}^{\infty} |2\pi k + 1|^{-2\alpha} C + \frac{1}{2\alpha - 1} \right\} \qquad (8.16)$$

$$= O(2^{-m(2\alpha-1)})$$

where C is a constant. By using (8.15) and (8.16) in (8.13) and observing that $2^{-m(2\alpha-1)} < 2^{-2\lambda m}$, we obtain the required inequality. $\qquad \square$

Corollary 8.2

Let $f \in H^\alpha$, let ϕ be as in Lemma 8.5, $\alpha > \lambda + \frac{1}{2}$; then the projection f_m of f onto V_m satisfies

$$\|f - f_m\|_\infty = O(2^{-m\lambda}).$$

This follows from the Schwarz inequality for Sobolev spaces. If $f \in H^\alpha$ and $g \in H^{-\alpha}$, then since $H^{-\alpha}$ is the dual space to H^α,

$$|\langle g, f \rangle|^2 = \left| \int \hat{g}(w) \overline{\hat{f}(w)} dw \right|^2 \leq \|g\|_{-\alpha}^2 \|f\|_\alpha^2,$$

and hence

$$|f(y) - f_m(y)| = |\langle f, \delta(\cdot - y) - q_m(\cdot, y) \rangle|$$
$$\leq \|f\|_\alpha \|\delta(\cdot - y) - q_m(\cdot, y)\|_{-\alpha} = O(2^{-m\lambda}).$$

8.6 Other partial sums of the wavelet expansion

The results of the last section may be given in terms of a partial sum of a wavelet expansion since

$$f_m(x) = \sum_{n=-\infty}^{\infty} a_{mn} 2^{m/2} \phi(2^m x - n)$$

$$= \sum_{k=-\infty}^{m} \sum_{n=-\infty}^{\infty} b_{kn} 2^{k/2} \psi(2^k x - n), \qquad (8.17)$$

where a_{mn} are the scaling coefficients and b_{kn} are the wavelet coefficients. In practice all of the infinite sums in (8.17) must be truncated, and partial sums of the form

$$f_{m,p,r,s}(x) = \sum_{k=-p}^{m} \sum_{n=-r}^{s} b_{kn} 2^{k/2} \psi(2^k x - n) \qquad (8.18)$$

must be considered.

We first consider the outer sums. Since

$$f_m(x) - f_{-p-1}(x) = \sum_{k=-p}^{m} \sum_{n=-\infty}^{\infty} b_{kn} 2^{k/2} \psi(2^k x - n),$$

we need merely find bounds on $f_m(x)$ as $m \to -\infty$. We already know, from the multiresolution analysis, that

$$\|f_m\|^2 \to 0 \qquad \text{as} \quad m \to -\infty,$$

but we don't know anything about the uniform convergence as $m \to -\infty$.

To show that it also converges to zero, we again use the Sobolev norm applied to the reproducing kernel

$$\|q_m(\cdot, y)\|_{-\alpha}^2$$

$$= 2^m \int \left(1 + 2^{2m} \xi^2\right)^{-\alpha} |\hat{q}(\xi, 2^m y)|^2 d\xi$$

$$= 2^m \int \left(1 + 2^{2m} \xi^2\right)^{-\alpha} |\hat{\phi}(\xi)|^2 |Z\phi(2^m y, \xi)|^2 d\xi. \qquad (8.19)$$

Since the Zak transform satisfies

$$|Z\phi(t, \xi)| \le \sum_{n} |\phi(t - n)|$$

and hence is uniformly bounded, (8.19) is dominated by

$$\|q_m(\cdot, y)\|_{-\alpha}^2 \leq 2^m \int_{-\infty}^{\infty} |\hat{\phi}(\xi)|^2 d\xi = (2\pi)2^m.$$

Thus for all $\alpha \geq 0$, we have the desired condition.

Corollary 8.3
  Let f, ϕ be as in Corollary 8.2; let $f_{m,p,r,s}(x)$ be given by (8.18); then

$$f_{m,p,r,s}(x) \rightarrow f(x) \qquad as \quad m, p, r, s \rightarrow \infty$$

uniformly on compact subsets of $\mathbb{R}$.

The remaining calculations are routine and will be omitted.

REMARK 8.1   The two convergence properties considered in this chapter, uniform convergence of the wavelet expansions for continuous functions and more rapid convergence for smoother functions, are not shared by the usual classical orthogonal systems. For example, the partial sums of the Fourier series of a continuous function do not necessarily converge. In fact there are continuous functions whose Fourier series diverge at a dense set of points [Z, p.298]. To overcome this lack of convergence, summability methods are frequently used instead. These give convergence for continuous functions but do not converge more rapidly for smooth functions [Z, p.122]. They can also be used to overcome Gibbs phenomenon. Unfortunately the usual summability methods do not apply to wavelet expansions and we must use another approach which gives us positive delta sequences. But first we look at Gibbs phenomenon. ∎

8.7   Gibbs phenomenon

As we saw in Chapter 4, the Fourier series of a function with a jump discontinuity exhibits Gibbs phenomenon [Gi], [Gi1]. That is, the partial sums overshoot the function near the discontinuity, and this overshoot continues no matter how many terms are taken in the partial sum. More precisely, if $f(x)$ is a piecewise Lipschitz continuous function with a positive jump discontinuity at $x = a$, then there is a sequence $x_n \to a^+$ such that $\lim\limits_{n \to \infty} s_n(x_n) > f(a^+)$. Here $s_n(x)$ is the sequence of partial sums of the Fourier series.

Gibbs phenomenon does not occur if the partial sums are replaced by the $\sigma_n(x)$, i.e., the average of the partial sums [Z, p.110]. Since the wavelet expansions have convergence properties similar to Cesàro summability, we might expect them not to exhibit Gibbs phenomenon. However, it has been shown in [Ke1], [S-V] that all standard wavelet expansionswavelet expansions do exhibit this overshoot phenomenon at the origin.

Lemma 8.6
[Ke1] The wavelet expansion for $\phi \in S_r$ with reproducing kernel $q(x,t)$ exhibits Gibbs phenomenon at 0 if there is an $a > 0$ such that

$$\int_0^\infty q(a,t)dt > 1 \tag{8.20}$$

(and/or an $a < 0$ such that $\int_{-\infty}^0 q(a,t)dt < 0$).

PROOF The proof involves the projection onto V_m of one of the Haar functions (Chapter 1). We take h to be the function

$$h(t) = \chi_{[0,1)}(t) - \chi_{[0,1)}(t+1)$$

and show that there is a sequence $\{x_n\}$ such that

$$\lim_{m \to \infty} h_m(x_m) > 1$$

if (8.20) is satisfied. Here h_m is the projection onto V_m, and

$$h_m(x) = \int q_m(x,t)h(t)dt$$

$$= \int_0^1 q_m(x,t) - q_m(x,-t)dt$$

$$= \int_0^{2^m} [q(2^m x, t) - q(2^m x, -t)]dt.$$

Let $x = 2^{-m}a$ where a is the value in (8.20). Then

$$h_m(2^{-m}a) = \int_0^{2^m} [q(a,t) - q(a,-t)]dt \to \int_0^\infty q(a,t) - q(a,t)dx$$

$$= 2 \int_0^\infty q(a,t)dt - \int_{-\infty}^\infty q(a,t)dt = 2 \int_0^\infty q(a,t)dt - 1 > 1.$$

This proves the first condition is sufficient to show the phenomenon for this particular function; the second is proved the same way. Any other piecewise Lipschitz continuous function with a jump discontinuity at 0 may be made continuous in a neighborhood of 0 by adding an appropriate multiple of $h(t)$. □

With this basic lemma it is possible to show that expansions in terms of all of the standard wavelet expansions exhibit Gibbs phenomenon. In fact Shim and Volkmer [S-V] have proved the following result

THEOREM 8.3

[S-V] Let ϕ be a continuous scaling function such that $\phi'(d) \neq 0$ for some dyadic rational number d and

$$\phi(x)| \leq C(1 + |x|)^{-\beta} \text{ for } x \in \mathbb{R}$$

with constants C and $\beta > 3$. Then the corresponding wavelet expansion shows Gibbs phenomenon on the right or the left at 0.

Most of the standard wavelets satisfy this condition. The two exceptions are the prototypes of Chapter 1.

The Haar wavelets do not and also do not exhibit Gibbs phenomenon. This can also be shown directly since $q_m(x,t)$ is a <u>positive</u> delta sequence in this case [Z, p.86].

For our other prototype, the Shannon wavelets, the decay condition is not satisfied. However the condition in the lemma is related to the overshoot for Fourier series (Chapter 4) since (8.20) becomes

$$\int_0^\infty \frac{\sin \pi(t-x)}{\pi(t-x)}dt = \frac{1}{\pi}\int_{-\pi x}^\infty \frac{\sin s}{s}ds$$

$$= \frac{1}{\pi}\int_0^\infty \frac{\sin s}{s}ds + \frac{1}{\pi}\int_{-\pi x}^0 \frac{\sin s}{s}ds.$$

If $x = 1$, then $\frac{1}{\pi}\int_{-\pi}^0 \frac{\sin s}{s}dx > \frac{1}{2}$ and therefore the condition (8.20) is satisfied ($\int_0^\infty \frac{\sin s}{s}ds = \pi/2$). Hence the Shannon expansions do exhibit Gibbs phenomenon.

In order to get around Gibbs phenomenon, one can sometimes use other approximations to $f \in V_m$ rather than the projection. In the case of the Shannon wavelet, the function $\sigma(x) = \frac{\sin^2 \pi x/2}{(\pi x/2)^2}$ belongs to V_0 and,

in fact, $f * \sigma \in V_0$ for each $f \in L^2(\mathbb{R})$. It follows (with a little work [Si])
that

$$\delta_m(x, y) := 2^m \sigma(2^m(x - y))$$

is a <u>positive</u> delta sequence, and hence the function

$$f^1_m(x) = \int_{-\infty}^{\infty} 2^m \sigma(2^m(x - y)) f(y) dy$$

is an approximation to $f(x)$ which does not show Gibbs phneomenon.
For wavelets with compact support, we use the results of the next section
on positive wavelets to get a positive delta sequence which avoids it.

8.8 Positive scaling functions

There are no continuous non-negative orthogonal scaling functions [J1].
In fact, it's not even clear that there is, in general, an $f \in V_0$ which is
≥ 0. The following construction shows there is.

8.8.1 A general construction

Let $\theta(t)$ be any continuous function on $\mathbb{R}$ with support in an interval
$[M, N]$. We say that $\theta(t)$ generates a partition of unity if

$$\sum_{n=-\infty}^{\infty} \theta(t - n) \equiv 1, \quad t \in \mathbb{R}. \tag{8.21}$$

This is a property shared by all orthogonal scaling functions of compact
support, in particular those due to Daubechies as well as the "Coiflets"
of Coifman and other "–lets". We define the Abel means of their series
(8.21) to be

$$\rho_r(t) := \sum_{n=-\infty}^{\infty} r^{|n|} \theta(t - n), \quad 0 < r \leq 1, \quad t \in \mathbb{R}. \tag{8.22}$$

This series converges uniformly on [0,1] since it is locally finite. Furthermore $\rho_r(t) \longrightarrow 1$ for $t \in [0, 1]$ as $r \to 1$ by the regular summability
property of Abel means. (If a series converges to S, it is also Abel
summable to S.)

The series also converges uniformly on any finite interval $[M, N]$ and therefore there is an $0 < r_0 < 1$ such that $\rho_r(t) \geq 1/2$ for $1 > r \geq r_0$, $t \in [M, N]$.

We denote by V the closed linear span of $\{\theta(t - n)\}_{n \in \mathbb{Z}}$ in $L^2(\mathbb{R})$. Then if in addition $\{\theta(t - n)\}$ is a Riesz basis of V and $a_n \in l^2$, we have

$$f(t) = \sum_{n=-\infty}^{\infty} a_n \theta(t - n) \in V.$$

This requirement will be met if $\widehat{\theta}(\omega)$ satisfies

$$0 < A \leq \sum_{k=-\infty}^{\infty} |\widehat{\theta}(\omega + 2k\pi)|^2 \leq B < \infty. \qquad (8.23)$$

The properties of $\rho_r(t)$ are given by the next lemma. Details may be found in [W-She4].

Lemma 8.7
Let $\theta(t)$ be a continuous function on $\mathbb{R}$ with compact support satisfying (8.21) and (8.23); let $V = CLS\{\theta(t - n), n \in Z\}$; then there is an $0 < r_0 < 1$, such that ρ_r given by (8.22) for $r_0 \leq r < 1$ satisfies
  (i) $\rho_r(t) \geq 0, t \in \mathbb{R}$
  (ii) $\rho_r \in V$.

8.8.2   Back to wavelets

From Lemma 8.7, we have an answer to the question posed at the start of this section.

Corollary 8.4
Let $\{V_m\}$ be a multiresolution of $L^2(\mathbb{R})$ associated with the scaling function $\phi(t)$ with compact support. Then there is an element $\rho \in V_m$ such that $\rho(t) \geq 0$.

We can say much more than this. In fact, we have

THEOREM 8.4 W-She4
Let $\rho_r(t) = \sum_n r^{|n|} \phi(t - n)$, where r is chosen so large that $\rho_r(t) \geq 0$, for $t \in \mathbb{R}$; then $\{2^{\frac{m}{2}} \rho_r(2^m t - n)\}_{n \in Z}$ is a Riesz basis of V_m; its dual

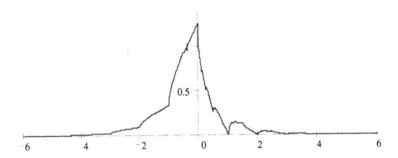

FIGURE 8.4
The summability function $\rho^r(x)$ for Daubechies wavelet $_2\phi(x)$,
$m = 0.$

basis is generated by $\widetilde{\rho}_r$, where

$$\widetilde{\rho}_r(t) = \frac{1}{2\pi(1-r^2)}[(1+r^2)\phi(t) - r\{\phi(t+1) + \phi(t-1)\}].$$

However, this biorthogonal system does not give us the positive kernel we need. Rather we use a modification which gives the desired properties. The kernel that gives us the approximation in V_0 to $f \in L^2$ is given by

$$k_r(t, s) = \left(\frac{1-r}{1+r}\right)^2 \sum_{n=-\infty}^{\infty} \rho_r(t-n)\rho_r(s-n), \qquad (8.24)$$

i.e.,

$$f_0^r(t) = \int_{-\infty}^{\infty} k_r(s, t)f(s)ds.$$

This kernel satisfies the conditions needed to generate a positive delta sequence $\{k_{r,m}\}$ where

$$k_{r,m}(s, t) = 2^m k_r(2^m s, 2^m t), \qquad m \in \mathbb{Z}. \qquad (8.25)$$

We summarize the properties of $k_{r,m}$ in the next proposition

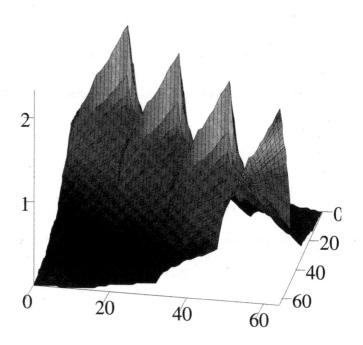

FIGURE 8.5
The positive delta sequence $k_{r,m}(x,y)$ for Daubechies wavelet
$_2\phi(x)$, $m = 0$.

Proposition 8.1
[W-She4] Let $k_{r,m}$ be as in (8.24), let $f \in L^1 \cap L^2(\mathbb{R})$; let

$$f_m^r(t) = \int_{-\infty}^{\infty} k_{r,m}(t,s) f(s) ds;$$

then $f_m^r \in V_m$ and $f_m^r \to f$ in the sense of $L^2(\mathbb{R})$. Furthermore,

(i) if $M_1 \le f(t) \le M_2$ for $t \in \mathbb{R}$, then $M_1 \le f_m^r(t) \le M_2$ for $t \in \mathbb{R}, m \in \mathbb{Z}$,

(ii) if $M_3 \le f(t) \le M_4$ for $t \in [a,b]$, then for each $\epsilon > 0, \delta > 0$, there is an m_0, such that for $t \in (a + \delta, b - \delta)$, $M_3 - \epsilon \le f_m^r(t) \le M_4 + \epsilon$, for $m \ge m_0$,

(iii) if f is continuous on (a,b), then for each $\epsilon > 0$, $f_m^r \to f$ uniformly on $[a + \epsilon, b - \epsilon]$.

Clearly, (ii) guarantees that $f_m^r(t)$ will not exhibit Gibbs phenomenon and oscillations will be controlled by the given ϵ. That is, the overshoot at discontinuities can be no more than ϵ, which is arbitrarily small.

If our function $f \in H^\alpha$, the Sobolev space, the usual projection wavelet estimator has error $O(2^{(1-\alpha)m})$ for sufficiently regular wavelets. This can be used to estimate the error for our new estimator (surprisingly).

Proposition 8.2
Let the scaling function $\phi \in C^2$, $f_m(x) = \int_{-\infty}^{\infty} q_m(t,s) f(s) ds$, the usual projection estimator, let $f \in H^r$, for $\alpha \ge 3$; then

$$f_m^r(x) - f_m(x) = O(2^{-2m})$$

uniformly in x as $m \to \infty$.

The proof is based on the fact that

$$\mathcal{F} f_m^r(\omega) = \frac{(1-r)^4}{(1 - 2r \cos 2^{-m}\omega + r^2)^2} \mathcal{F} f_m(\omega)$$

where $\mathcal{F}$ denotes the Fourier transform. The difference between these two ($\mathcal{F} f_m^r(\omega)$ and $\mathcal{F} f(\omega)$) is $O(2^{-2m}\omega^2)$. Hence for $\alpha \ge 3$, the rate of convergence of f_m^r to f is the same $O(2^{-2m})$.

8.9   Problems

1. For a <u>positive</u> delta sequence, the conditions are the same as for a quasi-positive (8.3) except that condition (i) is replaced by

$$(i)' \quad \delta_m(x,y) \geq 0, \quad x,y \in \mathbb{R}, \quad m \in \mathbb{N}.$$

Show that a positive delta sequence is also quasi-positive.

2. Show that the Fejer kernel is a positive delta sequence.

3. Let $f \in V_0$ for some r-regular scaling function ϕ and let $f(x) \geq 0$, $\hat{f}(2\pi k) = \delta_{0k}$. Show that

(a) $\int\limits_{-\infty}^{\infty} f(x)dx = 1$

(b) $\sum\limits_{n} f(x-n) = 1$

(c) $\sum\limits_{n} f^2(x-n) > 0 \quad x \in \mathbb{R}$

(d) $\sum\limits_{n} f(x-n)f(y-n) \leq C(1+|x-y|)^{-1}$.

4. Use the results from Problem 3 to show that

$$k_m(x,y) := 2^m \sum_{n} f(2^m x - n)f(2^m y - n)$$

is a positive delta sequence.

5. Show that the reproducing kernel $q_m(x,y)$ of V_m for the Haar wavelet is a positive delta sequence.

6. Let $k_m(x,y)$ be a positive delta sequence, let $m \leq f(x) \leq M$, $x \in \mathbb{R}$. Show that $f_m(x) = \int k_m(x,y)f(y)dy$ satisfies $m \leq f_m(x) \leq M$ (Lemma 8.2).

7. Show directly that the expansion of $\delta(x)$, in terms of 1-regular wavelets, converges to zero pointwise for all $x \neq 0$.

8. Show that if the tempered distribution f is equal to a continuous function $g(x)$ in some neighborhood of x_0, then f has the value $g(x_0)$ at x_0.

9. Show that for the Franklin wavelets, ϕ satisfies property Z_2 and for the Meyer wavelets with $\hat{\phi} \in C^\infty$, ϕ satisfies Z_λ for all $\lambda \geq 1$.

Chapter 9

A Shannon Sampling Theorem in Wavelet Subspaces

The classical Shannon sampling theorem given by the formula [Sh]

$$f(t) = \sum_{n=-\infty}^{\infty} f(nT) \frac{\sin \sigma(t - nT)}{\sigma(t - nT)}, \quad T = \pi/\sigma, \qquad (9.1)$$

holds for σ-band limited signals, i.e., continuous functions in $L^2(\mathbb{R})$ whose Fourier transform has support in $[-\sigma, \sigma]$. If σ is taken to be $\sigma = 2^m \pi$, and $\phi(t) = \sin \pi t / \pi t$ is the scaling function of Example 5, then (9.1) is a statement about the elements of the subspaces V_m in the multiresolution analysis. Many generalizations of this theorem have been proposed. They usually involve transforms other than the Fourier transform [B-S-S], [H], and can be put into a reproducing kernel Hilbert space setting [N-W].

The original theorem has widespread applications in communication theory since it enables one to recover an analog signal from its sampled values. It and many others can be imbedded in wavelet theory, which gives a unified approach to the subject [W5], [W11]. There are other advantages to this wavelet approach to sampling theorems as well. For example, the aliasing error may be given in terms of the wavelet coefficients [W11]. The "mother wavelet" $\psi(t)$ may be found in this case to be

$$\psi(t) = \frac{\sin \pi (t - \frac{1}{2}) - \sin 2\pi (t - \frac{1}{2})}{\pi (t - \frac{1}{2})}.$$

The translates of ψ, $\{\psi(t - n)\}$ are an orthonormal basis of W_0, the orthogonal complement of V_0 in V_1, which is composed of $f \in L^2(\mathbb{R})$ whose Fourier transform has support in $[-2\pi, -\pi] \cup [\pi, 2\pi]$.

In this chapter we shall turn this around and first ask if there always is a sampling theorem imbedded in any wavelet theory. That is, if we begin with any scaling function satisfying the required properties, is there a natural sampling function $S_n(t)$ that gives a sampling expansion for $f \in V_0$? Then, of course, by a change of scale, we have the same for all V_m. We shall see that there is indeed such a sampling expansion under very broad conditions [W11]. We shall first introduce another basis of the V_m spaces and use it to construct the sampling functions. This will work in some cases; in others we must extend V_m to the space T_m given in Chapter 5.

Next we consider a generalization to shifted sampling, i.e., we recover $f \in V_0$ from its values at the sampling points $t_n = a + n$. This approach, due to Janssen, works in other cases where an unshifted sampling function doesn't exist [J]. In particular, it works in spaces arising from splines of order 2.

We then consider Gibbs phenomenon for these sampling series. We give a condition under which Gibbs phenomenon arises and show that it is satisfied for many standard wavelets. However there are some, such as the sampling series associated with a Daubechies with 4 taps, which do not show Gibbs phenomenon even though it does occur for the series of orthogonal functions with the same MRA.

Finally we take up some elements of irregular sampling, which, however, is not as well developed for wavelet sampling in general as it is for the Shannon case.

9.1   A Riesz basis of $\mathbf{V}_m$

A Riesz basis of a separable Hilbert space H is a basis $\{f_n\}$ that is close to being orthogonal. That is, there is a bounded invertible operator that maps $\{f_n\}$ onto an orthonormal basis, [Y, p.31].

Each $f \in H$ has a representation

$$f = \Sigma a_n f_n.$$

The coefficients $a_n = \langle f, g_n \rangle$ where $\{g_n\}$ is the unique biorthogonal basis, which is also a Riesz basis, [Y, p.36].

We shall need an alternate basis of our spaces V_m different from that based on the scaling function. It will be based on the reproducing kernel

(Chapter 3, section 3)

$$q(x, t) = \sum_n \phi(x - n)\phi(t - n), \tag{9.2}$$

by taking one of the parameters equal to an integer (ϕ is assumed real).

Proposition 9.1
Let $\phi(t)$ be a scaling function in S_r such that

$$\hat{\phi}^*(w) := \sum_n \phi(n)e^{-iwn} \neq 0, \quad w \in \mathbb{R}; \tag{9.3}$$

then $\{q(t, n)\}$ is a Riesz basis of V_0, with biorthogonal Riesz basis $\{S_n(t)\}$.

PROOF Let T be the operator on V_0 that takes $\phi(x - n)$ into $q(x, n)$. Its Fourier transform induced operator $\hat{T}$, which maps the Fourier transform of $\phi(x - n)$ into that of $q(x, n)$, satisfies

$$\hat{T}\hat{\phi}(w) = \hat{q}(w, 0) = \sum_k \hat{\phi}(w)e^{-iwk}\phi(0 - k) = \overline{\hat{\phi}^*(w)}\hat{\phi}(w).$$

Since $\{\phi(k)\}$ is rapidly decreasing, $\hat{\phi}^*(w)$ is continuous and periodic and hence a bounded function. Hence $\hat{T}$ satisfies

$$\|\hat{T}\hat{\phi}\| \leq \|\hat{\phi}^*\|_\infty \|\hat{\phi}\| \tag{9.4}$$

and is therefore a bounded operator on $\hat{V}_0$. Since, by (9.3), $1/\hat{\phi}^*(w)$ is also a bounded function, $\hat{T}^{-1}$ is bounded as well and takes $\hat{q}(w, 0)e^{-iwn} = \hat{q}(w, n)$ into $\hat{\phi}(w)e^{-iwn}$. Since the Fourier transform is an isometry, T is also a bounded invertible operator, and hence $\{q(t, n)\}$ is a Riesz basis that must have a unique biorthogonal basis $\{S_n(t)\}$ [Y, p.32]. □

9.2 The sampling sequence in V_m

The biorthogonal Riesz basis of Proposition 9.1 turns out to be our sampling function as we shall see in this section. In addition we show that the aliasing error is given in terms of wavelet coefficients.

THEOREM 9.1

Let $f \in V_0$, let ϕ satisfy (9.3), and let $\{S_n(t)\}$ be the basis biorthogonal to $\{q(t,n)\}$; then $S_n(t) = S(t-n)$ where

$$\hat{S}(w) = \hat{\phi}(w)/\hat{\phi}^*(w), \qquad w \in \mathbb{R} \tag{9.5}$$

and

$$f(t) = \sum_n f(n)S(t-n), \qquad t \in \mathbb{R} \tag{9.6}$$

where the convergence is uniform on $\mathbb{R}$.

PROOF   Since $q(t,n) = q(t-n,0)$, it follows that $\{S_0(t-n)\}$ must also be a biorthogonal sequence, which because of uniqueness must be equal to $S_n(t)$. The expression (9.6) is just the expansion of f with respect to $\{S_n(t)\}$ (since $\int f(t)q(t,n)dt = f(n)$), which converges in the sense of $L^2(\mathbb{R})$. The calculation of $S(t) = S_0(t)$ follows at once by substituting $\phi(t)$ for $f(t)$ in (9.6), and then taking the Fourier transform,

$$\hat{\phi}(w) = \sum_n \phi(n)e^{-iwn}\hat{S}(w) = \hat{\phi}^*(w)\hat{S}(w).$$

This is solved for $\hat{S}(w)$ to get (9.5).

To show that (9.6) converges uniformly we use

$$\left| f(t) - \sum_{n=-N}^{N} f(n)S(t-n) \right|$$

$$= \left| \int_{-\infty}^{\infty} q(t,s) \left[f(s) - \sum_{n=-N}^{N} f(n)S(s-n) \right] ds \right|$$

$$\leq \left\{ \int |q(t,s)|^2 ds \right\}^{1/2} \left\| f - \sum_{n=-N}^{N} f(n)S_n \right\|.$$

Since $q(t,t) = \int |q(t,s)|^2 ds = \Sigma \phi^2(t-n)$ and hence is periodic, it must be uniformly bounded.   □

Several different types of errors may occur in this and other sampling theorems. They are truncation error, jitter error, round off error, and aliasing error. We shall treat only aliasing error, which usually refers to the error in the classical Shannon sampling theorem that arises when the signal is not σ-bandlimited but rather σ'-bandlimited for some $\sigma' > \sigma$. The natural analogy to this is when the function $f \in V_m$ rather than V_0 for some $m > 0$. In particular such error arises when $f \in V_1$.

Definition 9.1   *The <u>aliasing error</u> for $f \in V_1$ is the difference*

$$e(t) := f(t) - \sum_n f(n)S(t - n).$$

Proposition 9.2
Let $\{\psi(t - n)\}$ be the wavelet basis of W_0; then the aliasing error for $f \in V_1$ satisfies

$$|e(t)|^2 \leq C \sum_n |b_n|^2, \qquad t \in \mathbb{R}$$

where C is constant, and the $b_n = \int f(t)\overline{\psi(t - n)}dt$ are the wavelet co-efficients of f.

PROOF   Let $f = f_0 + f_1$ where $f_0 \in V_0$ and $f_1 \in W_0$; then $e(t) = f_1(t) - \Sigma f_1(n)S(t - n)$ since $f_0(t)$ is equal to its sampling series.
But we see that

$$|f_1(t)|^2 = \left| \sum_n b_n \psi(t - n) \right|^2 \leq \sum_n |b_n|^2 \sum_n |\psi(t - n)|^2$$

and

$$\left| \sum_n f_1(n)S(t - n) \right|^2 \leq \sum_n |f_1(n)|^2 \sum_n |S(t - n)|^2.$$

Moreover by the argument in Proposition 9.1 (applied to W_0 instead of V_0) we have

$$\sum_n |f_1(n)|^2 \leq C\|f_1\|^2 = C \sum_n |b_n|^2.$$

The proof is completed by observing that both $\psi(t)$ and $S(t)$ are continuous and rapidly decreasing and hence $\Sigma|\psi(t - n)|^2$ and $\Sigma|S(t - n)|^2$ are bounded for $t \in \mathbb{R}$.   □

9.3   Examples of sampling theorems

We return to the examples of Chapter 3 and try to construct sampling functions. For the first example, the <u>Haar system</u>, the reproducing kernel is $q(t, s) = \phi(t - [s])$ and hence the basis is $q(t, n) = \phi(t - n) = S_n(t)$, the sampling function. This and the Shannon case, Example 5, are the only ones among the examples of Chapter 3 for which this happens.

The <u>Franklin wavelets</u>, Example 2, have

$$\hat{\phi}^*(w) = \sum_k \phi(k)e^{-iwk} = \sum_k \hat{\phi}(w + 2\pi k) = \left(1 - \frac{2}{3}\sin^2\frac{w}{2}\right)^{-\frac{1}{2}}.$$

Thus we find

$$\hat{S}(w) = \hat{\phi}(w)/\hat{\phi}^*(w) = \left(\frac{\sin w/2}{w/2}\right)^2$$

whose inverse Fourier transform is the original "hat function", $S(t) = (1 - |t|)\chi_{(-1,1)}(t)$.

For <u>cubic splines</u>, Example 4, the space V_0 is the closed linear span of translates of the third order basic spline $\theta_3(t)$. The $\{\theta_3(t - k)\}$ are not orthogonal but may be orthogonalized in the standard way. However this is unnecessary since $\hat{S}(w) = \hat{\theta}(w)/\hat{\theta}^*(w)$ as well as (9.5).

Since $\theta_3(t)$ has support on $[0,4]$, only its values on the integers 1, 2, 3 are needed and are given by $\theta_3(1) = \theta_3(3) = 1/6$, $\theta_3(2) = 2/3$; therefore

$$\hat{\theta}_3^*(w) = \frac{1}{6}e^{-iw}\left[1 + 4e^{-iw} + e^{-2iw}\right],$$

which has no real zeros. The inverse $\hat{\theta}_3^{*-1}$ may be found by using the Laurent series of

$$\frac{6}{z^{-1}[1 + 4z^{-1} + z^{-2}]} = \sqrt{3}\sum_{n=0}^{\infty}(\sqrt{3} - 2)^{n+1}z^{n-1}$$

$$+\sqrt{3}\sum_{n=1}^{\infty}(\sqrt{3} - 2)^{n-1}z^{-n-1}.$$

From this we find that

$$S_3(t) = \sqrt{3}\sum_{n=0}^{\infty}\left(\sqrt{3} - 2\right)^{n+1}\theta_3(t - n + 1)$$

$$+\sqrt{3}\sum_{n=1}^{\infty}\left(\sqrt{3} - 2\right)^{n-1}\theta_3(t + n + 1).$$

In Example 3, the <u>Daubechies Wavelets</u>, the scaling function has support on $[0,3]$, and, since $\phi(j) = \sqrt{2}\Sigma c_k\phi(2j - k) = \sqrt{2}\Sigma c_{2j-k}\phi(k)$, it follows that $(\phi(0),\ \phi(1),\ \phi(2),\ \phi(3))^T$ is an eigenvector of the matrix

$C = \sqrt{2}[c_{2j-k}]$. Since $\phi(0) = \phi(3) = 0$, we need consider only the matrix

$$C = \begin{bmatrix} c_1 & c_0 \\ c_3 & c_2 \end{bmatrix} = \frac{1}{\nu^2 + 1} \begin{bmatrix} 1 - \nu & \nu(\nu - 1) \\ (\nu + 1)\nu & \nu + 1 \end{bmatrix}.$$

An eigenvector corresponding to eigenvalue 1 is $\phi(1) = \frac{\nu-1}{2\nu}$, $\phi(2) = \frac{\nu+1}{2\nu}$, which satisfies $\sum_k \phi(k) = 1$ as it must and

$$\hat{\phi}^*(w) = \frac{1}{2\nu} \left\{ (\nu - 1)e^{-iw} + (\nu + 1)e^{-2iw} \right\},$$

which has no zeros for $\nu \neq 0$. For $\nu < 0$ we see that

$$\hat{\phi}^{*-1}(w) = \frac{2\nu}{\nu - 1} \sum_{n=0}^{\infty} \left(\frac{1 + \nu}{1 - \nu} \right)^n e^{-i(n-1)w}$$

and hence

$$S(t) = \frac{2\nu}{\nu - 1} \sum_{n=0}^{\infty} \left(\frac{1 + \nu}{1 - \nu} \right)^n \phi(t - n + 1).$$

For $\nu > 0$, the series becomes

$$S(t) = \frac{2\nu}{\nu + 1} \sum_{n=0}^{\infty} \left(\frac{1 - \nu}{1 + \nu} \right)^n \phi(t + n + 2).$$

REMARK 9.1   The examples illustrate each of the three methods that can be used to find $\hat{\phi}^*(w)$. The first is the definition given by (9.3). If $\phi \in S_r$ for $r \geq 2$, then the periodic extension of $\hat{\phi}$ may also be used since the series

$$g(w) = \sum_{k=-\infty}^{\infty} \hat{\phi}(w + 2\pi k)$$

then converges uniformly. Its Fourier coefficients are

$$\frac{1}{2\pi} \int_0^{2\pi} \sum_k \hat{\phi}(w + 2\pi k)e^{-iwn} dw = \frac{1}{2\pi} \int_{-\infty}^{\infty} \hat{\phi}(w)e^{-iwn} dw = \phi(-n)$$

and hence

$$g(w) = \sum_n \phi(-n)e^{iwn} = \hat{\phi}^*(w).$$

We used this in Example 2, even though $r = 1$ in that case.   ∎

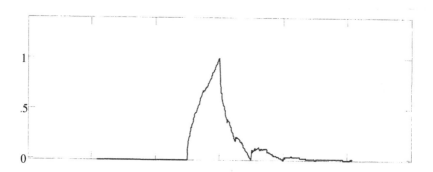

FIGURE 9.1
The sampling function for the Daubechies wavelet $_2\phi(t)$ with

$\gamma = -\frac{1}{\sqrt{3}}.$

In the case of Example 7, the sampling function is again easy to find. Since

$$\hat{\varphi}^2(w) = \int_{\pi-w}^{\pi+w} dP$$

satisfies $\hat{\varphi}(w) = 1$ for $|w| \leq 2\pi/3$ and has support in $\left[-\frac{4\pi}{3}; \frac{4\pi}{3}\right]$, it follows that

$$\hat{\varphi}^*(w) = \hat{\varphi}(w) + \hat{\varphi}(w - 2\pi) + \hat{\varphi}(w + 2\pi), \ |w| \leq 4\pi/3.$$

Hence $\hat{S}(w) = 1$ for $|w| \leq 2\pi/3$ and

$$\hat{S}(w) = \frac{\hat{\varphi}(w)}{\hat{\varphi}(w) + \hat{\varphi}(w - 2\pi) + \hat{\varphi}(w + 2\pi)}, \qquad w \in \mathbb{R}. \qquad (9.7)$$

If only the dilation equation is known, and only a finite number of the c_k's are not zero, then the procedure of Example 3 can be used. That is, we can look for the right eigenvector of $C = \sqrt{2}[c_{2j-k}]$ corresponding to the eigenvalue 1. Since $\sqrt{2}\Sigma c_{2j} = \sqrt{2}\Sigma c_{2j-1} = 1$ in most examples [D2, p.936], there is always a left eigenvector in these cases. Higher order Daubechies wavelets also appear to have sampling functions but no general proof seems to be known.

9.4   The sampling sequence in $\mathbf{T}_m$

In some cases $\hat{\phi}^*(w)$ has real zeros (see quadratic splines below). In these cases we can find a sampling sequence in T_m, the multiresolution subspace of S'_r in Chapter 5, even though there is none in V_m.

We begin with the same definition of the sampling function $S_n(t)$ in (9.5), but without the added condition that $\hat{\phi}^*(w) \neq 0$. Rather we require only that its zeros be isolated and of finite multiplicity $< r$ in $[-\pi, \pi]$. Then the function

$$\hat{\phi}^*(w) / \prod_j \sin \frac{(w - \alpha_j)}{2} e^{\frac{i(w - \alpha_j)}{2}} = \hat{\phi}^*(w)/f(w), \qquad (9.8)$$

where the α_i's are the zeros of $\hat{\phi}^*(w)$ (counted according to multiplicity), has no zeros and is periodic. Since $\hat{\phi}^*(w) \in C^\infty$, the reciprocal $f(w)/\hat{\phi}^*(w)$ of (9.8) is continuous provided the values at $w = \alpha_j$ are defined by continuity. The same is true for its derivatives. Hence it is in C^∞ and periodic, and its Fourier coefficients are rapidly decreasing.

The Fourier series of $e^{-iw/2}/\sin w/2$ may be found from the power series of

$$h(z) = \frac{-2i}{1 - z^2}$$

to be

$$\frac{e^{-iw/2}}{\sin w/2} \sim -2i \sum_{n=0}^{\infty} e^{iwn},$$

that converges in S'_r for $r > 0$. Each of the first order factors in $f^{-1}(w)$ therefore has a series that has bounded coefficients. If a factor is repeated k times the resulting series has coefficients $= O(|n|^k)$. The product of two such series has coefficients whose growth is the same order as the maximum order of the coefficient of each series. Hence $f^{-1}(w)$ has a Fourier series whose coefficients are $O(|n|^{r-1})$ and which therefore converges in S'_r.

Its product with a C^∞ function of period 2π again forms a Fourier series convergent in S'_r. Hence

$$\hat{\phi}^{*-1}(w) = \sum_n \alpha_n e^{-iwn}, \qquad \alpha_n = 0(|n|^r),$$

and the sampling function $S(t)$ as the inverse Fourier transform of

$$\hat{\phi}^{*-1}(w)\hat{\phi}(w)$$

is given by

$$S(t) = \sum_n \alpha_n \phi(t-n). \tag{9.9}$$

Hence $S(t) \in T_o$. Clearly $S(k) = \delta_{0k}$. To show the sampling series converges we observe that for $f \in T_0$ we have

$$\hat{f}(w) = \sum_n a_n \hat{\phi}(w) e^{-iwn} = \hat{a}^*(w)\hat{\phi}(w)$$

$$= \hat{a}^*(w)\hat{\phi}^*(w)\hat{S}(w),$$

the product of a periodic distribution with $\hat{S}(w)$. The inverse Fourier transform gives

$$f(t) = \sum_n b_n S(t-n)$$

with convergence in S'_r. For $f(t) \in S_r$, the convergence is also uniform, and we have

$$f(k) = \sum_n b_n S(k-n) = b_k.$$

Hence the sampling theorem (9.6) holds in T_0 under greater generality.

Example 4 also includes quadratic splines. Here $\theta_2(t)$ is the basic spline of order 2,

$$\hat{\theta}_2(w) = \left[\frac{1 - e^{-iw}}{iw} \right]^3, \qquad w \in \mathbb{R}.$$

It is a quadratic polynomial in $(n, n+1)$, $n \in \mathbb{Z}$, has support in $[0,3]$ and belongs to S_1. The periodic extension in $\hat{\theta}_2^*(w) = \frac{1}{2}e^{-iw}[1 + e^{-iw}]$, which has a zero at $w = \pm\pi$.

Its reciprocal converges in the sense of S'_1 and is used to calculate

$$S_2(t) = 2\sum_{n=0}^{\infty} (-1)^n \theta_2(t - n + 1).$$

Since $S_2(t) \notin V_0$, this space has no sampling theorem. But $S_2(t) \in T_0$, and, hence, for $f \in V_0 \cap S_1$, we have

$$f(t) := \sum_n f(n) S_n(t)$$

convergent in S_1' and uniformly pointwise convergent.

This setting enables us to treat sampling by means of a traditional engineering approach. A signal $f(t) \in T_0$ is sampled at the integers and an impulse train

$$f^*(t) = \sum_n f(n)\delta(t-n) \qquad (9.10)$$

is formed. This series converges in the sense of S_r'. The projection of $f^*(t)$ onto T_0 (actually a type of low-pass filter) gives

$$P_0 f^*(t) = \sum_n f(n)q(t-n,0)$$

where $q(t,u)$ is the reproducing kernel of V_0. Since $\hat{q}(w,0) = \hat{\phi}(w)\overline{\hat{\phi}^*(w)}$ and $\hat{S}(w) = \hat{\phi}(w)/\hat{\phi}^*(w)$, it follows that $\hat{q}(w,0) = |\hat{\phi}^*(w)|^2 \hat{S}(w)$. Hence by multiplying $\hat{P}f^*(w)$ by $|\hat{\phi}^*(w)|^{-2}$ and taking the inverse Fourier transform we recover our original function from the sampling series (9.6). This again requires $\hat{\phi}^*(w) \neq 0$.

9.5 Shifted sampling

An alternate approach to that of Section 9.4 is to sample at points $\{a+n\}$ and then ask if a sampling function exists for such points. The natural hypothesis to use here is the shifted version of (9.3); there is an $a \in [0,1)$ such that

$$(Z\phi)(a,w) = \sum_n \phi(a+n)e^{-iwn} \neq 0, \quad w \in \mathbb{R}.$$

This of course is the Zak transform of ϕ mentioned in Chapter 8. Janssen [J] showed that the B-spline wavelet of order 2 satisfies this condition for $a = \frac{1}{2}$, but, as we observed, it does not hold for $a = 0$.

In this case the sampling function $S_a \in V_0$ is given by

$$\hat{S}_a(w) := \frac{\hat{\varphi}(w)}{(Z\phi)(a,w)} = \hat{\varphi}(w) \sum_n \sigma_{n,a} e^{-iwn}. \qquad (9.11)$$

The cardinal series of $f \in V_0$ is then

$$f(t) = \sum_n f(a+n)S_a(t-n). \qquad (9.12)$$

The derivation and proof is similar to that in Section 9.2 for $a = 0$ and is omitted.

The question then arises, if several choices of a are available, which is the best? One possible criterion is the one that minimizes the aliasing error [J]. This is the error arising if $f \in V_1$ is approximated by the right side of (9.10). Since the error

$$e_{f,a}(t) = f(t) - f_0(t) + \sum_n (f(a+n) - f_0(a+n))S_a(t-n)$$

depends only on the difference between $f - f_0$ (f_0 is the projection on V_0) which belongs to W_0, we need only consider the $e_{f,a}$ for $f \in W_0$. We find that for $f \in W_0$,

$$\|e_{f,a}\|^2 = \|f\|^2 + \|\sum_n f(a+n)S_a(\cdot - n)\|^2$$

$$= \|f\|^2 + \sum_m \sum_n f(a+n)\overline{f(a+m)} \int_{-\infty}^{\infty} S_a(t-n)\overline{S_a(t-m)}dt$$

$$= \|f\|^2 + \sum_m \sum_n f(a+n)\overline{f(a+m)}\frac{1}{2\pi}\int_{-\infty}^{\infty} |\hat{S}_a(w)|^2 e^{i(m-n)w}dw$$

by Parseval's equality. But by (9.11) this last integral is

$$\frac{1}{2\pi}\int_{-\infty}^{\infty} |\hat{S}_a(w)|^2 e^{i(m-n)w}dw$$

$$= \frac{1}{2\pi}\int_{-\infty}^{\infty} |\hat{\varphi}(w)|^2 \left|\sum_k \sigma_{k,a} e^{-iwk}\right|^2 e^{i(m-n)w}dw$$

$$= \sum_k \sigma_{k-n,a}\overline{\sigma_{k-m,a}}$$

by the orthogonality of $\{\hat{\varphi}(w)e^{inw}\}$. Hence the error is

$$\|e_{f,a}\|^2 = \|f\|^2 + \sum_k \left|\sum_m \sigma_{k-m,a} f(m+a)\right|^2$$

$$= \|f\|^2 + \frac{1}{2\pi}\int_0^{2\pi} \frac{|Zf(a,w)|^2}{|Z\phi(a,w)|^2}dw \qquad (9.13)$$

since $Zf(a,w)$ has Fourier coefficients $f(n+a)$ and $\frac{1}{Z\phi(a,w)}$ has Fourier coefficients $\sigma_{n,a}$.

We now use the fact that $f \in W_0$ has a wavelet expansion as well,

$$f(t) = \sum_k \beta_k \psi(t - k).$$

The Zak transform of f then becomes

$$(Zf)(a, w) = \sum_k \beta_k e^{-iwk} (Z\psi)(a, w),$$

and the integral in (9.11) therefore satisfies

$$\frac{1}{2\pi} \int_0^{2\pi} \left| \frac{Zf(a, w)}{Z\phi(a, w)} \right|^2 dw = \frac{1}{2\pi} \int_0^{2\pi} \left| \frac{Z\psi(a, w)}{Z\phi(a, w)} \right|^2 \left| \sum_k \beta_k e^{-iwk} \right|^2 dw$$

$$\leq \sup_w \left| \frac{Z\psi(a, w)}{Z\phi(a, w)} \right|^2 \sum_k |\beta_k|^2.$$

Since $\sum |\beta_k|^2 = \|f\|^2$, the error satisfies

$$\|e_{f,a}\|^2 \leq \|f\|^2 \left(1 + \sup_w \left| \frac{Z\psi(a, w)}{Z\phi(a, w)} \right|^2 \right).$$

This leads to one of the criteria by Janssen [J] which is to choose a such that $\sup_w \left| \frac{Z\psi(a,w)}{Z\phi(a,w)} \right|$ is minimized. He found for the Daubechies wavelets that a value of $a \approx 0.37$ gives a value close to the minimum.

We have not remarked on the convergence of the various series considered here. For $\phi \in S_r$ this is no problem since $(Z\phi)(a, w) \in C^\infty$ in this case, and all coefficients are rapidly decreasing. The same conclusions hold under the weaker hypothesis [J] that ϕ is bounded and $\sum_n |\phi(t - n)|$ converges uniformly.

9.6 Gibbs phenomenon for sampling series

In Chapter 8 we showed that Gibbs phenomenon exists for orthogonal expansions for all reasonable wavelets. However, when we turn to sampling (interpolating) series rather than orthogonal series, few results are known. Recently Helmsberg [Hel] has shown Gibbs phenomenon occurs for Fourier interpolation. Shim [Si1] has shown it also exists for interpolating series in some wavelet subspaces for functions continuous on the right. But to our knowledge few other results involving interpolating series are known.

In this section we extend these results to other wavelet interpolating series following the procedure of [W-Si]. We shall show that it occurs for many of the standard wavelets, but not for all. We shall characterize it by a condition for interpolating series similar to that in [S-V] for orthogonal series.

We calculate an approximation to the amount of overshoot in certain cases. We then show that Gibbs phenomenon can be avoided by using an alternate interpolating series. For certain cases, notably for Franklin wavelets and Daubechies wavelets with 4 taps, it does not occur for interpolating series even though it does for the corresponding orthogonal series.

In order to study the Gibbs phenomenon, we require that f be piecewise continuous and in $L^2(\mathbb{R})$. We shall also suppose that a jump discontinuity be at a dyadic rational number, so that by translation we can take it to zero, which we do. The spaces V_m are not translation invariant for irrational translations in general. We shall also assume the jump is in the positive direction, i.e., that $f(0^+) > f(0^-)$. If there is a sequence $t_m \downarrow 0$ such that

$$f_m(t_m) \to \gamma^+ > f(0^+) \tag{9.14}$$

where

$$f_m(t) = \sum_n f(n2^{-m})S(2^m t - n) \tag{9.15}$$

then the sampling series (9.15) exhibits Gibbs phenomenon on the right hand side of 0 for the function f (and similarly on the left hand side). We shall simply say "Gibbs right" and "Gibbs left" for these two cases if they hold for any function with such a jump at 0. We shall see later that these are independent of the particular function.

There is a possible source of ambiguity in our series (9.15) at points of discontinuity. By changing the value of $f(0)$, we could change Gibbs right to Gibbs left and vice versa. This was avoided in [Si1] by assuming that $f(t) = f(t^+)$ for all $t \epsilon \mathbb{R}$. However this assumption is unnecessarily restrictive and by eliminating it, we can sometimes avoid Gibbs left or right. We shall however always suppose that

$$f(0^-) \leq f(0) \leq f(0^+)$$

to avoid pathological behavior.

9.6.1   The Shannon case revisited

The Shannon system, although it serves as a prototype, does not sat-isfy the hypotheses of the theorems about Gibbs phenomenon in [Si1] and [S-V]. The formulae however are rather simple and may be used to show directly that Gibbs occurs for both sampling series and orthogonal series. In this particular case the sampling function $S(t) = (sin\pi t)/\pi t = \phi(t)$, the orthonormal scaling function. However, the sampling approximation to a continuous function is not the same as the orthogonal projection since the coefficients need not be the same. Nonetheless both cases can lead to Gibbs phenomenon for functions with jump discontinuities at 0 and the overshoot calculated. Indeed in [S-V] it was shown that the overshoot is exactly the same as for Fourier series in the case of orthogo-nal approximations. We can also calculate it for sampling series [W-Si], but here we merely state the result

Proposition 9.3
[W-Si] Let h be the function given by

$$h(t) = \begin{cases} sgn\ t - t & ,0 < |t| \leq 1 \\ \alpha & ,0 = t \\ 0 & ,1 < |t| \end{cases} \quad ;$$

then the Shannon expansion of h exhibits Gibbs phenomenon on both the right and the left whatever the value of $h(0) = \alpha$.

9.6.2   Back to wavelets

In the last subsection we saw that for the Shannon system, the ex-istence of Gibbs phenomenon for a function with a jump discontinuity at zero holds whatever the value of the function at 0. In this section we attempt to get similar results for other wavelet sampling series. In [Si1], Gibbs phenomenon for these sampling series was studied under the hypothesis that $f(0) = f(0^+)$. As was seen by the example this is much too restrictive since Gibbs can occur for all choices of $f(0)$.

We shall require that the sampling function $S(t)$ satisfies the condi-tions implied by the usual conditions on the scaling function $\phi(t)$ in Chapter 3, from which $S(t)$ may be constructed. These do not involve differentiability, but we are still able to get local convergence results.

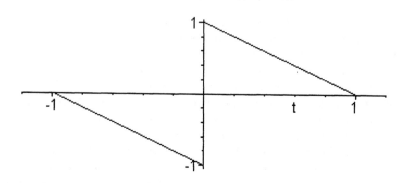

FIGURE 9.2
The function h of Proposition 9.3.

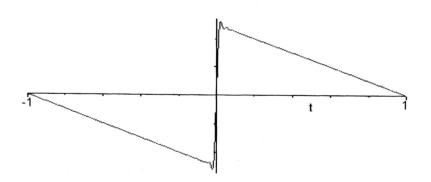

FIGURE 9.3
The partial sum of the Shannon series expansion.

These results should be compared to those for orthogonal series. The conditions in Theorem 3.1 will be analogous to the conditions found in [K1] for the orthogonal case, except that the integral in the latter case is replaced by a series in the former. In [S-V] it was shown that the integral condition for orthogonal wavelets is satisfied for all standard wavelets, in fact for all wavelets with continuous differentiable scaling functions that satisfy

$$\phi(x) = O(|x|^{-\beta}) \qquad |x| \to \infty$$

for some $\beta > 3$. We have been unable to obtain such a general result for the interpolating series. If there is such a result it would have to be more restrictive, since as we shall see, the interpolating series for $_2\phi(t)$ does not exhibit Gibbs while its orthogonal series does.

THEOREM 9.2
Let ϕ be a scaling function satisfying (9.3) and let S be the associated sampling function given by (9.5); let $f \epsilon L^2(\mathbb{R})$ be locally BV and continuous except for a jump discontinuity at zero where $f(0^-) \le f(0) \le f(0^+)$. Then the sampling series (9.15) exhibits Gibbs phenomenon on the right of 0 (respectively left of 0) if and only if

$$[f(0) - f(0^+)]S(a) > [f(0^+) - f(0^-)] \sum_{n=1}^{\infty} S(a+n) \qquad (9.16)$$

for some $a > 0$ (respectively

$$[f(0^-) - f(0)]S(a) > [f(0^+) - f(0^-)] \sum_{n=1}^{\infty} S(a-n)$$

for some $a < 0$).

The proof involves two lemmas.

Lemma 9.1
Let $f \epsilon L^\infty(\mathbb{R})$ be locally BV and continuous on $(-a, a), a > 0$; let f_m be the sum of the sampling series (9.15). Then for each $[-b, b] \subseteq (-a, a)$, $f_m \to f$ uniformly in $[-b, b]$.

The sampling approximation f_m is given by

$$f_m(x) = \sum_n f(n2^{-m})S(2^m x - n)$$

$$= \int_{-\infty}^{\infty} \sum_n S(2^m x - n)\delta(t - n2^{-m})f(t)dt$$

$$= \int_{-\infty}^{\infty} r_m(x,t)f(t)dt.$$

We find that the measure $r_m(x,t)$ satisfies

(i) $\int_{-\infty}^{\infty} r_m(x,t)dt = \sum_n S(2^m x - n) = 1$, $x \in \mathbb{R}$, $m \in \mathbb{Z}$
(ii) $\int_{-\infty}^{\infty} |r_m(x,t)|dt \leq \sum_n |S(2^m x - n)| \leq C < \infty$, $x \in \mathbb{R}$, $m \in \mathbb{Z}$
(iii) *For each* $\gamma > 0$,
$\int_{|x-t|\geq\gamma} |r_m(x,t)|dt \longrightarrow 0$, *as* $m \longrightarrow \infty$, *uniformly for* $x \in \mathbb{R}$.

It is clear that (i) and (ii) hold since $S(x) = O(|x|^{-2})$. To obtain (iii), we observe that

$$\int_{|x-t|\geq\gamma} |r_m(x,t)|dt \leq \sum_n |S(2^m x - n)| \int_{|x-t|\geq\gamma} \delta(t - n2^{-m})dt$$

$$= \sum_{|x-n2^{-m}|\geq\gamma} |S(2^m x - n)|$$

$$= \sum_{|2^m x-n|\geq 2^n \gamma} |S(2^m x - n)|$$

$$\leq \sum_{|2^m x-n|\geq\gamma 2^m} \frac{1}{|2^m x - n|^2 + 1}$$

$$\leq \sum_n \left(\frac{1}{|2^m x - n|^2 + 1}\right)^{2/3} \left(\frac{1}{|\gamma 2^m|^2 + 1}\right)^{1/3}$$

$$\leq C(2^{-m}\gamma^{-1})^{2/3}.$$

These three properties are all that is needed to prove the convergence since

$$f_m(x) - f(x) = \int_{-\infty}^{\infty} r_m(x,t)[f(t) - f(x)]dt$$

$$= \int_{|x-t|<\gamma} + \int_{|x-t|\geq\gamma} = I_1 + I_2.$$

Since f is continuous on $(-a, a)$, it is uniformly continuous on closed subintervals. For $\gamma < a - b$, we have

$$|I_1| \leq \int_{|x-t|<\gamma} |r_m(x,t)||f(x) - f(t)|dt.$$

Now given $\epsilon > 0$, choose γ such that $|f(x)-f(t)| < \epsilon$ for $|x-t| < \gamma < a-b$ and $x \in [-b, b]$. Then I_1 satisfies

$$|I_1| \leq \int_{-\infty}^{\infty} |r_m|(x,t)|dt\epsilon \leq C\epsilon$$

while I_2 satisfies, by (iii),

$$|I_2| \leq \int_{|x-t|\geq\gamma} |r_m(x,t)|2 \parallel f \parallel_\infty dt \leq \epsilon \text{ for } m \geq m_0.$$

Hence by first choosing γ and then m_0 we see

$$|f_m(x) - f(x)| \leq C\epsilon + \epsilon$$

for $m \geq m_0$, and $x \in [-b, b]$. □

We now can use a simpler standard function because of this lemma. We take h_α to be

$$h_\alpha(t) := \begin{cases} \text{sgn}\, t, & 0 < |t| \leq 1 \\ \alpha, & t = 0 \\ 0 & 1 \leq |t| \end{cases}.$$

We use h_α to get rid of the jump discontinuity of f at 0.

Lemma 9.2

Let g be given by

$$g(t) := \begin{cases} f(t) - f(0^+)h_\alpha(t), & t > 0 \\ 0, & t = 0 \\ f(t) + f(0^-)h_\alpha(t), & t < 0 \end{cases},$$

then $g_m(t) \to g(t)$ uniformly for $t\epsilon[-1/2, 1/2]$ as $m \to \infty$.

The proof of this lemma follows directly from Lemma 9.1 if we observe that $g(t)$ is continuous on $(-1, 1)$.

The value of $h_\alpha(t)$ at $t = 0$ did not enter into the definition of $g(t)$ in this lemma. However since it will turn out to be important, we define $h_\alpha(0) = \alpha$ to be the proportional value

$$\alpha = \frac{f(0) - \frac{f(0^+)+f(0^-)}{2}}{\frac{f(0^+)-f(0^-)}{2}}. \tag{9.17}$$

PROOF OF THE THEOREM. Let t_m be a positive sequence such that $t_m \to 0$ as $m \to \infty$. Then since $g_m(t_m) \to 0$, we need only consider $h_m(t_m)$ in studying Gibbs right. (Gibbs left is analogous.)

If Gibbs right exists at 0, then there is such a sequence $\{t_m\}$ such that $h_m(t_m) \to \gamma^+ > 1$, and hence

$$1 < h_m(t_m) = \left[\sum_{n=1}^{2^m} S(2^m t_m - n) + \alpha S(2^m t_m) - \sum_{n=1}^{2^m} S(2^m t_m + n) \right]$$

for $m \geq m_0$. We now take $a = 2^{m_0} t_{m_0}$ and obtain

$$1 < h_{m_0}(2^{-m_0} a) = \sum_{n=1}^{2^{m_0}} S(a - n) + \alpha S(a) - \sum_{n=1}^{2^{m_0}} S(a + n).$$

Moreover, by taking m_0 even larger if necessary we can deduce that

$$1 < \sum_{n=1}^{\infty} S(a - n) + \alpha S(a) - \sum_{n=1}^{\infty} S(a + n).$$

This condition is also sufficient for Gibbs right since the *rhs* is equal to $lim_{m\to\infty} h_m(a2^{-m})$. This inequality may be expressed by using the fact that $\sum_{n\in\mathbb{Z}} S(a - n) = 1$, as

$$1 < 1 - S(a) - 2 \sum_{n=1}^{\infty} S(a + n) + \alpha S(a)$$

or

$$(\alpha - 1)S(a) > 2 \sum_{n=1}^{\infty} S(a + n). \qquad (9.18)$$

By replacing α in (9.18) by the expression in (9.17), we obtain the first conclusion (9.16). The second is obtained by using the corresponding inequality for Gibbs left,

$$(\alpha + 1)S(a) < -2 \sum_{n=1}^{\infty} S(a - n). \qquad (9.19)$$

□

Corollary 9.1
Let S and f be as in the theorem and let $S(t) \geq 0$; then the sampling series (9.15) does not exhibit Gibbs phenomenon whatever the choice of $f(0)$ (satisfying $f(0^-) \leq f(0) \leq f(0^+)$).

PROOF Since $\alpha < 1$, the left side of (9.18) would be negative and the right positive for $S(t) \geq 0$. Hence the inequality cannot hold for any value of $a > 0$. Similarly (9.19) cannot hold for $\alpha > -1$. □

The piecewise linear spline with $S(t) = (1 - |t|)\chi_{[-1,1]}(t)$ satisfies the hypothesis of the Corollary and hence the sampling series does not exhibit Gibbs phenomenon. This is in contrast to the mean square wavelet approximation which does [S-V]. We shall see later that the same is true for the Daubechies wavelets with 4 taps [D].

REMARK 9.2 In the special case $\alpha = 1$, corresponding to f continuous on the right at 0, the condition for Gibbs right is $\sum_{n=1}^{\infty} S(a+n) < 0$. This can be expressed as

$$1 - \sum_{n=0}^{\infty} S(a - n) < 0$$

or

$$\sum_{n=0}^{\infty} S(a - n) > 1$$

which is the condition for Gibbs right in [Si1]. The condition (9.18) unfortunately is not easy to check. We next introduce a simpler sufficient condition for Gibbs right. It involves $\int_{-1}^{1} S(t)dt$, which $= 1$ for the linear spline case which has no Gibbs, but > 1 for the Shannon case which does.

Lemma 9.3
Let $S(t)$ be an even sampling function such that $S(t) > 0$ for $|t| < 1$ and

$$\int_{-1}^{1} S(t)dt = \gamma > 1;$$

let f and S satisfy the conditions of Theorem 9.2. Then there is a $\delta > 0$, such that if

$$f(0^+) - \delta < f(0) \leq f(0^+),$$

the sampling series exhibits Gibbs right at 0.

PROOF We use the well known fact that

$$\sum_{n=-\infty}^{\infty} S(t - n) = \int_{-\infty}^{\infty} S(t)dt = 1,$$

and let $S_\pm$ denote the continuous functions

$$S_\pm(t) = \sum_{n=1}^\infty S(t \pm n).$$

Then we have

$$\int_0^1 S_+(t)dt = \int_0^1 \sum_{n=1}^\infty S(t+n)dt = \int_1^\infty S(t)dt$$

and

$$\int_{-1}^0 S_-(t)dt = \int_{-\infty}^{-1} S(t)dt.$$

Hence by the symmetry of $S(t)$ we find that

$$1 = \int_{-\infty}^\infty S(t)dt = \int_{-1}^1 S(t)dt + 2\int_0^1 S_+(t)dt$$

and

$$\frac{1-\gamma}{2} = \int_0^1 S_+(t)dt.$$

By the mean value theorem there is an $a\epsilon(0,1)$ such that

$$\frac{1-\gamma}{2} = S_+(a) = \sum_{n=1}^\infty S(a+n).$$

The expression (9.18) then becomes

$$(\alpha-1)S(a) > 1-\gamma$$

or, since $S(a) > 0$,

$$\alpha > 1 - \frac{\gamma-1}{S(a)}. \tag{9.20}$$

This gives Gibbs right for $1 - \frac{\gamma-1}{S(a)} < \alpha \le 1$ for the standard function $h(t)$ which has $h(0) = \alpha$. The proof in the Theorem gives us the result for other functions. □

Corollary 9.2
Let $S(t)$ and $f(t)$ be as in the lemma, and let $f(t)$ be continuous on the right (resp. left) at 0. Then the sampling series exhibits Gibbs right (resp. left) at 0.

The result for Gibbs left follows from the symmetry.

REMARK 9.3   In many examples of wavelet systems, $S(t)$ is a convex function on [-1,1]. Since $S(0) = 1$, then $\int_{-1}^{1} S(t)dt > 1$ and the hypothesis holds.   ∎

Example 9.1
The Meyer wavelets have a scaling function $\phi(t)$ whose Fourier transform $\widehat{\phi}(w)$ has support on $[-\pi - \varepsilon, \pi + \varepsilon]$ for some $0 < \varepsilon \leq \frac{\pi}{3}$ and $\widehat{\phi}(w) = 1$ for $w \in [-\pi + \varepsilon, \pi - \varepsilon]$. The same conditions hold for $S(t)$ since $\widehat{S}(w) = \widehat{\phi}(w)/\widehat{\phi}^*(w)$. Thus it is possible to show that $\widehat{S}$ must be of the form

$$\widehat{S}(w) = \int_{w-\pi}^{w+\pi} h \qquad (9.21)$$

where h is some function ≥ 0 with support on $[-\varepsilon, \varepsilon]$ such that $\int h = 1$. We suppose that h and hence $\widehat{S}$ is symmetric. We may find $S(t)$ by using the inverse Fourier transform which gives us

$$S(t) = \frac{1}{2\pi} \int_{-\infty}^{\infty} \widehat{S}(w)e^{iwt} dw$$

$$= \frac{1}{2\pi} \int_{-\infty}^{\infty} \left(\int_{w-\pi}^{w+\pi} h \right) e^{iwt} dw$$

$$= \frac{1}{2\pi} \int_{-\infty}^{\infty} (h(w-\pi) - h(w+\pi)) \frac{e^{iwt}}{it} dw$$

$$= \frac{1}{\pi} \int_{-\infty}^{\infty} \frac{h(w)e^{i(w+\pi)t} - h(w)e^{i(w-\pi)t}}{2it} dw$$

$$= \int_{-\infty}^{\infty} h(w)e^{iwt} \frac{\sin \pi t}{\pi t} dw$$

$$= \frac{\sin \pi t}{\pi t} \int_{-\varepsilon}^{\varepsilon} h(w)e^{iwt} dw$$

$$= \frac{\sin \pi t}{\pi t} (1 + \int_{-\varepsilon}^{\varepsilon} h(w)(e^{iwt} - 1)dw). \qquad (9.22)$$

We already know that $\int_{-1}^{1} \frac{\sin \pi t}{\pi t} dt > 1$; in fact this is exactly the overshoot for Fourier series $\simeq 1.18$. Hence if we can show the last integral in (9.22) to be sufficiently small in magnitude, we will have shown that Gibbs phenomenon exists.

Let σ^2 denote the second moment of $h(w)$,

$$\sigma^2 = \int_{-\epsilon}^{\epsilon} w^2 h(w) dw.$$

Then we have

$$\int_{-1}^{1} S(t) dt = 2 \int_{0}^{1} \frac{\sin \pi t}{\pi t} 2 \int_{0}^{\epsilon} h(w) \cos wt \, dw \, dt$$

$$\geq 4 \int_{0}^{1} \frac{\sin \pi t}{\pi t} \int_{0}^{\epsilon} h(w) \left(1 - \frac{w^2 t^2}{2} \right) dw \, dt$$

since $\cos wt \geq 1 - \frac{w^2 t^2}{2}$ for $|wt| \leq \pi/3$. Furthermore the second integral satisfies

$$4 \int_{0}^{1} \int_{0}^{\epsilon} h(w) \frac{w^2}{2} t^2 \frac{\sin \pi t}{\pi t} dw \, dt = \sigma^2 \int_{0}^{1} \frac{t}{\pi} \sin \pi t \, dt = \frac{\sigma^2}{\pi^2}.$$

We can find a bound on σ^2 since

$$\sigma^2 = \int_{-\frac{\pi}{3}}^{\frac{\pi}{3}} w^2 h(w) dw \leq \left(\frac{\pi}{3} \right)^2 \int_{-\frac{\pi}{3}}^{\frac{\pi}{3}} h(w) dw$$

and hence

$$\int_{-1}^{1} S(t) dt \geq 1.179 - \frac{\pi^2}{\pi^2 \, 9} > 1,$$

i.e., Gibbs holds for all symmetric Meyer wavelets for functions continuous on the right or the left at 0.   □

Similar calculations can be used to show that the sampling series for certain Daubechies wavelet as well as spline based wavelet subspaces exhibit Gibbs phenomenon. (See [W-Si] for more details.) However we have

Example 9.2
The Daubechies wavelets with support on $[0, 3]$ are defined by the solution to the dilation equations

$$\phi(t) = \sqrt{2} \sum_{k=0}^{3} c_k \phi(2t - k) \tag{9.23}$$

where

$$c_0 = \nu(\nu - 1)/D$$
$$c_1 = (1 - \nu)/D$$
$$c_2 = (\nu + 1)/D$$
$$c_3 = \nu(\nu + 1)/D$$

and

$$D = \sqrt{2}(\nu^2 + 1), \quad \nu \in \mathbb{R}.$$

The standard case which has a vanishing first wavelet moment corresponds to $\nu = -\frac{1}{\sqrt{3}}$. The sampling function for $\nu < 0$ is given in Section 9.3 as

$$S(t) = \frac{2\nu}{\nu - 1} \sum_{n=0}^{\infty} \left(\frac{1 + \nu}{1 - \nu}\right)^n \phi(t - n + 1). \tag{9.24}$$

Since the $S(t)$ is not symmetric, we cannot use Lemma 9.2 but must use (9.16) directly. In order to do so, we must evaluate

$$\sum_{n=1}^{\infty} S(a + n).$$

We try $a = \frac{1}{2}$. Then

$$S(\frac{1}{2} - n) = \frac{2\nu}{\nu - 1} \sum_{k=0}^{\infty} \left(\frac{1 + \nu}{1 - \nu}\right)^k \phi(\frac{1}{2} - n - k + 1)$$

$$= \frac{2\nu}{\nu - 1} \sum_{j=n}^{1} \left(\frac{1 + \nu}{1 - \nu}\right)^{j-n} \phi(\frac{3}{2} - j). \tag{9.25}$$

We may evaluate $\phi(\frac{3}{2} - j)$ again by using (9.24). It gives us

$$\phi(\frac{1}{2}) = \sqrt{2} \sum c_k \phi(1 - k) = \sqrt{2} c_0 \phi(1)$$

$$= \frac{\sqrt{2}\nu(\nu - 1)}{\sqrt{2}(\nu^2 + 1)} \frac{(\nu - 1)}{2\nu} = \frac{(\nu - 1)^2}{2(\nu^2 + 1)}$$

$$\phi(\frac{3}{2}) = \sqrt{2} \sum c_k \phi(3 - k) = \sqrt{2}(c_1 \phi(2) + c_2 \phi(1))$$

$$= \left(\frac{1 - \nu}{\nu^2 + 1}\right)\left(\frac{\nu + 1}{2\nu}\right) + \left(\frac{1 + \nu}{\nu^2 + 1}\right)\left(\frac{\nu - 1}{2\nu}\right) = 0$$

$$\phi\left(\frac{5}{2}\right) = \sqrt{2}\sum c_k\phi(5-k) = \sqrt{2}c_3\phi(2)$$

$$= \frac{\nu(\nu+1)}{(\nu^2+1)}\frac{(\nu+1)}{2\nu} = \frac{(\nu+1)^2}{2(\nu^2+1)}$$

where $\phi(1)$ and $\phi(2)$ are also found from (9.24) and the relation $\phi(1) + \phi(2) = 1$. Hence we find by (9.25) that

$$S(\tfrac{1}{2}) = \frac{2\nu}{\nu-1}\left\{1\phi\left(\frac{3}{2}\right) + \left(\frac{1+\nu}{1-\nu}\right)\phi\left(\frac{1}{2}\right)\right\}$$

$$= \frac{2\nu}{\nu-1}\left\{\left(\frac{1+\nu}{1-\nu}\right)\frac{(\nu-1)^2}{2(\nu^2+1)}\right\} = -\frac{\nu(\nu+1)}{\nu^2+1}.$$

We also have

$$\sum_{n=1}^{\infty} S\left(\frac{1}{2}+n\right) = \sum_{n=1}^{\infty}\frac{2\nu}{\nu-1}\sum_{k=0}^{n+1}\left(\frac{1+\nu}{1-\nu}\right)^k\phi\left(\frac{3}{2}+n-k\right).$$

$$= \frac{2\nu}{\nu-1}\sum_{n=1}^{\infty}\left\{\left(\frac{1+\nu}{1-\nu}\right)^{n+1}\phi\left(\frac{1}{2}\right) + \left(\frac{1+\nu}{1-\nu}\right)^{n-1}\phi\left(\frac{5}{2}\right)\right\}$$

$$= \frac{2\nu}{\nu-1}\sum_{n=1}^{\infty}\left\{\left(\frac{1+\nu}{1-\nu}\right)^{n+1}\frac{(\nu-1)^2}{2(\nu^2+1)} + \left(\frac{1+\nu}{1-\nu}\right)^{n-1}\frac{(\nu+1)^2}{2(\nu^2+1)}\right\}$$

$$= \frac{\nu}{(\nu-1)(\nu^2+1)}\sum_{n=1}^{\infty}\left(\frac{1+\nu}{1-\nu}\right)^{n-1}2(1+\nu)^2$$

$$= \frac{(1+\nu)^2}{1+\nu^2}.$$

For $\nu < -1$, $S(\tfrac{1}{2}) < 0$, and hence (9.16) becomes $\alpha < -\tfrac{1}{\nu}$. We always have Gibbs right in this case. The case $\nu \geq -1$ is inconclusive.   ◻

9.7   Irregular sampling in wavelet subspaces

For the Shannon system, the standard result for irregular sampling is the "Kadec 1/4 theorem" which says that if sampled values are less than 1/4 from the integers, then there is a sampling sequence for π-bandlimited signals (the subspace V_0 for the Shannon wavelet), i.e.,

$$f(x) = \sum_n f(n + \delta_n) S_n(x)$$

provided that $\sup_n |\delta_n| < 1/4$.

Many attempts have been made to extend this theorem to other wavelet subspaces [Be],[W-L],[L1],[C-I-S],[S-Z]. We summarize a result below based on the following proposition which generalizes (9.5). (Recall that a family of functions $\{\phi_j\}_{j \in \mathbb{Z}}$ in a Hilbert space $\mathcal{H}$ is called a *frame* if there exist $A > 0, B < \infty$ such that $A\|f\|^2 \le \sum_{j \in J} |\langle f, \phi_j \rangle|^2 \le B\|f\|^2$ for any $f \in \mathcal{H}$. The constants A, B are called frame bounds. If only the right-hand inequality is satisfied for all $f \in \mathcal{H}$, then $\{\phi_j\}$ is called a Bessel sequence with bound B.)

Proposition 9.4

[S-Z] Let $\phi \in S_r$, and let V_0 be the closed subspace of $L^2(\mathbb{R})$ such that $\{\phi(\cdot - n) : n \in \mathbb{Z}\}$ is a frame for V_0 with bounds A and B. Put

$$E_\phi = \left\{ w \in \mathbb{R} : \sum_{n=-\infty}^{\infty} \left| \hat{\phi}(w + 2n\pi) \right|^2 > 0 \right\}.$$

Suppose that there exist two constants C_1 and $C_2 > 0$ such that

$$C_1 \, \chi_{E_\phi}(w) \le |Z_\phi(0, w)| \le C_2 \chi_{E_\phi}(w), \quad a.e. \tag{9.26}$$

where $Z_\phi(x, w)$ is the Zak transform of ϕ.

Then there is a frame $\{S(\cdot - n)\}$ for V_0 such that for any $f \in V_0$

$$f(x) = \sum_n f(n) S_n(x - n)$$

where the convergence is both in $L^2(\mathbb{R})$ and uniform on $\mathbb{R}$.

The following result in the paper by Sun and Zhou appears to be the most general [S-Z]. It is based on the theory of frames [Y] which is more general than Riesz bases.

THEOREM 9.3

[S-Z] Let the hypotheses be as in Proposition 9.4. Moreover, suppose that $\{\phi'(\cdot - n) : n \in \mathbb{Z}\}$ is a Bessel sequence with bound M. Suppose

that $\delta > 0$ and $\delta[2\delta] < AC_1^2/BM$. Then, for any sequence δ_n with $\sup_n |\delta_n| < \delta$, there exists a frame $\{S_n\}$ for V_0 such that for any $f \in V_0$

$$f(t) = \sum_n f(n + \delta_n)S_n(t)$$

where the convergence is both in $L^2(\mathbb{R})$ and uniform on $\mathbb{R}$.

Notice that for orthogonal scaling functions, the constants A and B are each equal to 1, while for sampling functions of the kind considered in section 9.2, $C_1 = C_2 = 1$, so that in either case the bounds for δ are simplified. In particular, for Meyer wavelets, all four constants are easy to estimate. They are $A = B = C_1 = 1$ and $C_2 = \sqrt{2}$.

9.8　Problems

1. Find the Riesz basis $\{q(x,n)\}$ of V_0 for the Daubechies wavelets with 4 taps (Example 3) as an explicit scaling function series.

2. For the Franklin wavelets, find a biorthogonal sequence $\{g(t-n)\}$ in V_0 to the Riesz basis $\{S(t-n)\}$ where $S(t)$ is the hat function. (Hint: Use the same trick as in orthogonalization and the Fourier transform $\sum_k \hat{S}(w + i\pi k)\overline{\hat{g}(w + 2\pi k)} = 1$.)

3. For the Franklin wavelets, shifted sampling is impossible for $a = \frac{1}{2}$. Show that the condition

$$\sum \phi(a + n)e^{-iwn} \neq 0$$

fails in this case. (Use the original B-spline $\theta_2(x)$ instead of ϕ.)

4. Show that for the B-spline $\theta_2(x)$ ($=$ hat function), $\theta_2(a) + \theta_2(a + 1) = 1$ for $0 < a < 1$. Use this to show that the criterion holds for such $a \neq \frac{1}{2}$ and hence there is a shifted sampling theorem.

5. Let $h(\xi) = 2\epsilon$, $|\xi| < \epsilon \leq \frac{\pi}{3}$ and $h(\xi) = 0$, $|\xi| \geq \epsilon$; let $\hat{S}(w)$ be given by (9.21). Show that the condition of Lemma 9.3 holds.

6. Find the orthogonal expansion of the Heaviside function with respect to the Franklin scaling function series and compare to the sampling series for the same function for m=0, 1, 2.

7. Find an upper bound on the aliasing error for $f \in V_2$ when the sampling series is in V_0.

8. Expand $f(t) = \frac{\sin \pi t/2}{t}$ in terms of the Shannon sampling theorem. Is $f \in V_0$? V_1?

9. Find the constants needed for Theorem 9.3 in the case of a Meyer wavelet with $h(\xi) = \frac{3}{\pi}(\frac{4\pi}{3} - \xi)$ for $|\xi| < \frac{\pi}{3}$, and $h(\xi) = 0$ for $|\xi| \geq \frac{\pi}{3}$.

Chapter 10

Extensions of Wavelet Sampling Theorems

Sampled values of a continuous function appear frequently in wavelet theory, even when the series is not of the form of (9.6). In fact, it is common to express the coefficients in the scaling function series at the finest scale m to be the sampled values, i.e.,

$$a_{n,m} \approx f(n2^{-m})2^{-m/2}$$

This is plausible (and proper) if the scaling function is strongly localized. In fact, it is just a one term Riemann sum of the coefficient integral.

We shall call series of the form

$$\sum_n f(n2^{-m})\phi(2^m t - n)$$

"hybrid series". While they do not converge to $f \in V_m$ as do the standard series, they do give an approximation of continuous functions as $m \to \infty$. We shall consider such series for both orthogonal $\phi(t)$ and positive scaling functions discussed in Chapter 8.

In some cases both the value of a function and that of the derivative are known (or can be estimated) at certain sampling points. This leads to an alternative version of the sampling theorem involving so-called Hermite interpolation. In wavelet subspaces, one can sometimes use the same approach, but it is more straightforward to work with "multiwavelets". These are systems in which the scaling function is a vector rather than a scalar of the form $\phi = [\phi_1, ..., \phi_n]^T$. We shall not explore them in detail, but rather consider only a few topics related to sampling.

A number of other topics related to sampling are also covered. We begin with a slight generalization of the usual sampling theorems applied to wavelet subspaces in which the scaling function serves as a sampling function but sampling occurs at double the rate [W7]. This enables us to have sampling series which converge at a much faster rate than (9.1). We then look at cases of wavelet subspaces in which the scaling function is also a sampling function. We discuss the work of Xia and Zhang [X-Z] and its extension to band limited scaling functions [W-L]. This leads to another orthogonal scaling-sampling function (besides the sinc function) given as a closed form expression. We also consider "hybrid sampling" in which sampling coefficients are used with orthogonal or other scaling functions which are themselves not sampling functions. We show that convergence is at a rate comparable to the orthogonal expansions. Finally we discuss "finite element multiwavelets" which lead to more general sampling series.

10.1 Oversampling with scaling functions

In the classical Shannon sampling theorem, the same sequence of functions is both orthonormal and a sampling sequence. This is not true for most wavelet subspaces in which the sampling functions and the orthonormal bases are different. However, by oversampling at double the rate the property of the Shannon wavelets can be extended to a much larger class, which includes the Meyer wavelets [W7]. This will be shown in this section.

More precisely, we shall look for $\phi(t)$ for which $\phi(2t-n)$ is a sampling function for V_0 by finding the properties it must satisfy. We then show that for the Meyer family, these properties are satisfied. Finally we show that the oversampling property together with another property characterize this family.

Accordingly, let $\phi(t) = 0\left((1+|t|)^{-1-\epsilon}\right)$ and $\widehat{\phi}(w) = 0\left((1+|w|)^{-1-\epsilon}\right)$, $\epsilon > 0$, and let $\{\phi(t-n)\}$ be an orthonormal basis of V_0 such that for each $f \in V_0$,

$$f(t) = \sum_n f\left(\frac{n}{2}\right)\phi(2t-n), \tag{10.1}$$

in the sense of $L^2(\mathbb{R})$. Then, in particular

$$\phi(t) = \sum_n \phi\left(\frac{n}{2}\right)\phi(2t-n)$$

and by taking Fourier transforms we find

$$\widehat{\phi}(w) = \frac{1}{2} \sum_n \phi\left(\frac{n}{2}\right) e^{-iwn/2} \phi\left(\frac{w}{2}\right)$$

$$= \sum_n \widehat{\phi}(w + 4\pi k) \widehat{\phi}\left(\frac{w}{2}\right).$$

The last equality is obtained by finding the Fourier coefficients of the 4π periodic function

$$\widehat{\phi}^*(w) := \sum_k \widehat{\phi}(w + 4\pi k),$$

which are exactly $\frac{1}{2}\phi\left(\frac{n}{2}\right)$. Thus, the Fourier transform of the dilation equation is

$$\widehat{\phi}(w) = \widehat{\phi}^*(w)\widehat{\phi}\left(\frac{w}{2}\right). \tag{10.2}$$

This is also sufficient for (10.1) to hold.

Lemma 10.1

Let ϕ be a scaling function such that $\phi(t) = O\left((1+|t|)^{-1-\epsilon}\right)$ and $\widehat{\phi}(w) = O\left((1+|w|)^{-1-\epsilon}\right)$ for $\epsilon > 0$; then (10.2) holds for ϕ if and only if (10.1) holds for all $f \in V_0$.

PROOF In order to show that (10.1) holds we must first show the series converges in the sense of $L^2(\mathbb{R})$. Since $\{\sqrt{2}\,\phi(2t - n)\}$ is orthonormal, we need only show that $\{f\left(\frac{n}{2}\right)\} \in \ell^2$. This is based on the fact that $f \in V_0 \subseteq V_1$, and therefore has an expansion convergent in $L^2(\mathbb{R})$ and, because of the decay property of $\phi(t)$, uniformly convergent on bounded sets,

$$f(t) = \sum_n a_{n,1} \sqrt{2}\,\phi(2t - n).$$

Thus we have

$$f\left(\frac{k}{2}\right) = \sum_n a_{n,1} \sqrt{2}\,\phi(k - n)$$

and by taking the discrete Fourier transform, we find

$$\sum_k f\left(\frac{k}{2}\right) e^{iwk} = \sqrt{2} \sum_n a_{n,1} e^{iwn} \sum_k \phi(k) e^{iwk}. \tag{10.3}$$

The right hand side is the product of a bounded function and an $L^2(-\pi, \pi)$ function. Hence the left hand side of (10.3) is in $L^2(-\pi, \pi)$ and $\{f\left(\frac{k}{2}\right)\} \in \ell^2$.

To show that it converges to $f(t)$ we use its expansion in V_0, which is

$$f(t) = \sum_n a_{n,0} \phi(t - n)$$

$$= \sum_n a_{n,0} \sum_j \phi\left(\frac{j}{2}\right) \phi(2t - j - 2n)$$

$$= \sum_n a_{n,0} \sum_k \phi\left(\frac{k}{2} - n\right) \phi(2t - k)$$

$$= \sum_k f\left(\frac{k}{2}\right) \phi(2t - k).$$

The interchange of the two series is justified since the inner series is a convolution of two ℓ^1 sequences. This is all we need since the last series converges in $L^2(\mathbb{R})$. □

We now turn to the Meyer type wavelets presented in Chapter 3. Their scaling functions ϕ were given by

$$\widehat{\phi}(w) = \left\{ \int_{w-\pi}^{w+\pi} h(\zeta) d\zeta \right\}^{1/2}, \tag{10.4}$$

where h is a positive distribution whose smallest support interval is $[-\epsilon, \epsilon]$, $0 \le \epsilon < \frac{\pi}{3}$.

Lemma 10.2
Let $\widehat{\phi}(w)$ satisfy (10.4); then it also satisfies (10.2)

PROOF   Since $\widehat{\phi}$ is taken to be the positive square root in (10.4), we need only show that

$$|\widehat{\phi}(w)|^2 = |\widehat{\phi}^*(w)|^2 \left|\widehat{\phi}\left(\frac{w}{2}\right)\right|^2. \tag{10.5}$$

Since $\widehat{\phi}$ has support on $[-\pi - \epsilon, \pi + \epsilon]$, it follows that $\widehat{\phi}^*$ has support on $\Lambda = \bigcup_k [-\epsilon + (4k - 1)\pi, \epsilon + (4k + 1)\pi]$. Thus on the support of $\widehat{\phi}\left(\frac{w}{2}\right)$, $\widehat{\phi} = \widehat{\phi}^*$ and (10.5) becomes

$$|\widehat{\phi}(w)|^2 = |\widehat{\phi}(w)|^2 \left|\widehat{\phi}\left(\frac{w}{2}\right)\right|^2.$$

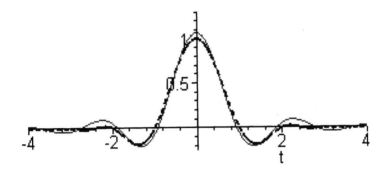

FIGURE 10.1
An example of the scaling function of a Meyer wavelet at scale m=0 and the sum of 5 terms of its sampling expansion in the next scale m=1.

Moreover, $\widehat{\phi}\left(\frac{w}{2}\right) = 1$ on $[-2\pi + 2\epsilon, 2\pi - 2\epsilon] \supseteq [-\pi - \epsilon, \pi + \epsilon]$, the support of $\widehat{\phi}$. Thus (10.5) holds. □

We can also go in the opposite direction. We begin with (10.2) and try to get (10.4).

Lemma 10.3

Let $\phi(t)$ be a scaling function satisfying the conditions of Lemma 10.1 and (10.2); let the support of $\widehat{\phi}$ be a bounded interval; then there is a distribution h, $\hat{h}(0) = 1$, with support in an interval of length $\leq 2\pi/3$ contained in $[-\pi, \pi]$ such that

$$\int_{w-\pi}^{w+\pi} h \geq 0 \quad \text{and} \quad \widehat{\phi}(w) = \left[\int_{w-\pi}^{w+\pi} h\right]^{1/2}.$$

PROOF We first observe that the support of $\widehat{\phi}$ must be a finite interval $[-a, b]$ where both a and b are positive. This follows from the fact that $\widehat{\phi}(0) = 1$. Since $\widehat{\phi}$ is continuous its support contains a neighborhood of the origin.

The support of $\widehat{\phi}^*$ is $\Lambda = \cup[-a + 4\pi k, b + 4\pi k]$ and hence if $b + a \geq 4\pi$, would be all of $\mathbb{R}$. But this is impossible since by (10.2) the support of $\widehat{\phi}\left(\frac{w}{2}\right)$ would also have to be $[-a, b]$. We can say much more since, by

(10.2),

$$[-a, b] = [-2a, 2b] \cap \Lambda.$$

Hence $-a + 4\pi \geq 2b$ and $b - 4\pi \leq -2a$, which may be expressed as

$$a + 2b \leq 4\pi, \ 2a + b \leq 4\pi,$$

which, in turn, may be added to obtain $a + b \leq 8\pi/3$.

On the other hand $2\pi \leq a + b$ since otherwise the orthogonality condition

$$\sum_k \left| \widehat{\phi}(w + 2\pi k) \right|^2 = 1, \qquad w \in \mathbb{R}$$

would be violated.

Since on the support of $\widehat{\phi}$, (10.2) becomes

$$\widehat{\phi}(w) = \widehat{\phi}(w)\widehat{\phi}\left(\frac{w}{2}\right),$$

it follows that $\widehat{\phi}\left(\frac{w}{2}\right) = 1$ on $[-a, b]$ or $\widehat{\phi}(w) = 1$ on $\left[-\frac{a}{2}, \frac{b}{2}\right]$.

We now define $h(w)$ to be

$$h(w) := \begin{cases} -\frac{d}{dw} \left| \widehat{\phi}(w + \pi) \right|^2, & 0 < w + \pi \\ 0, & w + \pi \leq 0 \end{cases}. \tag{10.6}$$

where the derivative is in general taken in the distribution sense.

It should be noticed that $h(w) = 0$ for $w < \pi - a$ or $w > b - \pi$. Furthermore, since $\left| \widehat{\phi}(w + \pi) \right|^2 + \left| \widehat{\phi}(w - \pi) \right|^2 = 1$ for $\pi - a < w < b - \pi$, it follows that

$$h(w) = \begin{cases} \frac{d}{dw} \left| \widehat{\phi}(w - \pi) \right|^2, & w - \pi < 0 \\ 0, & w - \pi \geq 0. \end{cases}$$

From these two expressions we deduce that

$$\left| \widehat{\phi}(w) \right|^2 = \int_{w-\pi}^{w+\pi} h(\zeta)d\zeta \geq 0, \qquad w \in \mathbb{R},$$

which may be rewritten as the conclusion of the Lemma, since the length of the support of h is $b + a - 2\pi \leq 2\pi/3$. $\square$

As a consequence of the hypothesis, $\left| \widehat{\phi}(w) \right|^2$ must be continuous but not necessarily differentiable, so that h is not necessarily a function but is always a tempered distribution. To be consistent we must take $\widehat{\phi}(w)$ to be the positive square root of $\left|\widehat{\phi}(w)\right|^2$. Its inverse Fourier transform $\phi(t)$ is not necessarily real. However, $\widehat{\phi}(w)$ is symmetric about $\frac{-a+b}{2}$ if $h(w)$ is symmetric about 0. Therefore, in this case $\phi(t)$ can be made real by shifting w by an amount $\frac{-a+b}{2}$. Then $h(w)$ satisfies the condition needed except for the positivity. We add this as a hypothesis to get

THEOREM 10.1
Let $\phi(t)$ be a real, symmetric scaling function such that

$$\phi(t) = O\left(1 + |t|\right)^{-1-\epsilon}, \widehat{\phi}(w) = O\left(1 + |w|\right)^{-1-\epsilon}, \epsilon > 0,$$

and $\widehat{\phi}$ is nonincreasing for $w > 0$; then $\phi(t)$ is a Meyer type scaling function if and only if
(i) $\phi(t)$ satisfies the double sampling property (10.1) and
(ii) the support of $\widehat{\phi}(w)$ is a bounded interval.

10.2   Hybrid sampling series

Hybrid sampling series arise when an orthogonal series is paired with sampling coefficients, i.e., if ϕ is an orthogonal scaling function, then a continuous function $f(t) \in L^1(\mathbb{R})$ is approximated by an element of V_m given by

$$f_m(t) = \sum_n f(n2^{-m})\phi(2^m t - n). \qquad (10.7)$$

If we begin with a function in V_m, this series does not recover the function as it would if the orthogonal expansion were used, but the sampling series (if it exists)

$$f_m^s(t) = \sum_n f(n2^{-m})S(2^m t - n)$$

does. Here the sampling function $S(t)$ is the one discussed in the last chapter.

We can show the convergence of $f_m(t)$ as $m \to \infty$ by showing that the difference

$$e_m(t) = f_m(t) - f_m^s(t) = \sum_n f(n2^{-m})(\phi(2^m t - n) - S(2^m t - n))$$

converges to 0. This is done by considering the Fourier transform

$$\widehat{e}_m(w) = \sum_n f(n2^{-m})[\widehat{\phi}(2^{-m}w) - \widehat{S}(2^{-m}w)]2^{-m}e^{-inw2^{-m}}$$

$$= \sum_n f(n2^{-m})[\widehat{\phi}(2^{-m}w) - \frac{\widehat{\phi}(2^{-m}w)}{\widehat{\phi}^*(2^{-m}w)}]2^{-m}e^{-inw2^{-m}}$$

where we have used (9.5) $\widehat{S}(w) = \frac{\widehat{\phi}(w)}{\widehat{\phi}^*(w)}$.

We can find a bound on this difference in the case of Meyer wavelets where we have $\widehat{\phi}(w) \geq 0$. Then we find that

$$\widehat{\phi}^*(w) = \sum_k \widehat{\phi}(w + 2\pi k) \geq \sum_n \widehat{\phi}^2(w + 2\pi k) = 1$$

and

$$(1 - \frac{1}{\widehat{\phi}^*(w)}) \leq \widehat{\phi}^*(w) - 1.$$

Thus we have, when $m = 0$,

$$|\widehat{e}_m(w)| \leq |F_m(w)||\widehat{\phi}(2^{-m}w)||\widehat{\phi}^*(2^{-m}w) - 1|, \qquad (10.8)$$

where

$$F_m(w) = \sum_n f(n2^{=m})e^{-inw2^{-m}}2^{-m}.$$

Since $F_m(w)$ is an approximation to the continuous Fourier transform of f, it can be shown to be dominated by a multiple of the L^1 norm of f. By using this and the mean value theorem, we can deduce that $\widehat{e}_m(w)$ converges to 0 uniformly. Furthermore, we obtain for $m = 0$

$$\int_{-\infty}^{\infty} |\widehat{e}(w)|dw \leq \| F \|_\infty \| \widehat{\phi} \|_\infty \int_{-\pi-\epsilon}^{\pi+\epsilon} |\widehat{\phi}^*(w) - 1|dw \qquad (10.9)$$

since $\widehat{\phi}$ is zero outside of $[-\pi - \epsilon, \pi + \epsilon]$. This gives us a bound on the L^1 norm of $\widehat{e}(2^{-m}w)$ obtained by replacing w by $2^{-m}w$ and enables us to deduce that e_m converges to 0 in the L^2 sense. Rather than explore this further, we consider similar series where the scaling function ϕ is replaced by the positive summability function ρ^r.

10.3 Positive hybrid sampling

If we relax the requirement that series be composed of orthogonal scaling functions but rather that it only be series in a Riesz basis of V_m, but with sampling coefficients, then we can get positive sampling series. To do so we use the positive summability functions introduced in Chapter 8. We repeat the main result.

THEOREM 10.2
Let $\rho^r(t) = \sum_n r^{|n|}\phi(t-n)$, where r is chosen so large that $\rho^r(t) \geq 0$, for $t \in \mathbb{R}$; then $\{2^{\frac{m}{2}}\rho^r(2^m t - n)\}_{n \in Z}$ is a Riesz basis of V_m; its dual basis is generated by $\widetilde{\rho}_r$, where

$$\widetilde{\rho}_r(t) = \frac{1}{2\pi(1-r^2)}[(1+r^2)\phi(t) - r\{\phi(t+1) + \phi(t-1)\}].$$

In this section we introduce another positive kernel based on the same summability function but now defined by

$$G^r(t,s) = \frac{1-r}{1+r}\sum_n \delta(s-n)\rho^r(t-n), \qquad (10.10)$$

where $\delta(t)$ is the delta distribution. It gives us an approximation $f_0^r \in V_0$ to a contiuous $f \in L^2$ where

$$f_0^r(t) = \int_{-\infty}^{\infty} G^r(s,t)f(s)ds.$$

The summability function $\rho^r(t)$, as defined in Theorem 10.2, is no longer compactly supported. A dilation equation relating $\rho^r(t)$ and a series of $\{\rho^r(2t-n)\}$ may be found but is now an infinite sum. However, the series is convergent rapidly [W-She4] and its Fourier transform is given by

$$\widehat{\rho}^r(w) = Q_r(w)\widehat{\phi}(w) = \frac{1-r^2}{1-2r\cos w + r^2}\widehat{\phi}(w) \qquad (10.11)$$

$$= \frac{1-r^2}{1-2r\cos w + r^2}m_0\left(\frac{w}{2}\right)\frac{1-2r\cos\frac{w}{2}+r^2}{1-r^2}\widehat{\rho}^r\left(\frac{w}{2}\right)$$

$$= \frac{1-2r\cos\frac{w}{2}+r^2}{1-2r\cos w + r^2}m_0\left(\frac{w}{2}\right)\widehat{\rho}^r\left(\frac{w}{2}\right).$$

Here, we use the identity $\widehat{\phi}(w) = m_0(\frac{w}{2})\widehat{\phi}(\frac{w}{2})$.

This (10.11) is the Fourier transformed version of the dilation equation for $\rho^r(t)$. By expanding the periodic factor in a Fourier series and taking the inverse Fourier transform, we can find the coefficients of the dilation equation in the time domain as follows:

$$a_k = \sum_{n=-\infty}^{\infty} \left(c_{k-2n} r^{|n|} \frac{1+r^2}{1-r^2} - c_{k-1-2n} \frac{r^{|n|+1}}{1-r^2} - c_{k+1-2n} \frac{r^{|n|+1}}{1-r^2} \right)$$

(10.12)

Notice that, for every fixed k, (10.12) contains only a finite number of terms. We then have the dilation equation:

$$\rho^r(t) = \sqrt{2} \sum_{k=-\infty}^{\infty} a_k \rho^r(2t - k).$$

Similarly the biorthogonal function $\widetilde{\rho}_r$ satisfies

$$\widehat{\widetilde{\rho}}_r(w) = \frac{1 - 2r\cos w + r^2}{1 - 2r\cos \frac{w}{2} + r^2} m_0(\frac{w}{2})\widehat{\widetilde{\rho}}_r(\frac{w}{2})$$

(10.13)

and hence

$$\widetilde{\rho}_r(t) = \sqrt{2} \sum_{k=-\infty}^{\infty} \widetilde{a}_k \widetilde{\rho}_r(2t - k),$$

with coefficients given by

$$\widetilde{a}_k = \sum_{n=-\infty}^{\infty} \left(c_{k-n} r^{|n|} \frac{1+r^2}{1-r^2} - c_{k-2-n} \frac{r^{|n|+1}}{1-r^2} - c_{k+2-n} \frac{r^{|n|+1}}{1-r^2} \right).$$

(10.14)

The pair of biorthogonal mother wavelets can be defined by

$$\xi^r(t) = \sqrt{2} \sum (-1)^k \widetilde{a}_{1-k} \rho^r(2t - k)$$

(10.15)

with its dual

$$\widetilde{\xi}_r(t) = \sqrt{2} \sum_{k=-\infty}^{\infty} (-1)^k a_{1-k} \widetilde{\rho}_r(2t - k)$$

(10.16)

as given in [C, p.263]. Then the Fourier transformed versions of (10.15) and (10.16) are

$$\widehat{\xi}^r(w) = e^{-\frac{w}{2}i} \frac{Q_r(\frac{w}{2} + \pi)}{Q_r(w)} \overline{m_0(\frac{w}{2} + \pi)} \widehat{\rho}^r(\frac{w}{2})$$

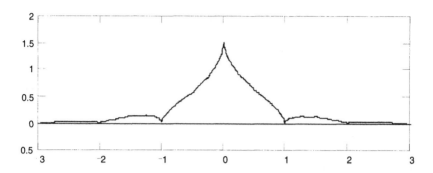

FIGURE 10.2
**The positive summability function for the Coiflet of degree 2
with r = 0.22.**

$$= \frac{(1 - 2r\cos w + r^2)(1 - r^2)}{(1 + r^2) - 2r^2\cos w}\widehat{\psi}(w)$$

and

$$\widehat{\widetilde{\xi}}_r(w) = \frac{(1 + r^2) - 2r^2\cos w}{(1 - 2r\cos w + r^2)(1 - r^2)}\widehat{\psi}(w),$$

respectively. Hence by taking the inverse Fourier transform and using
the Fourier series of the periodic factor, we have in the first case that

$$\xi^r(t) = \sum_n r^{|2n|}\psi(t - n) \tag{10.17}$$

$$-\frac{r}{1 + r^2}\sum_n r^{|2n|}[\psi(t - n - 1) + \psi(t - n + 1)].$$

Notice that both $\xi^r(t)$ and $\widetilde{\xi}_r(t)$ belong to the wavelet subspace W_0,
the space with basis $\{\psi(t - n)\}$ and each have as many vanishing moments as ψ has. Furthermore, it can be shown that each constitutes a
Riesz basis of W_0 [W-She1].

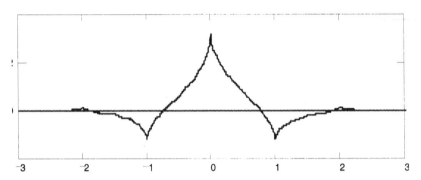

FIGURE 10.3
The dual of the positive summability function in Figure 10.2.

10.4   The convergence of the positive hybrid series

We assume that f is a uniformly continuous function in $L^1(\mathbb{R})$. We shall approximate it by a hybrid series in V_m given by

$$f_m^r(t) = \sum_n f(\frac{n}{2^m})\rho^r(2^m t - n)\frac{1-r}{1+r} \qquad (10.18)$$

whose convergence to f is what we are interested in. We investigate both conditions for uniform convergence and rate of convergence. This series can be expressed as

$$f_m^r(t) = \int_{-\infty}^{\infty} G_m^r(t,s)f(s)ds \qquad (10.19)$$

where G_m^r, the 2^m dilation of the kernel $G^r(t,s)$ is given by

$$G_m^r(s,t) = 2^m \sum_n \delta(s - \frac{n}{2^m})\rho^r(2^m t - n)\frac{1-r}{1+r}. \qquad (10.20)$$

The sequence of these $G_m^r(t,s)$ will be shown to constitute a positive delta sequence and therefore (10.18) will converge to $f(t)$ uniformly and will avoid Gibbs phenomenon (excessive oscillations). We can use the coefficients in (10.18) together with the decomposition algorithm to construct the coefficients at the coarser scales, but shall not do so here. See [W-She1] for details.

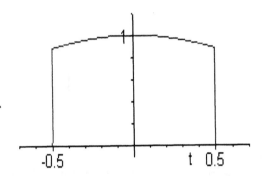

FIGURE 10.4

A non-negative function $f(x) = e^{-\frac{x^2}{2}} \chi_{[-1/2,1/2]}$.

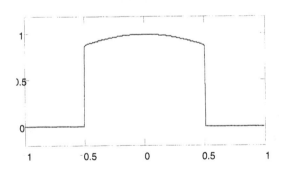

FIGURE 10.5
The positive hybrid series (m=4) using Coiflet of degree 2 for the function in Figure 10.4.

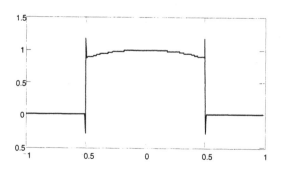

FIGURE 10.6
The hybrid sampling series for the function in Figure 10.4 ($m =$ 4) using Coiflet of degree 2.

The properties of a positive delta sequence needed for convergence (Section 8.2) are:

$$\begin{cases} (i)\ G_m^r(t,s) \geq 0 \\ (ii)\ \int_{-\infty}^{\infty} G_m^r(t,s)ds = 1 \\ (iii)\ \forall \gamma > 0,\ \int_{|t-s| \geq \gamma} G_m^r(t,s)ds \longrightarrow 0,\ \text{as } m \longrightarrow \infty. \end{cases}$$

$$(10.21)$$

The first inequality of (10.21) is obvious (but must be taken in the sense of distributions). The second inequality follows immediately from the fact that

$$\sum_n \rho^r(t-n)\frac{1-r}{1+r} \equiv 1.$$

In the case of (iii) we use the fact

$$\int_{|t-s| \geq \gamma} G_m^r(t,s)ds = \sum_{|n-2^m t| \geq 2^m \gamma} \rho^r(2^m t - n)\frac{1-r}{1+r}$$

$$= \sum_{|n-x| \geq 2^m \gamma} \rho^r(x-n)\frac{1-r}{1+r}.$$

But $\rho^r(x) = O(r^{|x|})$ [W-She4] and hence this series converges to 0 uniformly as $m \longrightarrow \infty$. Thus the three inequalities are satisfied and give us the following result whose proof is the same as for other positive delta sequences (Lemmas 8.1,8.2).

231231231231

231231231231231

Proposition 10.1

Let f be a function such that both f and $\hat{f} \in L^1(\mathbb{R})$, let $f^r_m \in V_m$ be given by (10.18), then

    (a) $f^r_m \longrightarrow f$ uniformly as $m \longrightarrow \infty$;

    (b) if $M_1 \leq f(t) \leq M_2$, then $M_1 \leq f^r_m(t) \leq M_2$.

We also have local convergence results. They are based on the same inequalities except we have to stay away from the end points of the interval of continuity.

Proposition 10.2

Let $f \in L^1(\mathbb{R})$ be piecewise continuous and be continuous in $[a, b]$; let $[\alpha, \beta] \subseteq (a, b)$; then

    (i) $f^r_m(t) \longrightarrow f(t)$ uniformly in $[\alpha, \beta]$;

    (ii) if $M_1 \leq f(t) \leq M_2$ in (a, b), then $\forall\, \epsilon > 0$, $\exists\, m_0$, such that

$$M_1 - \epsilon \leq f^r_m(t) \leq M_2 - \epsilon, \text{ for } t \in [\alpha, \beta],\ m > m_0.$$

The proof of these assertions is similar though a little harder and will be omitted. Compare to [Z, p.88].

In order to obtain the rate of convergence, we will compare $f^r_m(t)$ to the projection $f_m(t)$ of $f(t)$ onto V_m. We know its rate of convergence (see Chapter 8) provided that the original scaling function $\phi(t)$ has at least the first vanishing moment. That is, if ϕ satisfies $\int_{-\infty}^{\infty} t^k \phi(t)dt = 0$, $k = 1, 2, ..., [\lambda]$, and if $f \in H^\alpha$ for $\alpha > \lambda + \frac{1}{2}$; then

$$|f_m(t) - f(t)| = O(2^{-m\lambda}), \text{ uniformly in } \mathbb{R}.$$

We use this result to get our convergence rate for our positive hybrid series approximation. We have the following

THEOREM 10.3

[W-Sh1] Let $\phi(t)$ be a continuous orthogonal scaling function with compact support and MRA $\{V_m\}$; and with r chosen such that $\rho^r(t) \geq 0$. Then

    (i) if $f \in H^\alpha$ for $\alpha > \frac{3}{2}$, then $||f^r_m - f||_\infty = O(2^{-m})$.

    (ii) if in addition, $\int_{-\infty}^{\infty} t\phi(t)dt = 0$ and $f \in H^\alpha$, $\alpha > \frac{5}{2}$, then

$$||f^r_m - f||_\infty = O(2^{-2m}).$$

Additional vanishing moments, while they improve the convergence for the standard series approximation, do not improve the rate beyond that given in (ii). However this quadratic rate is still better than that arising in other classical approximations based on positive delta sequences which are linear at best [Z, p. 122].

10.5   Cardinal scaling functions

In the previous examples of Chapter 9, we considered wavelet subspaces in which the orthonormal basis is based on one function (a scaling function) and sampling on another (a cardinal function, i.e., a function f such that $f(n) = \delta_{0n}$). In none of these examples are the two functions the same except in the case of Shannon wavelets where $\phi(t)$ satisfies both $\phi(n) = \delta_{0n}$ and $\int_{-\infty}^{\infty} \phi(t)\phi(t-n)dt = \delta_{0n}$. In this case both the hybrid sampling of section 10.2 and the sampling series are the same.

However it is possible to construct new families of wavelets not previously considered that also satisfy both conditions, i.e., have a cardinal scaling function. One approach [X-Z] is based on the coefficients of the dilation equation

$$\phi(t) = \sum_k c_k \sqrt{2}\, \phi(2t - k). \tag{10.22}$$

The other [W-L] is based on the Fourier transform of $\phi(t)$, which satisfies

$$\widehat{\phi}(w) = m_0\left(\frac{w}{2}\right) \widehat{\phi}\left(\frac{w}{2}\right).$$

If $\phi(t)$ is to be a scaling function, then the coefficient $\{c_k\}$ must also satisfy (see Chapter 3)

$$\sum_k c_k c_{k-2n} = \delta_{0n}, \qquad n \in \mathbb{Z} \tag{10.23}$$

or equivalently

$$\left| m_0\left(\frac{w}{2}\right) \right|^2 + \left| m_0\left(\frac{w}{2} + \pi\right) \right|^2 = 1. \tag{10.24}$$

If it is to be a cardinal function then

$$\delta_{0n} = \sum_k c_k \sqrt{2}\, \delta_{0,2n-k} = \sqrt{2}\, c_{2n}, \quad n \in \mathbb{Z}, \tag{10.25}$$

or equivalently,

$$m_0\left(\frac{w}{2}\right) + m_0\left(\frac{w}{2} + \pi\right) = 1. \tag{10.26}$$

These conditions are necessary but not sufficient for $\phi(t)$ to be a cardinal scaling function. Necessary and sufficient conditions may be given in terms of the Fourier transform of $\phi(t)$:

$$\sum_k \left|\widehat{\phi}(w + 2\pi k)\right|^2 = 1, \tag{10.27}$$

$$\sum_k \widehat{\phi}(w + 2\pi k) = 1. \tag{10.28}$$

That this last condition (10.28) is equivalent to $\phi(n) = \delta_{0n}$ may be shown by taking the Fourier series of the left side, which converges in the sense of $L^2(0, 2\pi)$. Its Fourier coefficients are then seen to be $\phi(-n) = \delta_{0n}$, and the Fourier series is identically equal to 1.

Two other conditions needed are

$$\sum_k c_k = \sqrt{2}, \qquad \sum_k (-1)^k c_k = 0,$$

which correspond to

$$m_0(0) = 1, \qquad m_0(\pi) = 0.$$

The approach of Xia and Zhang uses the fact that the even coefficients are zero (by (10.25)), except for c_0 and hence

$$m_0(w) = \frac{1}{2} + \frac{1}{\sqrt{2}} \sum_k c_{2k+1} e^{iw(2k+1)}$$

$$= \frac{1}{2} + \frac{e^{iw}}{2} H(2w)$$

where

$$H(w) = \sqrt{2} \sum_k c_{2k+1} e^{iwk} = \sum_k h_k e^{iwk}.$$

The conditions (10.25) and (10.23) are converted into conditions involving $H(w)$. They are

$$|H(w)|^2 = \left|\sum_k h_k e^{iwh}\right|^2 = \left|\sum_{n\cdot}\left(\sum_k h_{n-k} h_k\right) e^{iwn}\right|^2$$

$$= \left|\sum_n \delta_{0n} e^{iwn}\right|^2 = 1, \tag{10.29}$$

and

$$H(0) = 1. \tag{10.30}$$

Now it is just a matter of finding $H(w)$ satisfying these two conditions and then checking that these necessary conditions are also sufficient. The simplest example is

$$H_1(w) = e^{iMw} \frac{1 + a\, e^{iw}}{a + e^{iw}} \tag{10.31}$$

where M is an integer and $0 < |a| < 1$. The Fourier series of $H_1(w)$ is

$$H_1(w) = e^{i(M-1)w}(1 + a\, e^{iw}) \sum_{n=0}^{\infty} (-a)^n e^{-inw}$$

$$= a\, e^{iMw} + \sum_{k=-\infty}^{M-1} (-a)^{M-k-1}(1 - a^2) e^{ikw}. \tag{10.32}$$

The coefficients can be converted back into c_k's and give us a Fourier series for $m_0(w)$ whose coefficients are exponentially decaying. A particular choice of a and M can be made to obtain an $m_0(w)$ corresponding to a moment condition $\int_{-\infty}^{\infty} t\psi(t)dt = 0$ (see Chapter 3). This is obtained by requiring $m'(\pi) = 0$, which leads to the condition

$$(2M + 3)a^2 + (4M + 2)a + 2M - 1 = 0$$

or

$$a = -\frac{2M - 1}{2M + 3}.$$

This lies between 0 and -1 for all $M \geq 1$ and is smallest when $M = 1$ and $a = -1/5$. This gives the most rapid convergence in (10.32).

THEOREM 10.4

[X-Z] Let $M = 1$ and $a = -1/5$ in $H_1(w)$ given by (10.32), let

$$m_0(w) = \frac{1}{\sqrt{2}}(1 + H_1(2w));$$

then $\phi(t)$ defined by

$$\widehat{\phi}(w) = \prod_{k=1}^{\infty} m_0(2^{-k}w)$$

is a continuous cardinal scaling function.

The proof may be found in [X-Z]. Other examples of cardinal scaling functions using a similar construction may be also found there. However, these are not bandlimited as are the Shannon scaling functions. It is shown in [W-L] that a family of cardinal scaling functions that are bandlimited and that are rapidly decreasing can also be constructed. They are based on the following lemma [W-L], which does not require that $\phi \in S_r$.

Lemma 10.4
Let $\phi(x) \in L^2(\mathbb{R})$. Then $\phi(x)$ is an orthonormal cardinal scaling function if

1. $\sum_k \left| \widehat{\phi}(w + 2\pi k) \right|^2 = 1$ *a.e.,*

2. *there exists a 2π-periodic and Lipschitz continuous function $m_0(w) \in L^2(-\pi, \pi)$ such that*

$$\widehat{\phi}(w) = m_0 \left(\frac{w}{2} \right) \cdot \widehat{\phi} \left(\frac{w}{2} \right) \quad a.e.,$$

3. $\widehat{\phi}(w)$ *is continuous at 0 and $\widehat{\phi}(0) = 1$, and*

4. $\sum_k \widehat{\phi}(w + 2\pi k) = 1$ *a.e.*

To construct our scaling function, we shall prove a necessary condition first. Then we will show in Theorem 10.5 that we have the desired scaling functions and the mother wavelets.

Lemma 10.5
If $\phi(t) \in (L^2 \cap L^1)(\mathbb{R})$ is an orthonormal cardinal scaling function with $\operatorname{supp} \widehat{\phi} \subseteq [-\pi - \epsilon, \pi + \epsilon]$ $(0 < \epsilon \le \frac{\pi}{3})$ and $\widehat{\phi}(w) = \overline{\widehat{\phi}(-w)}$, then $\widehat{\phi}(w)$ must have the form

$$\widehat{\phi}(w) = \frac{1}{2} + \frac{1}{2} e^{i\theta(w)}, \quad |w| < 2\pi$$

for some real function $\theta(w)$. Furthermore $\theta(w)$ must satisfy

(a)' $\theta(-w) = -\theta(w) + 2\pi k$,
(b)' $|\theta(w)| = (2k + 1)\pi$ for $|w| \ge \pi + \epsilon$,
(c)' $\theta(w) + \theta(2\pi - w) = (2k + 1)\pi$.

PROOF Since $\phi(n) = \delta_{no}$, it follows that (10.28) holds and the orthonormality of $\{\phi(t - n)\}$ implies that (10.27) holds.

The condition supp $\widehat{\phi} \subseteq [-\pi - \epsilon, \pi + \epsilon]$ ensures that the series in these equations reduces to two terms for each $w \in R$. In particular, for $0 \le w < 2\pi$, one has

$$\widehat{\phi}(w) + \widehat{\phi}(w - 2\pi) = 1 \quad \text{a.e.} \tag{10.33}$$

and

$$|\widehat{\phi}(w)|^2 + |\widehat{\phi}(w - 2\pi)|^2 = 1 \quad \text{a.e.} \tag{10.34}$$

The general solution to (10.34) and (10.35) must have the form

$$\widehat{\phi}(w) = \frac{1}{2} + \frac{1}{2}e^{i\theta(w)} \tag{10.35}$$

for some real function $\theta(w)$. Similarly (10.36) holds for $-2\pi < w \le 0$.

By using (10.36), (a)' follows from $\widehat{\phi}(w) = \overline{\widehat{\phi}(-w)}$, (b)' holds since supp $\widehat{\phi} \subseteq [-\pi - \epsilon, \pi + \epsilon]$, and (c)' holds because of (10.34) and (a)'. □

Next we shall choose $\theta(w)$ such that $\widehat{\phi}(w) = \frac{1}{2} + \frac{1}{2}e^{i\theta(w)}$ satisfies all conditions of Lemma 10.5. Let $\mathcal{F}$ be the set of all $\alpha \in L^1(\mathbb{R})$ such that

- $\alpha(w) \ge 0$,

- supp $\alpha \in [-\epsilon, \epsilon]$,

- $\alpha(-w) = \alpha(w)$ and

- $\{\int_{-\epsilon}^{\epsilon} \alpha(w)dw = \pi\}$ for some $0 < \epsilon \le \frac{\pi}{3}$.

Define, for such an $\alpha \in \mathcal{F}$,

$$\theta(w) := \int_{-w-\pi}^{w-\pi} \alpha(\xi)d\xi \tag{10.36}$$

and

$$\widehat{\phi}(w) := \frac{1}{2} + \frac{1}{2}e^{i\theta(w)}. \tag{10.37}$$

Then the corresponding $\phi(t)$ is the desired function for any $\alpha(w) \in \mathcal{F}$. In fact we have

THEOREM 10.5
For each $\alpha \in \mathcal{F}$, the function $\phi(t)$ defined by (10.36) and (10.37) is a real bandlimited orthonormal cardinal scaling function, and the corresponding mother wavelet $\psi(t)$ is given by

$$\psi(t) = 2\phi(2t - 1) - \phi\left(\frac{1}{2} - t\right).$$

PROOF Since $\alpha(w) \in \mathcal{F}$, it is easy to see that

$$\theta(w) = \int_{-w-\pi}^{w-\pi} \alpha(\xi) d\xi$$

satisfies
(i) $\theta(-w) = -\theta(w)$,
(ii) $|\theta(w)| = \pi$ for $|w| \geq \pi + \epsilon$,
(iii) $\theta(w) = 0$ for $|w| \leq \pi - \epsilon$,
(iv) $\theta(w) + \theta(2\pi - w) = \pi$ for $0 \leq 2\pi$.

Furthermore, it can be shown that the function $\hat{\phi}(w) = \frac{1}{2} + \frac{1}{2}e^{i\theta(w)}$ has the following properties:
(i)' $\hat{\phi}(w) = \overline{\hat{\phi}(-w)}$,
(ii)' $\hat{\phi}(w) = 0$ for $|w| \geq \pi + \epsilon$,
(iii)' $\hat{\phi}(w) = 1$ for $|w| \leq \pi - \epsilon$,
(iv)' $\hat{\phi}(w) + \hat{\phi}(w + 2\pi) + \hat{\phi}(w - 2\pi) = 1$ for $|w| < 2\pi$,
(v)' $|\hat{\phi}(w)|^2 + |\hat{\phi}(w + 2\pi)|^2 + |\hat{\phi}(w - 2\pi)|^2 = 1$ for $|w| < 2\pi$,
(vi)' $\hat{\phi}(w) = \sum_k \hat{\phi}(w + 4\pi k) \cdot \hat{\phi}\left(\frac{w}{2}\right)$ for $w \in \mathbb{R}$.

In fact, the conclusion (i)'–(iii)' follow from (i)–(iii) directly, and (iv)' and (v)' follow from (iv), by considering two cases $0 < 2\pi$ and $-2\pi < w \leq 0$. For the proof of (vi)', the real line $\mathbb{R}$ may be separated into three disjoint sets: $|w| \geq \pi + \epsilon$, $\pi - \epsilon \leq |w| \leq \pi + \epsilon$ and $|w| \leq \pi - \epsilon$.

We define as in the case of Meyer wavelets

$$m_0\left(\frac{w}{2}\right) := \sum_k \hat{\phi}(w + 4\pi k).$$

Then all conditions except (4) in Lemma 10.4 are satisfied because of (ii)', (iii)', (v)' and (vi)'.

As in the beginning of the proof of Lemma 10.5, it can be shown that $\phi(t)$ is a cardinal function by (iv)'. Obviously (i) implies that $\phi(t)$ is real.

Next one needs to calculate $\psi(t)$ by the general formula

$$\hat{\psi}(w) = e^{-i\frac{w}{2}} \overline{m_0\left(\frac{w}{2} + \pi\right)} \hat{\phi}\left(\frac{w}{2}\right).$$

Since $m_0\left(\frac{w}{2}\right) = \sum_k \hat{\phi}(w + 4\pi k)$ with supp $\hat{\phi} \in \left[-\frac{4\pi}{3}, \frac{4\pi}{3}\right]$ $\left(0 < \epsilon \leq \frac{\pi}{3}\right)$, $\overline{m_0\left(\frac{w}{2} + \pi\right)}\hat{\phi}\left(\frac{w}{2}\right)$ reduces to $\overline{\hat{\phi}(w + 2\pi) + \hat{\phi}(w - 2\pi)} \cdot \hat{\phi}\left(\frac{w}{2}\right)$. Therefore by $\widehat{\phi}(w) = \frac{1}{2} + \frac{1}{2}e^{i\theta(w)}$, one has

$$\hat{\psi}(w) = e^{-i\frac{w}{2}}\left[1 + \frac{e^{-i\theta(w+2\pi)} + e^{-i\theta(w-2\pi)}}{2}\right] \cdot \frac{1 + e^{i\theta(\frac{w}{2})}}{2}.$$

Furthermore it can be shown that

$$\hat{\psi}(w) = e^{-i\frac{w}{2}}\frac{1 + e^{i[\pi - \theta(w)]\chi(w)}}{2} \cdot \frac{1 + e^{i\theta(w)}}{2}$$

by using the properties (i)–(iv) of $\theta(w)$, where $\chi(w)$ is the characteristic function on $[-2\pi, 2\pi]$.

To complete the proof, one needs to show

$$\psi(t) = 2\phi(2t - 1) - \phi\left(\frac{1}{2} - t\right).$$

Since $\widehat{\phi}(w) = m_0\left(\frac{w}{2}\right)\widehat{\phi}\left(\frac{w}{2}\right)$ and $m_0\left(\frac{w}{2}\right)$ vanishes for those w whose $\widehat{\phi}\left(\frac{w}{2}\right)$ is not real (which is possible only on the interval $[2\pi - 2\epsilon, 2\pi + 2\epsilon]$), one has that

$$\widehat{\phi}(w) = m_0\left(\frac{w}{2}\right) \cdot \overline{\widehat{\phi}\left(\frac{w}{2}\right)}.$$

By (10.25) $m_0(w) + m_0(w + \pi) = 1$ and hence

$$\hat{\psi}(w) = e^{-i\frac{w}{2}}\overline{m_0\left(\frac{w}{2} + \pi\right)}\widehat{\phi}\left(\frac{w}{2}\right) = e^{i\frac{w}{2}}\overline{\left[1 - m_0\left(\frac{w}{2}\right)\right]}\widehat{\phi}\left(\frac{w}{2}\right)$$

$$= e^{-i\frac{w}{2}}\left[\widehat{\phi}\left(\frac{w}{2}\right) - \overline{m_0\left(\frac{w}{2}\right)}\widehat{\phi}\left(\frac{w}{2}\right)\right] = e^{-i\frac{w}{2}}\left[\widehat{\phi}\left(\frac{w}{2}\right) - \overline{\widehat{\phi}(w)}\right].$$

It follows, by taking the inverse Fourier transform, that $\psi(t) = 2\phi(2t - 1) - \phi\left(\frac{1}{2} - t\right)$. $\qquad\square$

Most scaling functions arising from this theorem are not in closed form except for the one case corresponding to constant α. This leads to the formula

$$\phi(t) = \frac{\sin \pi(1 - \beta)t + \sin \pi(1 + \beta)t}{2\pi t[1 + 2\beta t]}$$

for $\beta < 1/3$. This is one of the examples discussed in Chapter 3; it has the same properties as the sinc function but converges to zero more .rapidly as $|t| \to \infty$.

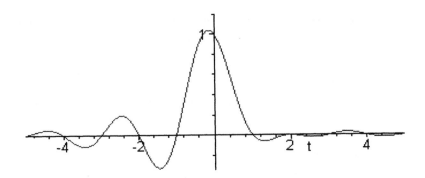

FIGURE 10.7
An example of cardinal scaling function of Theorem 10.5, type
two raised cosine wavelet.

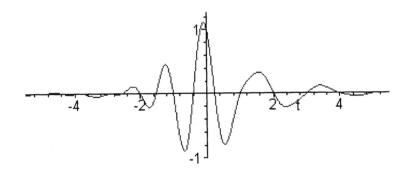

FIGURE 10.8
The mother wavelet for the scaling function in Figure 10.7.

10.6   Interpolating multiwavelets

In the previous sections we have seen that if we begin with any scaling function satisfying certain properties, then there is a sampling function $S(t)$ in the space V_0 such that for any f in V_0, $f(t) = \sum_n f(n)S(t-n)$. If we know the value of both the function and its derivative, we would expect to be able to sample f less frequently. That is exactly the case for Hermite Interpolation. For the Shannon example, we have the following theorem [Za, p. 75]:

THEOREM 10.6
Let f be π band limited, i.e., $f \in V_0$. Then,

$$f(t) = \sum_k (f(2k) + \frac{(t-2k)}{2} f'(2k))S^2(\frac{t-2k}{2}),$$

where

$$S(t) = \frac{\sin \pi t}{\pi t}.$$

This may be rewritten in the following form by observing that if $f \in V_1$, then $f(\frac{t}{2}) \in V_0$, and hence

$$f(\frac{t}{2}) = \sum_k (f(k) + \frac{(t-2k)}{2} f'(k))S(\frac{t-2k}{2})$$

or

$$f(t) = \sum_k f(k)S^2(t-k) + f'(k)(t-k)S^2(t-k).$$

Since S^2 has as its Fourier transform $\widehat{S}*\widehat{S}$ with support in $[-2\pi, 2\pi)$, $S^2 \in V_1$ and $tS^2 \in V_1$, as well, so this last expression becomes

$$f(t) = \sum_k f(k)s_1(t-k) + f'(k)s_2(t-k)$$

where $s_1(t) = S^2(t)$, $s_2(t) = tS^2(t)$ and

$$\begin{cases} s_1(n) = s_2'(n) = \delta_{0n} \\ s_2'(n) = s(n) = 0 \end{cases}, \quad n \in \mathbb{Z}. \qquad (10.38)$$

However, it is not clear at all that the same result holds for other wavelet systems. We first try to find, in general, conditions that the suitable s_1 and s_2 must satisfy, and then show that these conditions are satisfied for wavelets with certain properties.

The subject of multiwavelets was given its impetus by a number of recent papers by Geronimo, Hardin and Massopust (see, e.g., [G-H-M]) who constructed multiwavelet scaling functions from fractal interpolation functions. The concept of multiwavelets is similar to that of single wavelets except the space V_0 is now spanned by integer translates of n scaling functions $\phi_1(t), \cdots, \phi_n(t)$ instead of only one. In this section, we consider the problem of constructing multisampling functions in a given multiresolution analysis. This will give us a Hermite interpolation as observed in the Shannon example.

We begin with a single band-limited scaling function $\phi(t)$. The set of translates $\{\phi(t-n)\}$ need not be orthogonal, but will be assumed to be a Riesz basis of V_0 and to generate a multiresolution analysis $\{V_m\}$. We want to find sampling functions that have the Hermite interpolation property in these wavelet subspaces. The space we look at is V_1. We want to find s_1 and $s_2 \in V_1$ satisfying the two conditions (10.38). The properties we want will lead to equations [W-C]

$$a(w) \sum_k \widehat{\phi}(\frac{w}{2} + 2\pi k) + a(w + 2\pi) \sum_k \widehat{\phi}(\frac{w}{2} + 2\pi k + \pi) = 1;$$

$$a(w) \sum_k (w + 4\pi k)\widehat{\phi}(\frac{w}{2} + 2\pi k)$$
$$+ a(w + 2\pi) \sum_k (w + 4\pi k + 2\pi)\widehat{\phi}(\frac{w}{2} + 2\pi k + \pi) = 0$$

whose solutions exist when the scaling function satisfies certain conditions:

$\phi(t) \in H^r$, $r \geq 4$, and:
(a) $0 < a \leq |\Delta(w)| \leq b$ for $w \in \mathbb{R}$ for some constants a, b, where $\Delta(w)$ is the determinant of the system,
(b) there exists a constant c_1 such that

$$(t^2 + 1)^r |\phi^{(i)}(t)|^2 \leq c_1, \ i \leq r, \ t \in \mathbb{R},$$

(c) there is a constant c_2 such that either

$$0 < c_2 \leq |\sum \widehat{\phi}(w + 2\pi k)|, \ w \in \mathbb{R},$$

or

$$0 < c_2 \leq |\sum (2w + 4\pi k)\widehat{\phi}(w + 2\pi k)|, \ w \in \mathbb{R}.$$

Then we get

THEOREM 10.7

[W-C] Let $\phi(t)$ be a scaling function in $H^r, r \geq 4$, which satisfies (a), (b), (c). Then, there are multisampling functions s_1, s_2 in V_1 such that for any $f \in V_1$

$$f(t) = \sum f(n)s_1(t-n) + \sum f'(n)s_2(t-n), \quad \text{for} \quad t \in \mathbb{R} \quad (10.39)$$

where the convergence is in the sense of $L^2(\mathbb{R})$ and uniform. The s_1, s_2 are given by

$$\widehat{s_1}(w) = a(w)\widehat{\phi}(\frac{w}{2}),$$
$$\widehat{s_2}(w) = b(w)\widehat{\phi}(\frac{w}{2}),$$

where $a(w)$ and $b(w)$ are given by

$$\begin{cases} b(w) = -\dfrac{\sum \widehat{\phi}(\frac{w}{2}+(2k+1)\pi)}{i\Delta(w)} \\ a(w) = \dfrac{\sum \widehat{\phi}(\frac{w}{2}+2\pi k+\pi)(w+4\pi k+2\pi)}{\Delta(w)} \end{cases} \quad (10.40)$$

The proof may be found in [W-C], where it is also shown that the scaling functions considered in the last section satisfy the hypotheses of Theorem 10.7. The extension to higher order derivative sampling is conceptually straightforward, but the details have not been worked out to our knowledge.

10.7   Orthogonal finite element multiwavelets

In a paper by Strela and Strang [S-S], a new class of wavelet type functions are introduced which are an alternative to those discussed in the last section. They are based on the finite element scaling functions

$\phi_1(t), \phi_2(t), \ldots \phi_n(t)$ which generalize B-splines theory. They are polynomials of degree 2n−1 on [0,1] and [1,2] and zero elsewhere. They have $n-1$ continuous derivatives and satisfy the condition that

$$\Phi(n) = \delta_{1n} I, \qquad (10.41)$$

where

$$\Phi(t) = \begin{bmatrix} \phi_1(t) & \cdots \cdots & \phi_1^{(n-1)}(t) \\ \vdots & \vdots \;\; \vdots & \vdots \\ \phi_n(t) & \cdots \cdots & \phi_n^{(n-1)}(t) \end{bmatrix}.$$

For $n = 1$, $\phi_1(t) = (1 - |t - 1|)\chi_{[0,2]}(t)$ is the familiar hat function.

Example 10.1

For $n = 2$, we have, for $0 \le t \le 1$,

$$\phi_1(t) = (3t^2 - 2t^3), \quad \phi_2(t) = t^3 - t^2$$

and for $1 \le t \le 2$, we use the symmetry

$$\phi_i(2 - t) = \phi_i(t)(-1)^{i-1}, i = 1, 2.$$

▯

For higher values of n, the polynomials can also be worked out [S-S], but we shall restrict ourselves to detailed computation only in these two examples.

Each ϕ_i satisfies a dilation equation which may be expressed as

$$\phi(t) = \sum_k C_k \phi(2t - k) \qquad (10.42)$$

where $\phi(t) = [\phi_1(t), \ldots, \phi_n(t)]^T$ and C_k is an $n \times n$ matrix. Since

$$\phi^{(j)}(t) = \sum_k C_k 2^j \phi^{(j)}(2t - k)$$

it follows that

$$\Phi(t) = \sum_k C_k \Phi(2t - k)\Gamma \qquad (10.43)$$

where Γ is the diagonal matrix with 1,2, ..., 2^{n-1} on the main diagonal.

We may use the values at $t = \frac{m}{2}$ to find C_k. In fact

$$\Phi(\frac{m}{2}) = \sum_k C_k \Phi(m-k)\Gamma$$

$$= \sum_k C_k (\delta_{1,m-k}) I \cdot \Gamma$$

$$= C_{m-1}\Gamma, \qquad m \in \mathbb{Z}. \qquad (10.44)$$

Hence

$$C_0 = \Phi(\frac{1}{2})\Gamma^{-1}, \quad C_1 = \Phi(1)\Gamma^{-1} = I \cdot \Gamma^{-1} = \Gamma^{-1}, \qquad (10.45)$$

and

$$C_2 = \Phi(\frac{3}{2})\Gamma^{-1}.$$

Since ϕ has support on [0,2] all other $C_k's$ are 0.

10.7.1   Sobolev type norm

These scaling functions do not satisfy the usual orthogonality condition

$$\int \phi_i(t)\phi_i(t-n)dt = \delta_{0n}.$$

Rather the condition (10.40) leads to another type of orthogonality based on a Sobolev type inner product. We define

$$< f,g >_0 := \sum_{j=0}^{n-1} \sum_k f^{(j)}(k)\overline{g^{(j)}(k)} \qquad (10.46)$$

where f and g are in C_0^{n-1}. Then we have

$$< \phi_i, \phi_k(.-m) >_0 = \sum_{j=0}^{n-1} \sum_p \phi_i^{(j)}(p)\phi_k^{(j)}(p-m)$$

$$= \sum_{j=0}^{n-1} \phi_i^{(j)}(1)\phi_k^{(j)}(1-m)$$

$$= \sum_{j=0}^{n-1} \delta_{i,j+1}\delta_{k,j+1}\delta_{0,m}$$

$$= \delta_{i,k}\delta_{0,m}.$$

In terms of vectors this becomes

$$< \phi, \phi^T(. - m) >_0 = \delta_{0,m} I.$$

This can also be expressed as $\sum_k \Phi(k)\Phi^T(k - m) = \delta_{0,m} I$, where Φ is the matrix in (10.40).

We also introduce inner products based on a scale change

$$< f, g >_p := \sum_{j=0}^{n-1} \sum_k f^{(j)}(2^{-p}k)\overline{g^{(j)}(2^{-p}k)}, \ p \in \mathbb{Z}.$$

We do not have orthogonality of $\{\phi(t - m\}$ with respect to these inner products. In fact for $p = 1$, we have

$$< \phi, \phi^T(. - m) >_1 = \sum_k \Phi(\frac{k}{2})\Phi^T(\frac{k}{2} - m) = \sum_k C_{k-1}\Gamma^2 C_{k-1-2m}^T.$$

For $m = 1$ we have $< \phi, \phi^T(. - m) >_1 = C_2\Gamma^2 C_0^T$ while for $m = 0$ we have

$$< \phi, \phi^T(. - m) >_1 = C_0\Gamma^2 C_0^T + I^2 + C_2\Gamma^2 C_2^T.$$

We denote as usual by V_m the closed linear span in $L^2(\mathbb{R})$ of

$$\{\phi_1(2^m t - k), ..., \phi_n(2^m t - k)\},$$

with inner product $< , >_m$. This is equivalent to the usual L^2 inner product in V_m.

10.7.2 The mother multiwavelets

We now introduce a biorthogonal pair of wavelets $(\psi, \tilde{\psi})$, both of which belong to V_1. The first is in $V_0^\perp$ and is given by

$$\psi(t) = \sum_k D_k\phi(2t - k). \tag{10.47}$$

Since $\psi_i \in V_1$, we use its inner product to choose the D_k so that $\psi_i(t)$ is orthogonal to V_0 in the sense of V_1. We need

$$< \psi, \phi^T(. - m) >_1 = 0, \text{ for all } m \in \mathbb{Z}.$$

To get this we plug in the dilation equation (10.42)and (10.47) and use $<\phi(2t), \phi^T(2t - m) >_1 = \delta_{0,m}\Gamma^2$ to get

$$< \psi, \phi^T(. - m) >_1 = \sum_{k,j} D_k < \phi(2t - k), \phi^T(2t - 2m - j) >_1 C_j^T$$

$$= \sum_{k,j} D_k\Gamma^2\delta_{k,2m+j}C_j^T = \sum_j D_{2m+j}\Gamma^2 C_j^T = O, \qquad m \in Z. \quad (10.48)$$

These equations are simply, since D_m may be taken to be 0 for $m < 0$ and $m > 4$,

$$D_0\Gamma^2 C_0^T + D_1\Gamma^2 C_1^T + D_2\Gamma^2 C_2^T = O,$$
$$D_2\Gamma^2 C_0^T + D_3\Gamma^2 C_1^T + D_4\Gamma^2 C_2^T = O,$$

$$D_0\Gamma^2 C_2^T = O,$$
$$D_4\Gamma^2 C_0^T = O.$$

Since both C_0 and C_2 are nonsingular [S-S], it follows that $D_0 = O = D_4$. Hence we have only two equations

$$D_1\Gamma^2\Gamma^{-1} = -D_2\Gamma^2 C_2^T,$$
$$D_2\Gamma^2 C_0^T = -D_3\Gamma^2\Gamma^{-1},$$

since $C_1 = \Gamma^{-1}$. These have a solution obtained by taking $D_2 = C_1 = \Gamma^{-1}$. It is

$$D_m = (-1)^m\Gamma C_{3-m}^T\Gamma^{-1}, \; m \in Z.$$

Thus $\psi(t)$ may be given by

$$\psi(t) = -\Gamma C_2^T\Gamma^{-1}\phi(2t-1) + \Gamma^{-1}\phi(2t-2) - \Gamma C_0^T\Gamma^{-1}\phi(2t-3). \quad (10.49)$$

Clearly $\psi(t)$ has support on $[\frac{1}{2}, \frac{5}{2}]$ and

$$\psi^{(j)}(\frac{m}{2}) = \sum D_k\phi^{(j)}(m - k)2^j$$
$$= D_{m-1}2^j\phi^{(j)}(1), \; j = 0, 1, \cdots, n - 1,$$

or

$$\psi_j^{(i-1)}(m + \frac{3}{2}) = \delta_{i,j}\delta_{m,0}.$$

The other wavelet in the pair is defined simply as

$$\widetilde{\psi}(t) = \Gamma^{-1}\phi(2t - 2). \tag{10.50}$$

The two are related by:

Lemma 10.6

Let ψ be given by (10.47) and $\widetilde{\psi}$ by (10.50). Then for $i, j = 0, 1, \cdots, n-1, k \in \mathbb{Z}$.

(i) $\psi_{j+1}^{(i)}(k + \frac{3}{2}) = \widetilde{\psi}_{j+1}^{(i)}(k + \frac{3}{2}) = \delta_{i,j}\delta_{0,k}$,

(ii) $\psi_{j+1}^{(i)}(k) = 0$,

(iii) $< \psi_j(. - k), \widetilde{\psi}_j(. - m) >_1 = \delta_{km}$.

If $W_0 = CLS\{\psi_j(x - k)\}, \widetilde{W}_0 = CLS\{\widetilde{\psi}_j(x - k)\}$, then $W_0 \perp V_0, \widetilde{W}_0 \cap V_0 = \{0\}$, and $V_1 = \widetilde{W}_0 \dot{+} V_0$.

PROOF The three conditions $(i), (ii)$ and (iii) follow from the definitions and the conditions on ϕ in (10.41). The space W_0 has $\{\psi_j(x - k)\}$ as a Riesz basis and hence since $\psi_j(x-k) \in V_0^{\perp}$, it follows that $W_0 \perp V_0$. By (ii) each element $f \in \widetilde{W}_0$ must be zero for all $k \in Z$, together with its first $n - 1$ derivatives. If $f \in V_0$ as well then, since

$$f(t) = \sum_{j=1}^{n} \sum_{k} a_{kj}\phi_j(t - k),$$

$$a_{kj} = f^{(j-1)}(k + 1) = 0 \tag{10.51}$$

and hence $f \equiv 0$. Thus only the final statement remains to be proved. Indeed let $f \in V_1$ with an expansion

$$f(t) = \sum_{j=1}^{n} \sum_{k} a_{kj}^1 \phi_j(2t - k).$$

Then the coefficients are

$$a_{kj}^1 = f^{(j-1)}(\frac{k+1}{2})2^{1-j}.$$

We define $f_0 \in V_0$ to be the "projection" of f to V_0,

$$f_0(t) = \sum_{j=1}^{n} \sum_{k} f^{(j-1)}(k + 1)\phi_j(t - k). \tag{10.52}$$

We now have

$$f(t) - f_0(t) = \sum_{j=1}^{n} \left\{ \sum_k f^{(j-1)}(k + \tfrac{1}{2})2^{-j+1}\phi_j(2t - 2k) \right.$$

$$\left. + \sum_k f^{(j-1)}(k+1)[2^{-j+1}\phi_j(2t - 2k - 1) - \phi(t - k)] \right\}.$$

But we have

$$\Gamma^{-1}\phi(2t - 1) - \phi(t)$$
$$= \Gamma^{-1}\phi(2t - 1) - C_0\phi(2t) - \Gamma^{-1}\phi(2t - 1) - C_2\phi(2t - 2)$$
$$= -C_0\phi(2t) - C_2\phi(2t - 2)$$
$$= -C_0\Gamma\tilde{\psi}(t + 1) - C_2\Gamma\tilde{\psi}(t).$$

Therefore we have the representation

$$f_1(t) = f(t) - f_0(t) \tag{10.53}$$

$$= \sum_k \mathbf{f}^T(k + \tfrac{3}{2})\tilde{\psi}(t - k) - \sum_k \mathbf{f}^T(k + 1)C_0\Gamma\tilde{\psi}(t - k)$$

$$- \sum_k \mathbf{f}^T(k + 1)C_2\Gamma\tilde{\psi}(t - k - 1)$$

$$= \sum_k \left\{ \mathbf{f}^T(k + \tfrac{3}{2}) - \mathbf{f}^T(k + 2)C_0\Gamma - \mathbf{f}^T(k + 1)C_2\Gamma \right\} \tilde{\psi}(t - k).$$

Clearly $f_0 \in V_0$ and $f_1 \in \widetilde{W}_0$, and we have the desired decomposition $f = f_0 + f_1$. $\qquad\square$

The expression (10.53) can also be given in terms of $\psi(t - k)$ since by (10.47) we have

$$< \psi(t - k), \ f>_1 \tag{10.54}$$
$$= \ < D_1\phi(2t - 2k - 1, \ f) >_1 +$$
$$< D_2\phi(2t - 2k - 3, \ f) >_1 + D_3\phi(2t - 2k - 3, \ f) >_1$$
$$= D_1\Gamma\mathbf{f}(k + 1) + D_2\Gamma\mathbf{f}(k + \tfrac{3}{2}) + D_3\Gamma\mathbf{f}(k + 2)$$
$$= -\Gamma C_2^T\mathbf{f}(k + 1) + \mathbf{f}(k + \tfrac{3}{2}) - \Gamma C_0^T\mathbf{f}(k + 2)$$

in agreement with (10.53). The coefficients in the expansions of f_0, f_1, and f_2 lead to simple decomposition and reconstruction algorithms.

Lemma 10.7
Let $f \in V_1$, let f_0, f_1 be as in (10.52) and (10.53), and let

$$a^1_{kj} = <f, \phi_j(2t-k)>_1,$$
$$a^0_{kj} = <f, \phi_j(t-k)>_0,$$

and

$$b^0_{kj} = <f, \psi_j(t-k)>_0.$$

Then the decomposition algorithm is (in vector form)

$$\mathbf{a}^0_k = \Gamma \mathbf{a}^1_{2k+1},$$
$$\mathbf{b}^0_k = \Gamma(-C_2^T \Gamma \mathbf{a}^1_{2k+1} + \mathbf{a}^1_{2k+2} - C_0^T \Gamma \mathbf{a}^1_{2k+3})$$

while the reconstruction algorithm is

$$\mathbf{a}^1_{2k+1} = \Gamma^{-1} \mathbf{a}^0_k,$$
$$\mathbf{a}^1_{2k+2} = \Gamma^{-1} \mathbf{b}^0_k + C_2^T \Gamma \mathbf{a}^0_k + C_0^T \Gamma \mathbf{a}^0_{k+1}.$$

PROOF The first line of the decomposition algorithm follows from the fact that

$$a^1_{kj} = 2^{1-j} f^{(j-1)}(\frac{k+1}{2})$$

and

$$a^0_{kj} = f^{(j-1)}(k+1).$$

The second line is a restatement of (10.54) and the reconstruction algorithm is just a rearrangement of the same formulae. □

Example 10.2
We work out the details for $n = 1$ which are particularly simple. In that case we have $c_0 = 1/2 = c_2$, and $d_1 = -1/2 = d_3$. The decomposition algorithm is

$$-\frac{1}{2}a^1_{2k+1} + a^1_{2k+2} - \frac{1}{2}a^1_{2k+3}$$

while the reconstruction algorithm is

$$a^1_{2k+1} = a^0_k$$
$$a^1_{2k+2} = \frac{1}{2}a^0_k + b^0_k + \frac{1}{2}a^0_{k+1}.$$

In this case only the values of the signal $f(t)$ at the points $2^{-m}k$, $k \in Z$ at the finest scale of interest are needed.

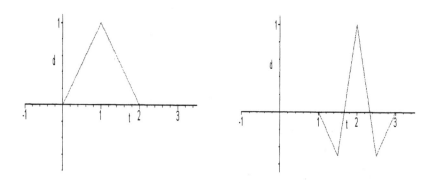

FIGURE 10.9
The scaling function and wavelet of interpolating multiwavelets
for n=1.

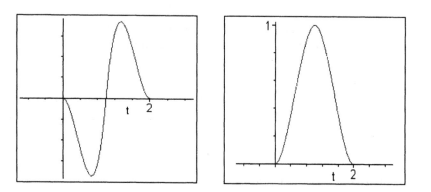

FIGURE 10.10
The two scaling functions of interpolating multiwavelets for
$n = 2.$

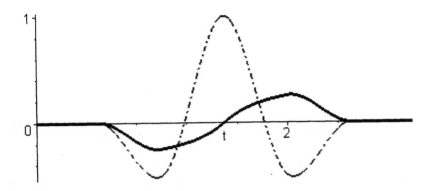

FIGURE 10.11
The first pair of wavelets of interpolating multiwavelets for n=2.

The scaling function and first wavelet are shown in Figure 10.9.

Example 10.3
For $n = 2$, the scaling function coefficients are matrices given by

$$C_0 = \Phi(1/2)\Gamma^{-1} = \begin{bmatrix} \frac{1}{2} & \frac{3}{4} \\ -\frac{1}{8} & -\frac{1}{8} \end{bmatrix}$$

and

$$C_3 = \begin{bmatrix} \frac{1}{2} & -\frac{3}{4} \\ \frac{1}{8} & -\frac{1}{8} \end{bmatrix}$$

while the first wavelet coefficients are

$$D_1 = \begin{bmatrix} -\frac{1}{2} & \frac{3}{8} \\ -\frac{1}{4} & \frac{1}{8} \end{bmatrix}, \quad D_3 = \begin{bmatrix} -\frac{1}{2} & -\frac{3}{8} \\ \frac{1}{4} & \frac{1}{8} \end{bmatrix}.$$

Then the decomposition algorithm becomes

$$\begin{bmatrix} a_{k,1}^0 \\ a_{k,2}^0 \end{bmatrix} = \begin{bmatrix} a_{2k+1,1}^1 \\ 2a_{2k+1,2}^1 \end{bmatrix},$$

$$\begin{bmatrix} b^0_{k,1} \\ b^0_{k,2} \end{bmatrix} = - \begin{bmatrix} -\frac{1}{2} & \frac{1}{4} \\ \frac{3}{2} & -\frac{1}{2} \end{bmatrix} \begin{bmatrix} a^1_{2k+1,1} \\ a^1_{2k+1,2} \end{bmatrix} + \begin{bmatrix} a^1_{2k+2,1} \\ 2a^1_{2k+2,2} \end{bmatrix}$$

$$- \begin{bmatrix} \frac{1}{2} & -\frac{1}{4} \\ \frac{3}{2} & -\frac{1}{2} \end{bmatrix} \begin{bmatrix} a^1_{2k+3,1} \\ 2a^1_{2k+3,2} \end{bmatrix}$$

and similarly for the reconstruction algorithm. The scaling functions and the first wavelets are shown in Figure 10.10 and Figure 10.11. □

These results can be generalized to scaling functions with support on other intervals, in particular on the real line. In that case it is possible to introduce band limit versions [W-Za], by the methods of section 10.6.

10.8   Problems

1. Show that the closed linear span of $\{\phi_1(t-k), .., \phi_n(t-k)\}$ in $L^2(\mathbb{R})$ is the same with the inner product $<,>_0$ as with the usual L^2 inner product. (If too messy, try it for n=1 first.)

2. Find a closed form expression for $\Phi(t)$ in (10.41) for the n=2 case. Use it to find the matrices C_0 and C_2 given in the text. Then find the matrices D_1 and D_3 and use them to find an expression for the first mother wavelets. (You can do this by hand, but it's easier with Maple.)

3. Let ϕ be the scaling function for a Meyer type wavelet and let $f(t) = \sin \pi t / \pi t$ be the sinc function. Show that $f(t) \in V_1$ by using Fourier transforms. Find the oversampling series (10.1) of f.

4. Let $g(t) = f(\sigma t / \pi)$ where f is the function of problem 3. For which values of σ is the usual sampling series

$$g(t) = \sum g(n)\phi(t-n),$$

where ϕ is as in Problem 3, valid?

5. Show directly that the raised cosine scaling function $\phi_1(t)$ given by example 8 of Chapter 3 satisifies the oversampling property (10.2) for $\beta = \pi/4$.

6. Let ϕ be a scaling function with support in the interval $[0,\mathrm{T}]$. Find the support of the dual basis $\tilde{\rho}_r$ of the of the positive scaling function given in Theorem 10.2.

7. Let $\alpha(w) = C(\pi/4 - |w|)\chi_{[-\pi/4,\pi/4]}(w)$. Show that there is a value of C for which (10.35) is an exponent leading to a valid cardinal scaling function in (10.36). Find the correct value of C and an expression for θ.

Chapter 11

Translation and Dilation Invariance in Orthogonal Systems

One of the reasons for taking orthogonal expansions of functions is that operations on the functions may be converted to simpler operations on the coefficients. Thus, in the example of the Sturm-Liouville systems, the differential operator is converted into multiplication by the eigenvalues.

Many of the operations that are useful in analysis are based on translation and dilation operators, the same operators that appear repeatedly in wavelet theory. In this chapter we study these two operators and their effect on the partial sums of the expansions. We first consider trigonometric systems and orthogonal polynomials and then a particular wavelet system in which everything works well. However, in most wavelet systems translation invariance fails, and we therefore consider a weaker concept that works more often. In fact, the wavelets that satisfy this "weak" translation property are characterized as a type of Meyer wavelet. The same ones also satisfy a weak dilation invariance property.

11.1   Trigonometric system

An important property of the trigonometric system of orthogonal functions is its invariance under arbitrary translations. Indeed if $f(t)$ is periodic 2π with Fourier series

$$f(t) \sim \sum_n c_n e^{int}, \tag{11.1}$$

then for each real α, $f(t-\alpha)$ has Fourier series

$$f(t-\alpha) \tilde{} \sum_n (c_n e^{-in\alpha}) e^{int}. \tag{11.2}$$

In particular, the number of terms in the partial sums of the first functions is invariant under translation, the moduli of the coefficients are the same $|c_n| = |c_n e^{-in\alpha}|$, and the translations in the space V_m spanned by $1, e^{\pm it}, \cdots, e^{\pm imt}$, are again in this space.

Things are not so nice when it comes to dilation since for arbitrary $\beta > 0$, $f(\beta t)$ is not even periodic. The only regularity obtained is when $\beta = p$, a positive integer. Then the partial sum $f_m(t)$ in V_m will be mapped into V_{pm}.

Other common operators arise from translations. For example, differentiation of $f(t)$, which comes from a limit of the difference quotient based on translations, leads to the corresponding Fourier series

$$f'(t) \tilde{} \sum_n in\, c_n\, e^{int},$$

and V_m maps into itself under this operation. The same is true for convolution. If g is periodic 2π and in $L^1(0, 2\pi)$, then the Fourier series of the convolution is

$$(f * g)(t) = \int_0^{2\pi} f(t-s)g(s)ds = 2\pi \sum_n c_n d_n e^{int},$$

and hence for $f_m \in V_m$,

$$f_m * g \in V_m.$$

11.2   Orthogonal polynomials

When we consider orthogonal polynomials, things are not quite so simple. If $f(t)$ has an orthogonal polynomial expansion

$$f(t) \sim \sum_{n=0}^{\infty} c_n p_n(t), \tag{11.3}$$

then

$$f(t - \alpha) \sim \sum_n c_n \sum_{k=0}^n a_{nk}^\alpha p_k(t)$$

$$= \sum_k \left(\sum_{n=k}^\infty c_n a_{nk}^\alpha \right) p_k(t), \qquad (11.4)$$

where a_{nk}^α are the expansion coefficients of $p_n(t - \alpha)$ in terms of p_k. If again we denote by V_m the space spanned by $p_0, p_1, \cdots, p_m$, then V_m is mapped into itself by translation. However, in the case of the entire series as opposed to the partial sums, there is no guarantee that the series of $f(t - \alpha)$ converges (or indeed that $f(t - \alpha)$ makes sense over the interval of orthogonality).

The same is true for arbitrary positive dilations. $p_n(\beta t)$ is again a polynomial of degree n and hence has an expansion in terms of p_0, $p_1, \cdots, p_n$. Thus, again V_m is mapped into itself by dilations. The differential operator maps V_m into V_{m-1} while the convolution operator maps V_m into V_m provided $g * p$ exists.

11.3 An example where everything works

The wavelet prototype II discussed in Chapter 1 has certain desirable properties with respect to dilation and translation. In particular, it has the translation invariance properties of Fourier series. If V_m is a subspace in the multiresolution analysis (MRA), then $T_\alpha f \in V_m$ whenever $f \in V_m$. This follows from the fact that for $f \in V_0$, say,

$$\hat{f}(w) = F(w)\hat{\phi}(w) \qquad (11.5)$$

where $F(w)$ is 2π periodic and $\hat{\phi}(w) = \chi_{[-\pi,\pi]}(w)$. The Fourier transform of $f(t-\alpha)$ is $e^{-iw\alpha} F(w)\hat{\phi}(w)$, which by taking its periodic extension and multiplying by $\hat{\phi}(w)$ can be put in the same form as (11.5).

The dilation $D_\beta f$ (defined by $(D_\beta f(x) = f(\beta x))$ transforms into $\hat{f}\left(\frac{w}{\beta}\right)\frac{1}{\beta}$. This will belong to $\hat{V}_o$ provided $0 < \beta \leq 1$, since then its support will be contained in $[-\pi, \pi]$. For dilations with $0 < \beta \leq 2$, $D_\beta f \in V_1$, and by induction $D_\beta f \in V_m$ whenever $0 < \beta \leq m$. Thus all positive dilations of $f \in V_m$ belong to one of the spaces in the MRA.

As a consequence of the translation property, all derivatives of $f \in V_m$ are in V_m and hence any differential operator with constant coefficients maps V_m into itself. This can be extended to operators with polynomial coefficients provided that $f \in V_m$ is sufficiently rapidly decreasing.

The convolution of $f \in V_m$ with any $L^1(\mathbb{R})$ function g is again in V_m since the Fourier transform $\hat{g}(w)$ is continuous, and the Fourier transform of the convolution is

$$\hat{g}(w)\hat{f}(w),$$

which has support in $[-\pi, \pi]$ and, since $\hat{g}$ is bounded, belongs to $L^2[-\pi, \pi]$. Thus, $\hat{g}(w)\hat{f}(w)$ may be put in the form (11.5). In this example, therefore, considerable analysis can be done without ever leaving the space V_m. Unfortunately, this is not typical of wavelets.

11.4   An example where nothing works

The wavelet prototype I of Chapter 1 is at the opposite extreme. In this case $\varphi(t)$ is the characteristic function of the unit interval. Clearly $\varphi(t - \alpha)$ where α is an irrational number does not belong to V_0. In fact, it belongs to <u>no</u> V_m in the associated MRA since $f \in V_m$ is piecewise constant with possible jumps at $2^{-m}n$, $n \in \mathbb{Z}$.

For dilations we have a similar negative result. The dilation $D_\beta \varphi$ belongs to none of the V_m when β is an irrational number.

We might expect the convolution of φ with some function g to be in V_0 (or at least in T_0, the subspace of S') in spite of the failure of translation invariance.

But $g * \varphi$ for $g \in L^1(\mathbb{R})$ is given by

$$\int_{-\infty}^{\infty} g(t)\varphi(x - t)dt = \int_{x-1}^{x} g(t)dt,$$

which is continuous. Since the only continuous functions in T_0 are constants (V_0 contains only the constant function which is identically zero but T_0 of Chapter 5 contains other constant functions), if this convolution is to be in T_0, then the derivative must be $0 = g(x) - g(x - 1)$ a.e. This makes g periodic and hence not in $L^1(\mathbb{R})$ unless it is zero.

One might ask which of these two examples is more typical. Unfortunately, it is the latter, since as shown by Madych [Md], the only scaling functions for which V_0 is translation invariant are those whose

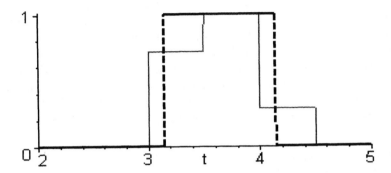

FIGURE 11.1
The approximation of the shifted scaling function by Haar series at scale m=1.

Fourier transform is a characteristic function of some set. The only such standard one is our prototype II.

11.5 Weak translation invariance

In this section we modify the definition of translation invariance. Rather than requiring that each subspace in a multiresolution analysis $\{V_m\}$ be invariant we shall require only that translations be in the next space up the ladder V_{m+1}. This is almost as good and includes many additional wavelet bases, in particular the Meyer wavelets [W8].

We first study weakly translation invariant MRA in general and find necessary and sufficient conditions for them. With an added connectivity condition, these are exactly the MRA of Meyer type wavelets. These are shown to be invariant as well under other operators including dilation, differentiation, and convolution in the next section.

Definition 11.1   Let $\{V_m\}$ be a multiresolution analysis of $L^2(\mathbb{R})$. Then $\{V_m\}$ is said to be *weakly translation invariant* if the translation operator, T_α, $\alpha \in \mathbb{R}$, maps V_m into V_{m+1}.

Any translation invariant MRA will be weakly translation invariant since $V_m \subseteq V_{m+1}$. This includes the Shannon wavelets $\left(\phi(t) = \frac{\sin \pi t}{\pi t}\right)$ but no other nonpathological cases [Md]. In common with the Shannon wavelets, the scaling function associated with the other translation invariant MRA's are not in $L^1(\mathbb{R})$ and thus give poor time localization.

The prototype of the weakly translation invariant MRA is the family of Meyer wavelets [L-M] for which $\phi(t)$ is rapidly decreasing. These are quite useful in some applications, such as $1/f$ processes, in which the frequency is known only over a finite range.

Proposition 11.1

Let $\{V_m\}$ be a weakly translation invariant MRA of $L^2(\mathbb{R})$ with scaling function ϕ; let $\Omega = \text{supp } \hat{\phi}$. Then

    (a) $\Omega \subset 2\Omega$

    (b) $\Omega \bigcap \{\Omega + 4\pi k\} \simeq \Phi$, $k \in \mathbb{Z}$, $k \neq 0$

    (c) $\displaystyle\bigcup_{k \in \mathbb{Z}} \{\Omega + 2\pi k\} \simeq \mathbb{R}$

    (d) $\displaystyle\bigcup_{j \in \mathbb{Z}} (2^j \Omega) \simeq \mathbb{R}$

    (e) There is a distribution (or function) h with support on $\{\Omega + \pi\} \bigcap \{\Omega - \pi\}$ such that

$$\frac{d}{dw} \left|\hat{\phi}(w)\right|^2 = \begin{cases} +h(w + \pi), w \in \Omega - \{\Omega + 2\pi\} \\ -h(w + \pi), w \in \Omega - \{\Omega - 2\pi\} \end{cases}$$

    (f) $\left|\hat{\phi}(w)\right| = 1$, $w \in \Omega - \{\Omega + 2\pi\} - \{\Omega - 2\pi\}$.

PROOF   Since $\hat{\phi}(w) = m_0\left(\frac{w}{2}\right)\hat{\phi}\left(\frac{w}{2}\right)$, where m_0 is a 2π-periodic function and $\text{supp } \hat{\phi}\left(\frac{w}{2}\right) = 2\Omega$, (a) clearly follows. For (b), we first observe that $\widehat{T_\alpha \phi}(w) = e^{-iw\alpha}\hat{\phi}(w)$, and since $T_\alpha \phi \in V_1$ by the weak translation invariance,

$$\widehat{T_\alpha \varphi}(w) = m_\alpha\left(\frac{w}{2}\right)\hat{\phi}\left(\frac{w}{2}\right),$$

where m_α is also 2π periodic. Hence we have

$$\left(m_\alpha\left(\frac{w}{2}\right) - e^{-iw\alpha}m_0\left(\frac{w}{2}\right)\right)\hat{\phi}\left(\frac{w}{2}\right) = 0$$

and by shifting w by $4\pi k$ we find

$$m_\alpha\left(\frac{w}{2}\right) - e^{-(w+4\pi k)\alpha}m_0\left(\frac{w}{2}\right) = 0, \quad w \in \{2\Omega + 4\pi k\},$$

since the support of $\hat{\phi}\left(\frac{w+4\pi k}{2}\right)$ is $\{2\Omega + 4\pi k\}$.

Since $\{\Omega + 4\pi k\} \subset \{2\Omega + 4\pi k\}$ by (a) we have that

$$\left(e^{-i(w+4\pi k)\alpha} - e^{-iw\alpha}\right) m_0\left(\frac{w}{2}\right) = 0, \quad w \in \Omega \bigcap\{\Omega + 4\pi k\}.$$

Moreover, $m_0\left(\frac{w}{2}\right) \neq 0$ for $w \in \Omega$ and by the periodicity of m_0 is also nonzero for $w \in \{\Omega + 4\pi k\}$. Hence either $e^{-i4\pi k\alpha} = 1$ or else Ω and $\{\Omega + 4\pi k\}$ do not intersect for $k \neq 0$ and (b) follows.

The proof of (c) follows immediately from the orthogonality condition

$$\sum_k \left|\hat{\phi}(w + 2\pi k)\right|^2 = 1. \tag{11.6}$$

If (d) does not hold, then there is set A of finite positive measure such that $A \subseteq (\mathbb{R} - \cup_j 2^j\Omega)$. Let $\hat{f}(w) = \chi_A(w)$; then $f \in L^2(\mathbb{R})$, $\hat{f} = 0$ on $2^m\Omega$ for each $m \in \mathbb{Z}$, and $f_m = P_m f = 0$ where P_m is the projector on V_m. This contradicts the fact that $\bigcup V_m$ is dense in $L^2(\mathbb{R})$.

For conditions (e) and (f) we use the fact that (11.6) contains at most two nonzero terms for each $w \in \mathbb{R}$, which follows from (b). We define $h(w)$ as the distribution $\in \mathcal{S}'$ given by

$$h(w) = \frac{d}{dw}|\hat{\phi}^2(w - \pi)| \text{ for } w \in \{\Omega + \pi\} \setminus \{\Omega + 3\pi\}, \tag{11.7}$$

and 0, otherwise, and then use the fact that

$$\left|\hat{\phi}^2(w + \pi)\right| + \left|\hat{\phi}^2(w - \pi)\right| = 1$$

for $w \in (\{\Omega + \pi\} \bigcup\{\Omega - \pi\}) \setminus (\{\Omega + 3\pi\} \bigcup\{\Omega - 3\pi\})$.

Hence

$$\frac{d}{dw}\left|\hat{\phi}(w + \pi)\right|^2 = -h(w), \quad w \in \{\Omega - \pi\} \setminus \{\Omega - 3\pi\}. \tag{11.8}$$

The support of $h(w)$ is contained in $\{\Omega + \pi\}$ by (11.7) and in $\{\Omega - \pi\}$ by (11.8). Their intersection is not empty by (11.6). The remaining conclusion in (e) follows by shifting (11.7) and (11.8).

The conclusion (f) also follows easily from (11.6). □

This result should be compared to those in [Md], section 3, in which translation invariance is considered. In particular, it is shown there that $|\hat{\phi}|$ is the characteristic function of Ω and hence is not continuous. Therefore, $\phi \notin L^1(\mathbb{R})$. In our case $\hat{\phi}$ can be taken to be C^∞ with compact support and hence ϕ will be rapidly decreasing.

When Ω is a connected set, then the results can be made more exact. We assume also that Ω is symmetric about the origin. It can be made so by a simple shift of w so that nonsymmetric results are easily obtained from the symmetric ones.

Corollary 11.1

Let $\{V_m\}$ *be a MRA as in Proposition 11.1 and suppose* $\Omega = supp \, \widehat{\phi}$ *is a symmetric connected set. Then there is an* $0 \leq \epsilon \leq \frac{\pi}{3}$ *and a distribution* h, *with support in* $[-\epsilon, \epsilon]$ *such that*

(i) $|\widehat{\phi}(w)|^2 = \int\limits_{w-\pi}^{w+\pi} h,$

(ii) $supp \, \widehat{\phi} \subseteq [-\pi - \epsilon, \pi + \epsilon],$

(iii) $|\widehat{\phi}(w)| = 1, \quad |w| \leq \pi - \epsilon.$

PROOF   Since Ω is connected it must be an interval $[a, b]$ which by (d) must contain a neighborhood of 0. It must also be bounded by (b) and in fact $b \leq a + 4\pi$, or

$$b - a \leq 4\pi. \tag{11.9}$$

Similarly by (c) it must be bounded below by

$$2\pi \leq b - a. \tag{11.10}$$

The intersection of $\{\Omega + \pi\}$ and $\{\Omega - \pi\}$ is $[a + \pi, b - \pi]$, which cannot be empty by (11.10), and the set on which $|\widehat{\phi}(w)| = 1$ is $[b - 2\pi, a + 2\pi]$ by (f).

Since Ω is symmetric $a = -b$ and ϵ may be taken as $\epsilon = b - \pi$. Then conclusions (ii) and (iii) follow. Clearly $0 \leq \epsilon \leq \pi$ by (11.9) and (11.10). We must show that $\epsilon \leq \frac{\pi}{3}$.

To do so, we return to the dilation equation. The support of $\widehat{\phi}\left(\frac{w}{2}\right)$ is $[-2\pi - 2\epsilon, 2\pi + 2\epsilon]$; hence $m_0\left(\frac{w}{2}\right)$ must be zero for $\pi + \epsilon < |w| \leq 2\pi + 2\epsilon$. Since it is periodic 4π as a function of w, it follows that in every interval of length 4π, $m_0\left(\frac{w}{2}\right) = 0$ on a set of measure $\geq \pi + \epsilon$. In particular, on the interval $[-2\pi, 2\pi]$ the support of $m_0\left(\frac{w}{2}\right)$ (or of $\widehat{\phi}(w)$) is of length $\leq 3\pi - \epsilon$. Since Ω has length $2\pi + 2\epsilon$, it follows that $2\pi + 2\epsilon \leq 3\pi - \epsilon$ or $\epsilon \leq \frac{\pi}{3}$. The conclusion (i) then follows from (e). $\square$

It is clear that h need not necessarily be positive. Take for example

$$|\widehat{\phi}(w)|^2 = \begin{cases} 1, & |w| \leq \pi - \varepsilon \\ 1/4, & \pi - \varepsilon < |w| \leq \pi \\ 3/4, & \pi < |w| \leq \pi + \varepsilon \\ 0, & \pi + \varepsilon < |w| \end{cases}$$

then $h(w) = \frac{3}{4}\delta(w + \epsilon) - \frac{1}{2}\delta(w) + \frac{3}{4}\delta(w - \epsilon)$. This may also be made into a smooth version by approximating the δ's by a smooth function.

However, since $|\widehat{\phi}(w)|^2 \geq 0$, the mean of h over any interval of length 2π must be nonnegative. With this added condition the conclusions of the Corollary are sufficient as well as necessary.

Proposition 11.2
Let the function $\phi(t)$ be given by

$$\widehat{\phi}(w) = \left\{ \int_{w-\pi}^{w+\pi} h \right\}^{1/2}$$

where h is a distribution with support in $\left[-\frac{\pi}{3}, \frac{\pi}{3}\right]$ such that $h * \chi_{[-\pi,\pi]}$ is a nonnegative bounded measurable function, and $\int_{-\infty}^{\infty} h = 1$. Then $\phi(t)$ is a scaling function and its associated MRA $\{V_m\}$ is weakly translation invariant.

PROOF We first consider some of the properties of $|\widehat{\phi}(w)|^2$. It is clear that

(i) $\operatorname{supp}|\widehat{\phi}(w)|^2 \subseteq \left[-\dfrac{4\pi}{3}, \dfrac{4\pi}{3}\right]$,

(ii) $|\widehat{\phi}(w)| = 1$ for $|w| \leq \dfrac{2\pi}{3}$,

(iii) $\displaystyle\sum_k |\widehat{\phi}(w + 2\pi k)|^2 = \int_{-\infty}^{\infty} h = 1$ a.e.

In order to show that ϕ is a scaling function we must show that the dilation equation holds. If we define

$$m_0\left(\frac{w}{2}\right) := \sum_k \widehat{\phi}(w + 4\pi k)$$

then $m_0\left(\frac{w}{2}\right) = 0$ for $\frac{4\pi}{3} < |w| < \frac{8\pi}{3}$ and hence

$$\widehat{\phi}(w) = m_0\left(\frac{w}{2}\right)\widehat{\phi}\left(\frac{w}{2}\right)$$

holds. Since by condition (iii), ϕ is orthogonal to its translates as well, it is a scaling function.

Now let $f \in V_0$ and $f_\alpha = T_\alpha f$ be its translate by an amount α. Then

$$\hat{f}_\alpha(w) = \sum_n a_n e^{-iwn} e^{-iw\alpha} \widehat{\phi}(w)$$

$$= F(w) e^{-iw\alpha} \widehat{\phi}(w)$$

$$= F(w) e^{-iw\alpha} \theta(w) \widehat{\phi}(w)$$

where $\theta(w) \in C^\infty$, supp $\widehat{\phi}(w) \subseteq$ supp $\theta(w) \subseteq (-2\pi, 2\pi)$, and $\theta(w) = 1$ for $w \in$ supp $\widehat{\phi}(w)$. We form the 4π periodic function

$$G_\alpha(w) := \sum_{k=-\infty}^{\infty} e^{-i(w+4\pi k)\alpha} \theta(w + 4\pi k).$$

Then $\hat{f}_\alpha$ becomes

$$\hat{f}_\alpha(w) = F(w) G_\alpha(w) \widehat{\phi}(w) = F(w) G_\alpha(w) m_0 \left(\frac{w}{2}\right) \widehat{\phi}\left(\frac{w}{2}\right)$$

where $F G_\alpha m_0 \left(\frac{\cdot}{2}\right) \in L^2(-2\pi, 2\pi)$ and is periodic 4π. Hence if its Fourier series is

$$F(w) G_\alpha(w) m_0 \left(\frac{w}{2}\right) \sim \sum_{n=\infty}^{\infty} b_n e^{-iwn/2},$$

then

$$f_\alpha(t) = \sum_{n=\infty}^{\infty} b_n 2\varphi(2t - n), \quad \{b_n\} \in \ell^2.$$

Thus $f_\alpha \in V_1$, and $\{V_m\}$ is a weakly translation invariant MRA.   □

These MRA's have many other nice properties (see below). Some of the examples have been discussed in Chapter 3 where they were generated by Meyer type wavelets (Example 7), which are slightly less general in that $h \geq 0$.

Lemarie and Meyer [L-M] defined a wavelet whose scaling function [D, p.137] turned out to be

$$\widehat{\phi}(w) = \begin{cases} 1, & |w| \leq \frac{2\pi}{3} \\ \cos[\frac{\pi}{2} \nu(\frac{3}{2\pi}|w| - 1)], & \frac{2\pi}{3} < |w| \leq \frac{4\pi}{3} \\ 0, & \frac{4\pi}{3} < |w| \end{cases}$$

where $\nu(x)$ is a C^k function satisfying

$$\nu(x) = \begin{cases} 0, & x \le 0 \\ 1, & 1 \le x \end{cases}$$

and $\nu(x) + \nu(1-x) = 1$. (See Example 6 of Chapter 3.)

This can be put into the form of Proposition 10.3 by finding, for $w \in \left[\frac{2\pi}{3}, \frac{4\pi}{3}\right]$

$$h(w - \pi)$$

$$= -\frac{d}{dw}\cos^2\left[\frac{\pi}{2}\nu\left(\frac{3}{2\pi}w - 1\right)\right]$$

$$= 2\cos\left[\frac{\pi}{2}\nu\left(\frac{3}{2\pi}w - 1\right)\right]\sin\left[\frac{\pi}{2}\nu\left(\frac{3}{2\pi}w - 1\right)\right]\frac{\pi}{2}\nu'\left(\frac{3}{2\pi}w - 1\right)\frac{3}{2\pi}.$$

If $\nu(x) \in C^1$ at least, then $\nu'\left(\frac{3}{2\pi}w - 1\right)$ has support in $\frac{2\pi}{3} \le w \le \frac{4\pi}{3}$ and hence $h(w)$ has support in $\left[-\frac{\pi}{3}, \frac{\pi}{3}\right]$. By a little algebraic manipulation we find

$$h(w) = \frac{3}{4}\sin\left[\pi\nu\left(\frac{3w}{2\pi} + \frac{1}{2}\right)\right]\nu'\left(\frac{3w}{2\pi} + \frac{1}{2}\right), \quad |w| \le \frac{\pi}{3}.$$

The $\widehat{\phi}(w)$ can be made to be in C^∞ by an appropriate choice of $h(w)$.

This may be done by taking $e(x) = \exp(1/(x^2 - 1))\chi_{[-1,1]}(x)$. Then $h(w) = C_\epsilon e\left(\frac{w}{\epsilon}\right)$, where $C_\epsilon = (\epsilon \int e(x)dx)^{-1}$, satisfies the required conditions. This is the familiar regularization kernel of distribution theory. It may also be used to generate other $h(w)$. Indeed let h_0 be any non-negative function in L^1 with support on $[-\epsilon, \epsilon]$ and $\int h_0 = 1$. Then $h_1 = h * h_0$ has support on $[-2\epsilon, 2\epsilon]$ and belongs to C^∞. Since $\widehat{\phi}^2(w) = (\chi_{[-\pi,\pi]} * h)(w)$, the appropriate new $\widehat{\phi}$ may be given by

$$\widehat{\phi}_1^2 = \chi_{[-\pi,\pi]} * h_1 = \chi_{[-\pi,\pi]} * h_0 * h = \widehat{\phi}_0^2 * h,$$

where $\widehat{\phi}_0$ is the scaling function of h_0, since convolution is commutative and associative.

11.6 Dilations and other operations

We cannot expect V_0 to be invariant under all dilations. If it were, then all V_m would be the same as V_0 and $\bigcup_m V_m$ could not be dense in

$L^2(\mathbb{R})$. Our definition of weak dilation invariance will therefore be

Definition 11.2   *Let $\{V_m\}$ be a multiresolution analysis of $L^2(\mathbb{R})$. Then $\{V_m\}$ is said to be <u>weakly dilation invariant</u> if for each $\alpha \in \mathbb{R}$ there is an $m \in \mathbb{Z}$ such that the dilation operator D_α maps V_0 into V_m.*

Again the Haar wavelets and the ones based on splines are not weakly dilation invariant, but the Shannon wavelets are. This follows from the fact that $\widehat{D_\alpha\phi}(w) = \frac{1}{\alpha}\hat{\phi}\left(\frac{w}{\alpha}\right) = \frac{1}{\alpha}\chi_{[-\alpha\pi,\alpha\pi]}(w)$ which is clearly a periodic multiple of $\hat{\phi}\left(\frac{w}{2}\right)$ for $1 < \alpha < 2$. Hence $D_\alpha\phi \in V_1$ for such α. Similar results hold for the Meyer type wavelets.

Lemma 11.1
Let $\phi(t)$ be the scaling function of a Meyer type wavelet for some h_ϵ with support on $[-\epsilon, \epsilon]$, $\epsilon \leq \pi/3$. Then $D_\alpha\phi \in V_1$ if $\phi \in V_0$ for $1 \leq \alpha \leq 2(\pi - \epsilon)/(\pi + \epsilon)$.

PROOF The proof is straightforward since $\hat{\phi}\left(\frac{w}{\alpha}\right)$ has support in $[-\alpha(\pi + \epsilon), \alpha(\pi+\epsilon)]$. Thus $D_\alpha\phi \in V_1$ if $\hat{\phi}\left(\frac{w}{\alpha}\right) = m_\alpha\left(\frac{w}{2}\right)\hat{\phi}\left(\frac{w}{2}\right)$ where $m_\alpha\left(\frac{w}{2}\right)$ is 4π periodic. This is possible if the support of $\hat{\phi}\left(\frac{w}{\alpha}\right)$ is contained in the set where $\hat{\phi}\left(\frac{w}{2}\right) = 1$, i.e., $[-2\pi + 2\epsilon, 2\pi - 2\epsilon]$. A translation of $\hat{\phi}\left(\frac{w}{\alpha}\right)$ by $2\pi k$, $k \neq 0$, will then have its support outside of the support of $\hat{\phi}\left(\frac{w}{2}\right)$. Hence the 4π periodic extension of $\hat{\phi}\left(\frac{w}{\alpha}\right)$ will give us $m_\alpha\left(\frac{w}{2}\right)$. In terms of α this is

$$\alpha(\pi + \epsilon) \leq 2\pi - 2\epsilon,$$

which gives us our conclusion.   □

Corollary 11.2
Let $\{V_m\}$ be the MRA of a Meyer type wavelet with $\epsilon < \frac{\pi}{3}$, then $\{V_m\}$ is weakly dilation invariant.

PROOF By the Lemma, if $\epsilon < \frac{\pi}{3}$, then $D_\alpha\phi \in V_1$ for $\alpha < \alpha_\epsilon$, where $\alpha_\epsilon = 2\left(\frac{\pi - \epsilon}{\pi + \epsilon}\right) > 2\frac{2\pi/3}{4\pi/3} = 1$. By a change of scale it follows that $D_\alpha\phi_1 \in V_2$ for $\phi_1 \in V_1$. Hence, for $\phi \in V_0$, $D_{2\alpha}\phi \in V_2$ and it follows that $D_{2^m\alpha}\phi \in V_{m+1}$, $m \geq 0$. Since $2^m\alpha$ must eventually be greater than 1, the conclusion follows.   □

We can combine the two results on dilation and translation to get

Corollary 11.3

Let $\{V_m\}$ be the MRA of a Meyer type wavelet $\psi(t)$ with $\epsilon < \frac{\pi}{3}$, let $\alpha, \beta \in \mathbb{R}$; then $\psi(\alpha t - \beta) \in V_m$ for some m.

Other operators associated with translation are difference operators, differential operators and convolution operators. The invariance of these operators can be based on Proposition 11.2, but stronger results can be obtained directly using the same techniques.

Proposition 11.3

Let $\{V_m\}$ be the MRA of a Meyer type wavelet; let $P(\Delta)$ $(P(D))$ be a linear difference operator (differential operator) with constant coefficients; let H be the operator of convolution with $h \in L^1(\mathbb{R})$. Then for each $f \in V_m$, $P(\Delta)f$, $P(D)f$, and Hf belong to V_{m+1}.

The proof is along the same lines as in the previous results and will be omitted.

11.7 Problems

1. Show directly that for the Shannon wavelets, the space V_m is invariant under differentiation.

2. Recall from Chapter 7 that the Green's function of the differential operator $D^2 y$ with boundary conditions $y(0) = y(1) = 0$ is given by

$$g(x,t) = \begin{cases} x(1-t), \, 0 \le x \le t \le 1 \\ t(1-x), \, 0 \le t \le x \le 1 \\ 0, \qquad \text{otherwise} \end{cases}$$

and hence is in $L^1(\mathbb{R})$.

Show therefore that the solution to $D^2 y = f$, $f \in V_0$ satisfying the boundary conditions is also in V_0 for the Shannon wavelet case.

3. For the Franklin wavelets, show by example that neither translation invariance nor weak translation invariance holds for V_0.

4. Show directly that the weak translation invariant property holds for the Meyer wavelets generated by $h(\xi) = 1/2\epsilon$, $|\xi| < \epsilon \leq \frac{\pi}{3}$, $h(\xi) = 0$, $\epsilon \leq \xi$.

5. For the Franklin wavelets, show that the weak dilation invariance property does not hold.

6. Show the derivative of the hat function $\theta(t) = (1 - |t|)\chi_{[-1,1]}(t)$ is not continuous and hence does not belong to any of the spaces V_m for the Franklin wavelets.

7. Find the approximation to the characteristic function of the interval $[\pi, \pi + 1]$ by the scaling function series for the Haar wavelet at the scale $m = 5$.

Chapter 12

Analytic Representations Via Orthogonal Series

The analytic representation of functions or distributions on the real line is usually given by a Cauchy type formula (Chapter 2, Section 4) [Br] but in some cases may also be given by an orthogonal series. This is evident for periodic functions and distributions for which trigonometric series may be used [W15]. For nonperiodic functions the Hermite series can be used, but it leads to analytic representations that are no longer of the same form [W16]. For functions with compact support, series of Legendre polynomials lead to analytic representations involving Legendre functions of the second kind [W17].

The more natural approach for arbitrary functions on $\mathbb{R}$ seems to be one involving wavelets [W18]. The analytic representation preserves dilations and translations. Since these are the operators used to define wavelets, their analytic representatives also are wavelets. This is also true for harmonic functions that may be obtained from analytic representations. The corresponding "harmonic wavelets" also satisfy the same defining relations.

In this chapter, we first present elements of the analytic representation theory involving the trigonometric, Hermite, and Legendre systems. We then consider the wavelet approach in more detail. Both functions and tempered distributions are considered since they lead to slightly different theories. Some wavelets are already entire functions, in which case it is unnecessary to use analytic representations. In these cases we characterize the expansions of functions analytic in a strip in terms of their wavelet coefficients.

12.1   Trigonometric series

We have seen in Chapter 4 that each tempered distribution of period 2π has a Fourier series convergent in the sense of $\mathcal{S}'$. The analytic representation of such distributions is most easily obtained by splitting up the Fourier series.

Let $f(x)$ be a periodic distribution of period 2π with Fourier series (see Chapter 4)

$$\sum_{n=0}^{\infty} a_n \cos nx + b_n \sin nx.$$

Consider the series

$$\sum_{n=0}^{\infty} (a_n - ib_n)e^{inz} \quad \text{and} \quad \sum_{n=0}^{\infty} (a_n + ib_n)e^{-inz}.$$

The first converges for all z in the upper half-plane, the second for all z in the lower half-plane since the a_n and b_n, as Fourier coefficients of a distribution, grow no faster than a power of n. Also both series converge in the sense of distributions for z on the real axis. Clearly their sum also does and, in fact, since it is just twice the Fourier series again, converges to $2f(x)$. Thus the function

$$f^+(z) = \sum_n \frac{a_n - ib_n}{2} e^{inz}$$

for $\operatorname{Im} z > 0$, and

$$f^-(z) = \sum_n \frac{-a_n - ib_n}{2} e^{-inz}$$

for $\operatorname{Im} z < 0$, is an analytic representation of $f(x)$ since

$$\lim_{\epsilon \to 0} \{ f^+(x + i\epsilon) - f^-(x - i\epsilon) \}$$

$$= \lim_{\epsilon \to 0} \left\{ \sum_n \frac{a_n - ib_n}{2} e^{in(x+i\epsilon)} + \sum_n \frac{a_n + ib_n}{2} e^{-in(x-i\epsilon)} \right\}$$

$$= \lim_{\epsilon \to 0} \left\{ \sum_n (a_n \cos nx + b_n \sin nx) e^{-n\epsilon} \right\}$$

$$= \sum_n (a_n \cos nx + b_n \sin nx) \lim_{\epsilon \to 0} e^{-n\epsilon} = f(x).$$

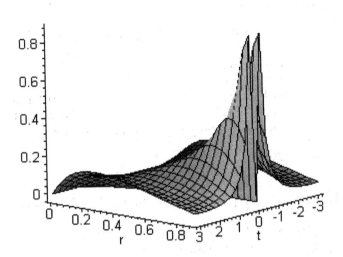

FIGURE 12.1
The kernel $Kr(t)$ given in Lemma 12.1 with $m = 2$.

Of course, the limits in the preceding computations must be taken in the sense of S' in which case the interchange of limits is justified.

These Fourier series of distributions are not convergent in the usual pointwise sense, but they may be pointwise summable at certain points. In fact we shall see that at points at which the distribution has a value in the sense of Lojasiewicz [Lo] (see Section 8.2) the Fourier series is Abel summable to that value. We now recall (see Chapter 4) that the Abel means are

$$f(r, x) := \frac{1}{\pi} \int_{-\pi}^{\pi} f(x + t) P_r(t) dt,$$

where P_r, the Poisson kernel, is given by

$$P_r(t) = \frac{1 - r^2}{2(1 - 2r \cos t + r^2)},$$

whence we may prove the statement about Abel summability by using the following.

Lemma 12.1

Let $P_r(t)$ be the Poisson kernel. Then

$$K_r(t) = ((-t)^m P_r^{(m)}(t))/\pi m! \rightarrow \delta(t) \text{ as } r \rightarrow 1^- \text{ for } t \in (-\pi, \pi).$$

Moreover $K_r(t)$ is a quasi-positive type delta family.

PROOF   It is well known that $P_r(t) \rightarrow \pi\delta(t)$ in the distribution sense on $(-\pi, \pi)$ as $r \rightarrow 1^-$. Both differentiation and multiplication by t are continuous operations with respect to distribution convergence. Hence

$$\frac{(-t)^m P_r^{(m)}(t)}{\pi m!} \rightarrow \frac{(-t)^m \delta^{(m)}(t)}{m!} = \delta(t) \quad \text{as } r \rightarrow 1^-.$$

The kernel $K_r(t)$ is a quasi-positive delta family if it satisfies

(i) $\int_{-\pi}^{\pi} K_r(t)dt = 1,$

(ii) $\int |K_r(t)|dt \leq M$, a constant,   (12.1)

(iii) $K_r(t) \rightarrow 0$ uniformly on $[-\pi,-\delta]$ and $[\delta,\pi]$ for each $\delta > 0$.

(See Chapter 8 where a slightly more general definition is used.)

To prove (i) we use the fact that $(1/\pi)\int_{-\pi}^{\pi} P_r(t)dt = 1$ and integrate by parts m times. The integrated terms are all bounded and what we have left is

$$\frac{1}{\pi} \int_{-\pi}^{\pi} \frac{(-t)^m P_r^{(m)}(t)}{m!} dt.$$

To prove (ii) we use the inequality $|t^{m+1} P_r^{(m)}(t)| \leq K_m$, a constant independent of r and t, which may be proved inductively. We then integrate

$$\int_{-\pi}^{\pi} |K_r(t)| \, dt = 2 \int_0^{\pi} t^m |P_r^{(m)}(t)| \, dt$$

$$= \pm 2 \sum_{k=1}^{m+1} \int_{\tau_{k-1}}^{\tau_k} (-1)^k t^m P_r^{(m)}(t)dt$$

by parts m times. Here τ_k are the zeros of $P_r^{(m)}$ for $k = 1, \cdots, m$ and $\tau_0 = 0, \tau_{m+1} = \pi$.

The integrated terms are each of the form

$$\pm 2 \sum_k \left\{ \tau_k^{n+1} P_r^{(n)}(\tau_k) - \tau_{k-1}^{n+1} P_r^{(n)}(\tau_{k-1}) \right\}$$

which, by the preceding inequality, are bounded. The left-over integral is exactly

$$\pm 2 \sum_{k=1}^{m+1} \int_{\tau_{k-1}}^{\tau_k} (-1)^k P_r(t) dt,$$

which is bounded by the fact used in the proof of (i).

Finally, we prove (iii) by the inequality

$$|P_r^{(m)}(t)| \leq \frac{A_m(1-r)}{t^{2m+2}},$$

which follows from the same inequality as in the case $m = 0$ (see [Z, p.97]). $\qquad\square$

THEOREM 12.1

The Fourier series of a distribution f with value γ at x_0 is Abel summable to γ at x_0.

PROOF To prove this we must show that $f(r, x_0) \to \gamma$ as $r \to 1^-$, where

$$f(r, x) = \frac{1}{\pi} \int_{-\pi}^{\pi} f(x+t) P_r(t) dt.$$

If we assume, as we may without loss of generality, that the Fourier series of f has constant terms 0, we may write $f = D^m F$, where F is a periodic and a continuous function. Moreover, by picking m large enough we may assume

$$\frac{F(x)}{(x-x_0)^m} \to \frac{\gamma}{m!} \quad \text{as} \quad x \to x_0.$$

The Abel means, by integrating by parts (the integrated terms are the 0-distribution), may be expressed as

$$f(r, x_0) = \frac{(-1)^m}{\pi} \int_{-\pi}^{\pi} F(t+x_0) P_r^{(m)}(t) dt$$

$$= \frac{m!}{\pi} \int_{-\pi}^{\pi} \frac{F(t+x_0)(-t)^m P_r^{(m)}(t)}{t^m \, m!} dt.$$

The first part of the lemma seems to suggest that the limit as $r \to 1^-$ should be

$$m! \int_{-\pi}^{\pi} \frac{F(t+x_0)}{t^m} \delta(t) dt = \lim_{t \to 0} m! \frac{F(t+x_0)}{t^m} = \gamma$$

and indeed would prove this statement if $F(t+x_0)/t^m$ were differentiable sufficiently often. Since it is not, we must use the fact that $K_r(t)$ is a kernel of quasi-positive type.

For such kernels $\int f K_r \to f(0)$ under the hypothesis that f is continuous at 0 (see Chapter 8). Hence our conclusion follows. □

The relation between the Abel means $f(r,x)$ and the analytic representation $f^\pm(z)$ of f is clear if we look at their respective series. In fact for $z = re^{ix}$, $f(r,x) = f^+(z) - f^-(\overline{z})$ so that any statement about Abel summability can be translated into a statement about analytic representation.

Corollary 12.1

Let f be a periodic distribution. If f has a value γ at x, then
$$f^+(x+i0) - f^-(x-i0) = \gamma.$$

12.2   Hermite series

In the case of trigonometric Fourier series the analytic representation f^+ is distinguished by having its negative Fourier coefficients equal to zero. If we move to functions in $L^2(\mathbb{R})$, these analytic representations belong to the Hardy space $H^2(\mathbb{C}^+)$, but the Hermite functions do not. Hence the (f^+) cannot be characterized in terms of vanishing coefficients.

In order to determine which sequences of coefficients do belong to functions in $H^2(\mathbb{C}^+)$, we use the auxiliary sequences $\{\alpha_{nk}\}$ where

$$\alpha_{nk} := \int_0^\infty h_n h_k, \qquad n,k = 0,1,\cdots.$$

Then $\sum i^n \alpha_{nk} h_n$ is the Hermite series of a function in $H^2(\mathbb{C}^+)$ since its Fourier transform vanishes on $(-\infty,0)$. By using the properties of the Hermite functions we may derive the following properties of the α_{nk}:

$$\begin{aligned}&\text{(i) } \alpha_{nk} = \tfrac{1}{2}\delta_{nk}, \quad n+k \text{ even},\\ &\text{(ii) } \sum_{n=0}^\infty \alpha_{nk}\alpha_{nl} = \alpha_{kl}.\end{aligned} \qquad (12.2)$$

These furnish the basic tools to prove

Proposition 12.1

Let $\sum c_n h_n$ be the Hermite series expansion of a function $f \in L^2(-\infty,$ $\infty)$; let

$$a_n = \sum_{k=0}^{\infty} i^{n-k} \alpha_{nk} c_k, \quad n = 0, 1, \cdots,$$

$$\left(b_n = \sum_{k=0}^{\infty} (-i)^{n-k} \alpha_{nk} c_k, \quad n = 0, 1, \cdots \right);$$

then $\sum a_n h_n (\sum b_n h_n)$ is the expansion of a function in $H^2(\mathbb{C}^+)$ $(H^2(\mathbb{C}^-))$. Moreover $c_n = a_n + b_n$, and if f itself $\in H^2(\mathbb{C}^+)$, $c_n = a_n$.

PROOF The proof involves using the infinite matrices $A = [i^{n-k} \alpha_{nk}]$ and $\overline{A} = [(-i)^{n-k} \alpha_{nk}]$ as operators on ℓ^2. They have the properties that $A + \overline{A} = I$ and $A\overline{A} = 0$. From these and the fact that A maps ℓ^2 into sequences that are the Hermite coefficients of H^2 functions, i.e., $H^2(\mathbb{C}^+)$ or $H^2(\mathbb{C}^-)$, the conclusions follow. □

Unfortunately, when we look at convergence of Hermite expansions of H^2 functions, we cannot in general extend the convergence on the real axis to the complex plane. In order to imitate the convergence of Fourier series of analytic representations, we need to introduce new functions associated with the Hermite functions. One such family of functions consists of projections of Hermite functions.

Since the Hermite functions themselves are not in H^2 and since Hermite series of H^2 functions do not behave very well off the real axis, it is better to consider other series instead. Thus, we shall consider the projections of the Hermite functions on H^2 and the behavior of series of the projections in the complex plane. These projections $\{g_k\}$ may be given by the functions

$$g_k(t) = \sum_{n=0}^{\infty} (i)^{n-k} \alpha_{nk} h_n(t), \quad k = 0, 1, 2, \cdots. \tag{12.3}$$

Then if $f \in H^2(\mathbb{C}^+)$ with Hermite expansion $\sum a_n h_n$ we see that

$$f = \sum_n a_n h_n = \sum_n \left\{ \sum_k (i)^{n-k} \alpha_{nk} a_k \right\} h_n = \sum_n \left\{ \int f \overline{g_n} \right\} h_n$$

by Parseval's equality since by the symmetry of α_{nk}, the Hermite coefficients of $\overline{g_n}$ are $i^{n-k} \alpha_{nk}$. Thus we find that $a_n = \int f h_n = \int f \overline{g_n}$ for

$f \in H^2(\mathbb{C}^+)$ and similarly $\int f h_n = \int f g_n$ for $f \in H^2(\mathbb{C}^-)$. Whenever $f \in L^2$ we can rewrite the results of Proposition 12.1 as

$$\int f h_n = \int f \overline{g_n} + \int f g_n,$$

whence it follows that

$$\overline{g_n} + g_n = h_n \quad a.e.$$

Thus all $f \in L^2$ may be written as

$$f = \sum c_n h_n = \sum c_n g_n + \sum c_n \overline{g_n}$$

where $\sum c_n g_n \in H^2(\mathbb{C}^+)$. These $\{g_n\}$ may be shown, by using the properties of the α_{nk}, to satisfy

$$g_k(x) = \frac{1}{\sqrt{2\pi}} \int_0^\infty e^{ixt} (-1)^k h_k(t) dt,$$

$$g_k(z) = \frac{1}{2\pi i} \int_{-\infty}^\infty \frac{h_k(x)}{x - z} dx, \quad \operatorname{Im} z > 0.$$

Using these properties we may attack the problem of convergence in the complex plane.

THEOREM 12.2

Let $f \in L^2(-\infty, \infty)$ with Hermite series expansion $\sum c_n h_n$; let

$$f^+(z) = \frac{1}{2\pi i} \int_{-\infty}^\infty \frac{f(x)}{x - z} dx, \quad \operatorname{Im} z > 0$$

· *be the projection of f on $H^2(\mathbb{C}^+)$; then $\sum c_n g_n(z)$, g_n given by (12.3),*

  (i) *converges absolutely for $\operatorname{Im} z > 0$,*

  (ii) *converges uniformly to $f^+(z)$ in any region for which $\operatorname{Im} z \geq \epsilon > 0$,*

  (iii) *converges in the sense of $L^2(-\infty, \infty)$ for each fixed $y \geq 0$, and*

$$\sum c_n \{\overline{g}_n(z) + g_n(z)\}$$

converges to a harmonic function whose boundary value on $(-\infty, \infty)$ are given by $f(x)$.

PROOF The first conclusion follows from Bessel's inequality since the $g_n(z)$ are just the Hermite coefficients of $(2\pi i)^{-1}(x-z)^{-1}$. Hence we see that

$$\left|\sum_{n=0}^{N} c_n g_n(z)\right|^2 \leq \sum |c_n|^2 \sum |g_n(z)|^2 \leq \|f\|_2^2 \int_{-\infty}^{\infty} \frac{1}{4\pi^2 |x-z|^2}\,dx.$$

To prove the uniform convergence we consider the equality

$$\left|\sum_{n=0}^{N} c_n g_n(z) - f^+(z)\right| = \frac{1}{2\pi}\left|\int_{-\infty}^{\infty} \frac{\sum_{n=0}^{N} c_n h_n(x) - f(x)}{x-z}\,dx\right|$$

$$\leq \left\|\sum_{n=0}^{N} c_n h_n - f\right\|_2 \left\{\frac{1}{2\sqrt{\pi y}}\right\}$$

from which our assertion follows immediately.

Finally to show L^2 convergence we use Plancherel's identity twice to calculate

$$\int_{-\infty}^{\infty} \left|\sum_{n=0}^{N} c_n g_n(x+iy) - f^+(x+iy)\right|^2 dx$$

$$= \int_{0}^{\infty} e^{-2yt} \left|\sum_{n=0}^{N} c_n h_n(t)(-i)^n - \hat{f}(t)\right|^2 dt$$

$$\leq \int_{-\infty}^{\infty} \left|\sum_{n=0}^{N} c_n h_n(t)(-i)^n - \hat{f}(t)\right|^2 dt$$

$$= \left\|\sum_{n=0}^{N} c_n h_n - f\right\|_2^2 \to 0 \text{ as } N \to \infty.$$

Here we have used the fact that

$$f^+(x+iy) = \frac{1}{2\pi i}\int_{0}^{\infty} e^{i(x+iy)t} \hat{f}(t)dt,$$

which is another consequence of Plancherel's identity. The last statement follows immediately from the others. $\square$

We note that with the replacement of h_n by g_n the convergence of the series is improved off the real axis. However it is still not good enough to include a number of cases of interest. We know that each function of polynomial growth has a Hermite series with coefficients

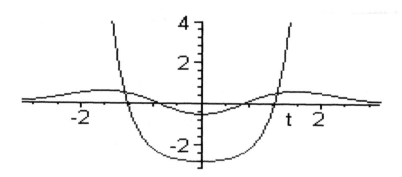

FIGURE 12.2
The Hermite function $h_2(x)$ and the real part of the Hermite

function of the second kind $\tilde{h}_2(x + i \cdot 0)$.

$O(n^p)$ convergent to it in the sense of tempered distributions. Also each such function has an analytic representation to which we should like the series with the g_n to converge with some sort of uniformity. Therefore, in order to improve the convergence we introduce another related sequence of functions $\{\tilde{h}_n\}$, the Hermite functions of the second kind. These are more closely analogous to the functions $\{e^{inz}\}$ used with Fourier series.

   These Hermite functions of the second kind $\tilde{h}_n$ are solutions to the Hermite equation

$$(z^2 - D^2)w = (2n + 1)w, \qquad n = 0, 1, \cdots,$$

which vanish at $i\infty$ when z is in the upper half-plane and at $-i\infty$ when z is in the lower half-plane. The functions were studied in [W16] where it is shown that they satisfy the same recurrence formulae as the Hermite functions themselves, as well as the following formulae:

(i) $\tilde{h}_n(z) = h_n(z) \int_{i\infty}^{z} \frac{2}{h_n^2}$, $\text{Im } z > 0$,

(ii) $\tilde{h}_n(z)h_k(z) = -\int_{-\infty}^{\infty} \frac{h_n(x)h_k(x)}{x-z}dx$, $\text{Im } z \neq 0, \, n \geq k$,   (12.4)

(iii) $\tilde{h}_n(x + i0) - \tilde{h}_n(x - i0) = -2\pi i h_n(x)$.

THEOREM 12.3
Let f be a continuous real valued function of polynomial growth with Hermite series expansion $\sum c_n h_n$; then

(i) $\sum(ic_n/2\pi)\tilde{h}_n(z)$ *converges uniformly in compact subsets of either the upper or lower open half-plane,*

(ii) $\sum(ic_n/2\pi)\{\tilde{h}_n(z) - \tilde{h}_n(\bar{z})\}$ *converges in the upper half-plane to a real harmonic function* $u(z)$,

(iii) $\displaystyle\lim_{y\to 0} u(x+iy) = f(x)$ *uniformly on compact sets of* $\mathbb{R}$.

PROOF We have seen in Chapter 2 that a function of polynomial growth belongs to $\mathcal{S}'$, the space of tempered distributions; convergence in $\mathcal{S}'$ is equivalent to weak convergence, i.e., $\sum c_n h_n$ converges as $\sum c_n \langle h_n, \varphi \rangle$ converges for each $\varphi \in \mathcal{S}$. Also an element f of $\mathcal{S}'$ has a Hermite expansion $\sum c_n h_n$ which converges to f and whose coefficients satisfy $c_n = O(n^p)$ for some integer p. Now by property (ii)

$$\sum \frac{ic_n}{2\pi}\tilde{h}_n(z) = \sum \frac{ic_n}{2\pi} \int_{-\infty}^{\infty} \frac{h_n(t)\exp(z^2/2)\exp(-t^2/2)}{z-t}dt$$

$$= \exp(z^2/2) \sum c_n \frac{1}{2\pi i} \int_{-\infty}^{\infty} \frac{h_n(t)}{t-z}\exp(-t^2/2)dt$$

$$= \exp(z^2/2) \sum c_n \langle \phi_z, h_n \rangle$$

where $\phi_z(t) = \exp(-t^2/2)/(t-z) \in \mathcal{S}$. Therefore this series converges pointwise for Im $z \ne 0$.

To show that this convergence is in fact uniform on compact sets we use the fact that $h_n = (x^2 - D^2)^{p+2}h_n/(2n+1)^{p+2}$, where p is the nonnegative integer associated with the asymptotic behavior of the coefficients. Then we have

$$\sum_n c_n \langle \phi_z, \frac{(x^2-D^2)^{p+2}}{(2n+1)^{p+2}} h_n \rangle = \sum_n \frac{c_n}{(2n+1)^{p+1}}\langle (x^2-D^2)^{p+2}\phi_z, h_n \rangle.$$

The series $\sum(c_n/(2n+1)^{p+2})$ converges absolutely and $\langle (x^2-D^2)^{p+2}\phi_z, h_n \rangle$ is uniformly bounded on compact sets avoiding the real axis.

The convergence in (ii) follows from that in (i), and the limit function will clearly be harmonic. We need only show that it is real. This can be done by showing $i(\tilde{h}_n(z) - \tilde{h}_n(\bar{z}))$ to be real. But we have, by (12.4), with $k = 0$,

$$i(\tilde{h}_n(z) - \tilde{h}_n(\bar{z}))$$

$$= -i \int_{-\infty}^{\infty} h_n(t)\exp(-t^2/2)\left\{ \frac{\exp(z^2/2)}{t-z} - \frac{\exp(\bar{z}^2/2)}{t-\bar{z}} \right\}dt$$

whence, since the expression in brackets is purely imaginary, it follows that it is.

The last conclusion of Theorem 12.3 is a consequence of the fact that $u(z)\exp(-x^2/2)$ is the Poisson integral representation of $f(t)\exp(-t^2/2)$. Hence $u(z)\exp(-z^2/2) \to f(t)\exp(-t^2/2)$ uniformly on compact sets and hence our conclusion. □

12.3 Legendre polynomial series

A function or distribution with support in $[-1,1]$ may be given by a series of Legendre polynomials

$$f(x) = \sum_{n=0}^{\infty} a_n P_n(x)\chi(x)$$

where $\chi(x)$ is the characteristic function of $[-1,1]$. The P_n, as we have seen in Chapter 6, satisfy the differential equation

$$((1-x^2)y')' + n(n+1)y = 0.$$

Since this is a second order differential equation, it must have a second linearly independent solution Q_n which we take to be

$$Q_n(z) = 2^{-n-1}\int_{-1}^{1}(1-t^2)^n(z-t)^{-n-1}dt, \quad \operatorname{Im} z \neq 0$$

[Sz, p.74]. This solution is related to the analytic representation since

$$\frac{i}{\pi}Q_n(z) = \frac{1}{2\pi i}\int_{-1}^{1} P_n(t)\frac{1}{t-z}dt \quad . \tag{12.5}$$

Thus it appears we have the following

Proposition 12.2
Let $\{a_n\}$ be a sequence of numbers such that $a_n = O(n^p)$ for some integer $p \geq 0$; then the series $\sum a_n P_n \chi$ converges to a tempered distribution g and the series $\sum a_n \frac{i}{\pi}Q_n(z)$ converges to a holomorphic function $g^{\pm}(z)$ for $\operatorname{Im} z \neq 0$, which is the analytic representation of g.

Details of the proof are omitted. They may be found in [W17].

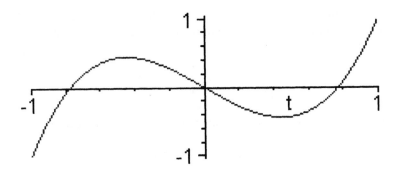

FIGURE 12.3
A Legendre polynomial $(n = 3)$.

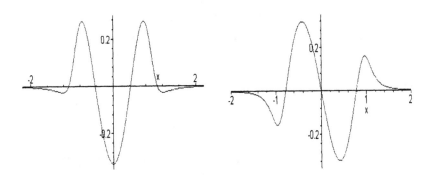

FIGURE 12.4
The real and imaginary parts of the analytic representations
at y=2 for the Legendre polynomial in Figure 12.3.

12.4   Analytic and harmonic wavelets

The operator that maps a function or distribution into its analytic representation is linear and commutes with certain other operators. Indeed, if $A^{\pm}(f(x)) = f^{\pm}(z)$, then

$$\frac{d}{dz}f^{\pm}(z) = A^{\pm}\left(\frac{d}{dx}f(x)\right)(z).$$

The same is true for the dilation operator D_{α} and translation operator T_{β}, i.e.,

$$f^{\pm}(\alpha z) = A^{\pm}(D_{\alpha}f(x))(z)$$

and

$$f^{\pm}(z - \beta) = A^{\pm}(T_{\beta}f(x))(z).$$

This is obvious in the case of L^2 functions and can easily be shown for distributions by the "integration of parts" device.

The operator A^+ maps $L^2(\mathbb{R})$ into the Hardy space $H^2(\mathbb{C}^+)$ consisting of all functions in the upper half-plane $\mathbb{C}^+$ that satisfy

$$\sup_{y>0} \int_{-\infty}^{\infty} |f(x + iy)|^2 dx < \infty.$$

This is clear if we use the Fourier transform definition,

$$f^+(x + iy) = \frac{1}{2\pi} \int_0^{\infty} \hat{f}(w)e^{iwx}e^{-wy}dw,$$

since by Parseval's equality

$$\int_{-\infty}^{\infty} |f^+(x + iy)|^2 dx = \frac{1}{2\pi} \int_0^{\infty} |\hat{f}(w)e^{-wy}|^2 dw$$

$$\leq \frac{1}{2\pi} \int_0^{\infty} |\hat{f}(w)|^2 dw \leq \frac{1}{2\pi}\|f\|^2. \quad (12.6)$$

This inequality also shows that A^+ is a bounded operator whose range is all of $H^2(\mathbb{C}^+)$.

It is also possible to show that

$$\begin{aligned}
&(i) \ \ f^+(x + iy) \to f^+(x + i0) \qquad a.e.,\\
&(ii) \ \|f^+(\cdot + iy) - f^+(\cdot + i0)\| \to 0,
\end{aligned}$$

as $y \to 0$ [S-W, p. 90], so that $H^2(\mathbb{C}^+)$ may be considered a closed subspace of $L^2(\mathbb{R})$.

If we assume that the scaling function $\phi \in \mathcal{S}_r$ as usual, then it certainly is in $L^2(\mathbb{R})$ and has an associated analytic representation $\phi^{\pm}(z)$. The translations of ϕ map into those of ϕ^+, and the subspace V_0 maps into a closed subspace V_0^+ of $H^2(\mathbb{C}^+)$.

Since $\{\phi(x-n)\}$ is an orthonormal basis of V_0, it follows by (12.5) that

$$\|f^+(z) - \Sigma a_n \phi^+(z-n)\|^2 \leq \frac{1}{2\pi}\|f - \Sigma a_n \phi(\cdot - n)\|^2, \qquad (12.7)$$

and hence V_0^+ is the closed linear span of $\{\phi^+(z-n)\}$. Unfortunately $\{\phi^+(z-n)\}$ is not necessarily orthogonal, since for $n \neq 0$

$$\int_{-\infty}^{\infty} \phi^+(x+i0-n)\overline{\phi^+(x+i0)}dx = \frac{1}{2\pi}\int_0^{\infty} |\hat{\phi}(w)|^2 e^{-iwn}dw \neq 0,$$

but it is a <u>basis</u>. Indeed, we need merely show the uniqueness of the expansion $\Sigma a_n \phi^+(z-n)$ to conclude this. This in turn follows if $\hat{\phi}(w) \neq 0$ a.e. on an interval of length 2π in $[0, \infty)$ since if

$$\sum_n a_n \phi^+(z-n) = 0,$$

then

$$\sum_n a_n e^{iwn}\hat{\phi}(w)\chi_{[0,\infty)}(w) = 0.$$

Thus it follows that $\Sigma a_n e^{iwn} = 0$ a.e. in some interval of length 2π. Hence by the uniqueness theorem for Fourier series $a_n = 0$, $n \in \mathbb{Z}$.

The dilation subspaces V_m also map onto subspaces V_m^+ of $H^2(\mathbb{C}^+)$, which are dilations of V_0^+. This gives

Proposition 12.3
Let $\phi \in \mathcal{S}_r$ be a scaling function such that $\hat{\phi}(w) \neq 0$ a.e. on an interval of length 2π in $[0, \infty)$ with associated multiresolution analysis $\{V_m\}$ of $L^2(\mathbb{R})$; then $V_m^+ = A^+ V_m$, $m \in \mathbb{Z}$ satisfies

$$\cdots \subset V_{-1}^+ \subset V_0^+ \subset V_1^+ \subset \cdots \subset V_m^+ \subset \cdots \subset H^2(\mathbb{C}^+),$$

$$\cap_m V_m^+ = \{0\}, \qquad \overline{\cup_m V_m^+} = H^2(\mathbb{C}^+),$$

and $\{\phi^+(2^m x - n)\}$ is a basis of V_m^+.

The same thing can be done for the wavelets themselves; the analytic representation of the mother wavelet is

$$\psi^{\pm}(z) = \frac{1}{2\pi i} \int_{-\infty}^{\infty} \frac{\psi(x)}{x - z} dx$$

and the <u>analytic wavelets</u> $\psi_{mn}^{\pm}$ are obtained by dilation and translation of $\psi^{\pm}$. The subspace W_m with orthogonal basis $\{2^{m/2}\psi(2^m x - n)\}$ is mapped onto W_m^+ with basis $\{\psi^+(2^m z - n)\}$. Moreover $W_m^+ \cap V_m^+ = \{0\}$. Suppose, on the contrary for $m = 0$, that $g^+ \in W_0^+ \cap V_0^+ \neq 0$. Then

$$\hat{g}(w) = \sum_n a_n e^{inw} \hat{\psi}(w) = \sum_n b_n e^{inw} \hat{\phi}(w) \quad \text{a.e.,} \ w > 0 \qquad (12.8)$$

for some $\{a_n\}$ and $\{b_n\}$ not zero. Since

$$\hat{\phi}(w) = m_0 \left(\frac{w}{2}\right) \hat{\phi}\left(\frac{w}{2}\right),$$

and

$$\hat{\psi}(w) = e^{-iw/2} \overline{m_0 \left(\frac{w}{2} + \pi\right)} \hat{\phi}\left(\frac{w}{2}\right),$$

it follows that in some interval of length 4π,

$$\alpha(w) e^{-iw/2} \overline{m_0 \left(\frac{w}{2} + \pi\right)} = \beta(w) m_0\left(\frac{w}{2}\right), \quad \text{a.e.}$$

(where $\alpha(w) = \Sigma a_n e^{iwn}$, $\beta(w) = \Sigma b_n e^{iwn}$). But both sides of (12.8) are periodic (4π) and hence equal a.e. in $\mathbb{R}$. Thus (12.7) holds a.e. in $\mathbb{R}$ and its inverse Fourier transform $g(x) \in W_0 \cap V_0$ and must be zero.

The same calculation works for V_m^+, and hence we may express V_{m+1}^+ as the nonorthogonal direct sum

$$V_{m+1}^+ = V_m^+ \oplus W_m^+.$$

This enables us to rewrite one conclusion of Proposition 11.8 as

$$\bigoplus_m W_m^+ = H^2(\mathbb{C}^+).$$

By the same argument as in V_m^+, $\{\psi^+(2^m x - n)\}_{n \in \mathbb{Z}}$ is a basis of W_m^+, but $\{\psi_{mn}^+\}$ is not a basis of all of $H^2(\mathbb{C}^+)$.

Indeed let $f \in L^2(\mathbb{R})$ be such that $\hat{f}(w) = 0$, $w > 0$ but $\|f\| \neq 0$. Then

$$f = \Sigma b_{mn} \psi_{mn}$$

where not all b_{mn} are zero. But

$$0 = f^+ = \Sigma b_{mn} \psi_{mn}^+,$$

and hence we do not obtain the uniqueness needed.

We can, by an identical procedure, obtain a multiresolution analysis of $H^2(\mathbb{C}^-)$, the Hardy space of analytic functions in the lower half-plane. Moreover we can express each $f \in L^2(\mathbb{R})$ as a linear combination of elements of $H^2(\mathbb{C}^+)$ and $H^2(\mathbb{C}^-)$, which are orthogonal, i.e.,

$$L^2(\mathbb{R}) = H^2(C^+) \bigoplus H^2(C^-) = \bigoplus_m (W_m^+ \oplus W_m^-).$$

By taking $f^+(z) - f^-(\bar{z})$ we are able to find the harmonic function in the upper half-plane with boundary values given by the L^2 function f. It can be expressed in terms of the analytic wavelet expansion as

$$u(x,y) = \sum_{mn} b_{mn}[\psi_{mn}^+(x+iy) - \psi_{mn}^-(x-iy)]$$

$$= \sum_{n=-\infty}^{\infty} a_{mn}[\phi_{mn}^+(x+iy) - \phi_{mn}^-(x-iy)]$$

$$+ \sum_{k=m}^{\infty}\sum_{n=-\infty}^{\infty} b_{kn}[\psi_{kn}^+(x+iy) - \psi_{kn}^-(x-iy)].$$

Thus one of the attractive features of wavelet theory, the decomposition algorithm, is present here also. The finest scale of interest is chosen and the series

$$u_m(x,y) = \sum_n a_{mn}[\phi_{mn}^+(x+iy) - \phi_{mn}^-(x-iy)]$$

is used to approximate $u(x,y)$. The error in going to the next finer scale is given by the wavelet terms

$$\sum_n b_{m-1,n}[\psi_{m-1,n}^+(x+iy) - \psi_{m-1,n}^-(x-iy)].$$

Since the a_{mn} and b_{mn} are the same as before they are still related by (3.20).

12.5   Analytic solutions to dilation equations

We have seen in Chapter 5 that dilation equation (3.3) with rapidly decreasing coefficients

$$\phi(t) = \sqrt{2}\sum_k c_k \phi(2t - k)$$

always has a distribution solution in $\mathcal{S}'$. By taking the analytic representation of the solution we obtain a function $\phi^+(z)$, which satisfies (3.3) in the upper half-plane, and similarly in the lower half-plane.

By analogy to the last section we can define subspaces T_m^+ (T_m^-) consisting of analytic functions in the upper (lower) half-plane. For example, we have

$$T_0^+ := \left\{ f(z) = \sum_n a_n \phi^+(z - n) \mid a_n = O(|n|^p) \right\}.$$

However, we cannot do much with this since we cannot characterize $\overline{\cup_m T_m}$ (Section 5.2) except that it is a subspace of $\mathcal{S}'$.

Example 12.1
Let $c_0 = \sqrt{2}$, $c_k = 0$, $k \neq 0$; then the solution to the dilation equation in $\mathcal{S}'$ is $\phi(t) = \delta(t)$. The corresponding analytic solution in the upper half-plane is $\phi^+(z) = -\frac{1}{2\pi i z}$. If $f \in T_m$, i.e., if $f = \Sigma a_n \delta(2^m t - n)$ then

$$f^+(z) = -\frac{1}{2\pi i}\sum_n a_n (2^m z - n)^{-1}.$$

▯

The dilation equation (3.3) also has a solution composed of harmonic functions $u(x, y)$ in the upper half-plane, i.e.,

$$u(x, y) = \sqrt{2}\sum_k c_k u(2x - k, y).$$

This is again given by $u(x, y) = \phi^+(x + iy) - \phi^-(x - iy)$.

The construction of $\phi^+(x)$ and $u(x, y)$ can proceed in two ways. One is first to construct the solution in $\mathcal{S}'$ by starting with $\phi_0(t) = \delta(t)$ and letting

$$\phi_n(t) = \sqrt{2}\Sigma c_k \phi_{n-1}(2t - k), \qquad n = 1, 2, \cdots \qquad (12.9)$$

and then finding the analytic representation of the limit in $\mathcal{S}'$. The other is to start with $\phi_0^+(z) = -1/2\pi i z$ and use (12.9) to find $\phi_n^+(z)$ directly and then taking the pointwise limit in the upper half-plane. The latter is more amenable to computation.

12.6 Analytic representation of distributions by wavelets

In order to represent an element of $\mathcal{S}'$ by series of analytic and harmonic wavelets, we impose conditions on the scaling function again and assume $\phi \in \mathcal{S}_r$. The spaces T_m^+ and T_m^- are now composed of analytic functions in the upper and lower half-plane respectively whose boundary functions are continuous functions of polynomial growth (Chapter 5).

Since $\overline{\cup T_m} = \mathcal{S}_r'$ we might expect to obtain analytic representations in terms of wavelets for each $f \in \mathcal{S}_r'$,

$$f^+(z) = \sum_{n=-\infty}^{\infty} a_n \phi^+(z - n)$$

$$+ \sum_{m=0}^{\infty} \sum_{n=-\infty}^{\infty} b_{mn} 2^{m/2} \psi^+(2^m z - n),$$

but the first series may not converge so we proceed differently.

The analytic representation represents a continuous map from $\mathcal{S}_r'$ to a corresponding space of analytic functions, i.e., if $f_m \to f$ in $\mathcal{S}_r'$ then $f_m^+(z) \to f^+(z)$ uniformly on bounded subsets of the upper half-plane. This of course requires that the proper analytic representation be chosen. One such choice is as follows: any $f \in \mathcal{S}_r'$ (or $\mathcal{S}'$) may be given as a finite order derivative of a continuous function of polynomial growth, say $F(x)$

$$f = D^k F.$$

An analytic representation of F is given by

$$F^+(z) = \frac{1}{2\pi i} \int_{-\infty}^{\infty} \frac{F(x)}{x - z} \left(\frac{z^2 + 1}{x^2 + 1}\right)^p dx$$

where p is sufficiently large to assure that $F(x)/(x^2+1)^p \in L^2(\mathbb{R})$. Then $f^+(z)$ is defined by

$$f^+(z) = D_z^k F^+(z).$$

The approximation to f from the space T_m is the function

$$f_m(x) = (f, q_m(\cdot, x)).$$

Thus if the analytic representation is taken as above

$$f_m^+(z) \to f^+(z).$$

We may also express f_m as

$$f_m = f_0 + f_m - f_0 = f_0 + \sum_{k=0}^{m-1} \sum_{n=-\infty}^{\infty} b_{kn} \psi_{kn},$$

and if the inner series converges we have

$$f_m^+(z) - f_0^+(z) = \sum_{k=0}^{m-1} \sum_{n=-\infty}^{\infty} b_{kn} \psi_{kn}^+(z) + g_m(z) \qquad (12.10)$$

where $g_m(z)$ is an entire function.

Lemma 12.2
Let $b_n = O(|n|^{r-\epsilon})$, $\epsilon > 0$; then

$$\sum_{n=-\infty}^{\infty} b_n \psi^+(z-n)$$

converges uniformly on compact subsets of the open upper half-plane.

PROOF   The proof is based on the moment property (Chapter 3)

$$\int_{-\infty}^{\infty} x^p \psi(x) dx = 0, \qquad p = 0, 1, \cdots, r.$$

Then

$$z^p \psi^+(z) = \frac{1}{2\pi i} \int_{-\infty}^{\infty} \frac{z^p - x^p}{x - z} \psi(x) dx + \frac{1}{2\pi i} \int_{-\infty}^{\infty} \frac{x^p \psi(x)}{x - z} dx$$

$$= \frac{1}{2\pi i} \int_{-\infty}^{\infty} \frac{x^p \psi(x)}{x - z} dx \qquad (12.11)$$

holds for $p \leq r + 1$. Hence $|z^{r+1}\psi^+(z)|$ is uniformly bounded for Im $z \geq \epsilon > 0$, and the conclusion follows. Thus if the coefficients in (12.10) satisfy the hypothesis, the series converges. $\quad\square$

Lemma 12.3
Let $f \in \mathcal{S}'_p$, $\phi \in \mathcal{S}_r$, $p < r$; then the wavelet coefficients of f satisfy

$$|b_{mn}| = |(f, \psi_{mn})| \leq C(|n| + 2^m)^{p+\frac{1}{2}}. \tag{12.12}$$

PROOF We use the fact that each $f \in \mathcal{S}'_p$ is of the form

$$f = D^p[(x^2 + 1)^j \mu]$$

for some integer j and finite measure μ on $\mathbb{R}$ [Ze, p.109].
   Also $\psi(x)$ satisfies the inequality

$$|\psi^{(k)}(x)| \leq \frac{C_j}{(1 + |x|)^j}, \qquad k = 0, 1, \cdots, r, \;\; j \geq 0,$$

since $\psi \in \mathcal{S}_r$. Hence after integrating by parts p times we obtain

$$\begin{aligned}
|b_{mn}| &\leq \int_{-\infty}^{\infty} (1 + x^2)^j |\psi_{mn}^{(p)}(x)| d|\mu| \\
&\leq \int_{-\infty}^{\infty} (1 + |x|)^{2j} \frac{C_{2j} 2^{m/2+pm}}{(1 + |2^m x - n|)^{2j}} d|\mu| \\
&\leq (1 + |n|2^{-m})^{2j} C_{2j} 2^{m/2+pm} \int_{-\infty}^{\infty} d|\mu|.
\end{aligned}$$

The inequality (12.12) follows by taking $2j = p + \frac{1}{2}$. $\quad\square$
   We now combine the two lemmas to obtain the convergence we need.

Lemma 12.4
Let $f \in \mathcal{S}'_p$, $\phi \in \mathcal{S}_r$, $p < r$, $b_{mn} = (f, \psi_{mn})$; then the series

$$\sum_{m=0}^{\infty} \sum_{n=-\infty}^{\infty} b_{mn} \psi_{mn}^+(z)$$

converges uniformly on the half-plane Im $z \geq 1$.

PROOF   That the inner series converges is immediate, but we have to be a little more careful with the outer series. By (12.11) we find that

$$|(2^m z - n)^{r+1} \psi_{mn}^+(z)| \leq \frac{1}{2\pi} \int \frac{|x^{r+1} \psi(x)| 2^{m/2} dx}{|x - 2^m z + n|}$$

$$\leq \frac{1}{2\pi} 2^{-m/2} \int |x^{r+1} \psi(x)| dx$$

for Im $z \geq 1$. Hence the inner series satisfies

$$\sum_n |b_{mn} \psi_{mn}^+(z)|$$

$$\leq C' \sum_n \frac{(|n| + 2^m)^{p+\frac{1}{2}}}{|2^m z - n|^{r+1}} 2^{-m/2}$$

$$\leq C' \sum_n \left\{ \frac{(|n| + 2^m)^2}{(n - 2^m \operatorname{Re} z)^2 + 2^{2m}} \right\}^{\frac{p}{2}+\frac{1}{4}} 2^{-m/2} \left[(n - 2^m \operatorname{Re} z)^2 + 2^{2m} \right]^{\frac{p}{2}-\frac{r}{2}-\frac{1}{4}}$$

since $|2^m z - n|^2 = (n - 2^m \operatorname{Re} z)^2 + (2^m \operatorname{Im} z)^2$.

In order to get a bound on the expression in parentheses we observe that

$$\frac{(a+1)^2}{(a-b)^2 + 1} \leq (b+1)^2 + 1 \qquad \text{for } a, b > 0.$$

By taking $a = |n| 2^{-m}$ and $b = \pm \operatorname{Re} z$ we obtain a bound that depends only on z. The series is thus dominated by the remaining factor

$$2^{-m/2} \sum_{n=-\infty}^{\infty} \left[(n - 2^m \operatorname{Re} z)^2 + 2^{2m} \right]^{\frac{p}{2}-\frac{r}{2}-\frac{1}{4}}$$

$$\leq 2^{-m/2} C'' \sum_{n=-\infty}^{\infty} (n^2 + 2^{2m})^{-3/4} \leq C''' 2^{-m/2}.$$

Thus the terms in the outer series are dominated by a geometric series and converge uniformly on the half-plane.   □

We now return to the analytic representation. By taking the limit in (12.10) as $m \to \infty$, we find

$$f^+(z) = f_0^+(z) + \sum_{k=0}^{\infty} \sum_{n=\infty}^{\infty} b_{kn} \psi_{kn}^+(z) + g_\infty(z), \qquad (12.13)$$

where $g_\infty(z) = \lim\limits_{m\to\infty} g_m(z)$ is also an entire function. We can drop this g_∞ in (12.13) since an analytic representation plus an entire function is again an analytic representation.

THEOREM 12.4
Let $f \in S'_p$, $\phi \in S_r$, $p < r$, and let $b_{mn} = \langle f, \psi_{mn} \rangle$, $m = 0, 1, 2, \cdots$, $n = 0, \pm1, \pm2, \cdots$, be the wavelet coefficients of f. Then an analytic representation of f_m, the projection of f on T_m, is given by

$$f_m^\pm(z) = f_0^\pm(z) + \sum_{k=0}^{m} \sum_{n=-\infty}^{\infty} b_{kn} \psi_{kn}^\pm(z) \qquad (12.14)$$

and $f_0^\pm(z)$ is any analytic representation of f_0. Also $f_m^\pm(z)$ converges to an analytic representation of f for Im $z \geq 1(+)$ or Im $z \leq -1(-)$ as $m \to \infty$.

We have only worked out the convergence for f^+ but that for f^- is parallel. Also we have only shown the convergence for $|\text{Im } z| \geq 1$, but we may continue the sum of the series analytically to the upper (lower) half-plane.

In case f is an ordinary function in L^2, then the results are straightforward since the analytic representation may be given directly by the scaling function series

$$f_m^\pm(z) = \sum_{n=-\infty}^{\infty} a_{mn} \phi_{mn}^\pm(z) \qquad (12.15)$$

which is uniformly convergent for $|\text{Im}(z)| > \epsilon > 0$.

12.7 Wavelets analytic in the entire complex plane

In the previous sections we have been concerned primarily with the analytic representation of wavelets which are not themselves analytic functions. But certain types of wavelets are already analytic and in fact are entire functions. These include the Meyer wavelets mentioned in Chapter 2, example 7.

The Fourier transforms of the associated scaling functions of the Meyer wavelets have compact support in an interval $[-\frac{4}{3}\pi, \frac{4}{3}\pi]$. Hence the scaling functions themselves are entire functions of exponential type. The same is true of series of their translates.

Proposition 12.4
Let $\phi(x)$ be a Meyer-type scaling function such that $\hat{\phi} \in C^{p+2}$ and let $a_n = O(|n|^p), p > 0$, then

$$f_0(z) = \sum_{n=-\infty}^{\infty} a_n\phi(z-n), \quad z \in \mathbb{C} \tag{12.16}$$

is an entire function of exponential type whose restriction to the real axis is a function of polynomial growth.

The proof follows by taking the Fourier transform to get

$$\hat{f}_0(w) = \sum_{n=-\infty}^{\infty} a_n e^{iwn}\hat{\phi}(w).$$

The series $\sum_{n=-\infty}^{\infty} a_n e^{iwn}$ converges to a periodic distribution for which $\hat{\phi}$ is a multiplier. The product has compact support in S' and hence by the Schwartz-Paley-Wiener theorem its inverse Fourier transform is an entire function of exponential type with polynomial growth on the real axis [Yo]. □

If we turn to the wavelet expansions rather than the scaling functions we get other results and can in fact characterize the analytic functions by their expansion coefficients. The Fourier transform of the mother wavelet has its support in the union of the intervals $[-\frac{8}{3}\pi, -\frac{2}{3}\pi] \cup [\frac{2}{3}\pi, \frac{8}{3}\pi]$. The full wavelet expansion

$$f(x) = \sum_{k=-\infty}^{\infty} \sum_{n=-\infty}^{\infty} b_{kn}\psi_{kn}(x) \tag{12.17}$$

converges to an entire function for certain values of the coefficients.

Proposition 12.5
[Za-W] Let ψ be a Meyer type wavelet; let $\{b_{mn}\}$ be a sequence of complex numbers such that

$$\sum_{n=-\infty}^{\infty} |b_{mn}| \le C\exp(-\gamma 2^{(\alpha+1)m}) \quad \text{for some } \alpha, \gamma, C > 0.$$

Then $f(x)$ given by (12.17) is an entire function.

The proof involves finding bounds on the derivatives of $f(x)$ given by derivatives of the series(12.17).

It is also possible to characterize the coefficients of these series when the function is analytic in a strip about the real axis.

THEOREM 12.5
[Za-W] Let ψ be a Meyer type wavelet; let $\{b_{mn}\}$ be a sequence of complex numbers and let $f(x)$ be given by (12.17). Then $f(z)$ is analytic in the strip $\text{Im}|f(z)| < d$ if and only if the coefficients satisfy

$$\sum_{n=-\infty}^{\infty} |b_{mn}| \leq \varepsilon_m \exp(-\frac{16}{3}\pi d2^m) \quad \text{for some } \varepsilon_m > 0, m \in \mathbb{Z},$$

where $\sum_{m=-\infty}^{\infty} \varepsilon_m < \infty$.

The proof of this, which is rather complicated, may be found in [Za-W].

12.8 Problems

1. Find the Fourier series of $\delta'^*(x)$ (the periodic extension of $\delta'(x)$), and use it to find its analytic representation.

2. Show that the analytic representation in Problem 1 satisfies the conclusion of Corollary 11.3 for $x \neq 0$, $x \in (-\pi, \pi)$, $\gamma = 0$.

3. Find the analytic representation as well as the Hermite series of $\delta'(x)$, $x \in \mathbb{R}$.

4. Show that for Problem 3 the series for a_n, $\sum_{k=0}^{\infty} i^{(n-k)} c_k \alpha_{nk}$ diverges for $n = 0$. (The exact values of c_k and α_{ok} may be found by using the recurrence formulae for the Hermite functions. For α_{ok} first use $\int_0^{\infty}(2k+1)h_k h_o = \int_0^{\infty}(x^2 - D^2)h_k h_o$ and integrate by parts twice.)

5. Find the analytic representation $\phi^\pm(z)$ of the scaling function of the Haar wavelet and show directly that $\phi^\pm(z - n)$ is the analytic representation of $\phi(x - n)$.

6. Find the analytic representation $\phi^\pm(z)$ of the scaling function of the Shannon wavelet by using its Fourier transform. Find a sequence $\{a_n\} \neq 0$ such that $\sum\limits_n a_n \phi^+(z - n) = 0$. (Use the Fourier coefficients of a function that has support in $(-\pi, 0)$.)

7. Repeat Problem 6 for the Meyer wavelets (except you will not find a closed form for $\phi^\pm$).

8. Show that the series

$$\sum a_n \phi^+(z - n)$$

diverges when ϕ is the Shannon scaling function, and a_n are the coefficients of $F(x) = x$.

Chapter 13

Orthogonal Series in Statistics

Orthogonal series form an important tool in many branches of statistics. They are used in probability density estimation, in spectral density estimation of time series, in linear and nonparametric regression, and in Bayes and empirical Bayes estimation, among others.

We can only present a sampling of these applications. We choose several for which wavelets can also be used. We shall concentrate on methods used in probability density estimation and only treat the others (threshold methods, regression, time series, mixtures) briefly.

The density estimation problem consists of trying to estimate properties of a probability density function $f(x)$ from an independent sample $X_1, X_2, \cdots, X_N$ drawn from a distribution with this density. For example, the mean and variance of the sample may be used to estimate the mean and variance of $f(x)$. If the sample is large enough, the density itself can be estimated. One way of doing so is by means of the empiric distribution

$$f^*(x) := \frac{1}{N} \sum_{i=1}^{N} \delta(x - X_i). \qquad (13.1)$$

This is an unbiased estimator for continuous $f(x)$. This means that the expected value of $f^*(x)$ is $f(x)$, which is immediate from the properties of expected values,

$$E(f^*(x)) = \frac{1}{N} \sum_{i=1}^{N} E(\delta(x - X_i))$$

$$= \frac{1}{N} \sum_{i=1}^{N} \int_{-\infty}^{\infty} \delta(x - y_i) f(y_i) dy_i = \frac{1}{N} \sum_{i=1}^{N} f(x) = f(x).$$

The indefinite integral of $f^*(x)$ is the empiric distribution function

$F^*(x)$. By the law of the iterated logarithm $F^*(x) \to F(x)$, the distribution function given by the integral of $f(x)$, almost surely (almost everywhere with respect to the measure induced by $f(x)$) [Ki]. The rate of convergence is $O\left(\left(\frac{\log \log N}{N}\right)^{1/2}\right)$.

Unfortunately the estimator (13.1) is very irregular even when $f(x)$ is very smooth. In order to obtain a smooth estimator a number of different procedures have been introduced (and widely studied). The simplest is the histogram estimator composed of piecewise constant functions. It is more regular than $f^*(x)$ but still not smooth; in terms of wavelet subspaces it is defined as

$$\tilde{f}_m(x) := \frac{1}{N} \sum_{i=1}^{N} q_m(X_i, x) \tag{13.2}$$

where $q_m(x, y)$ is the reproducing kernel of V_m in prototype 1. The scaling function $\phi(x) = \chi_{[0,1)}(x)$, and $q(x,y) = \phi(x - [y])$, in this case. Since $\phi(2^m X_i - [2^m x])$ counts the number of X_i's in the interval $[k2^{-m}, (k+1)2^{-m})$ containing x, this is the same as the usual definition of histogram with nodes at $2^{-m}k$.

Other traditional smoother estimators are based on kernels [P], on Fourier series [K-T], on splines [Wh], and on general delta-sequences [W-B]. Some of these estimators converge in the mean square sense for all continuous densities, and some converge more rapidly for C^∞ densities, but none do both. Smooth estimators based on wavelets do both. This comes out of the properties of delta sequences of wavelets discussed in Chapter 8. Wavelet estimators are also more amenable to non-linear threshold methods in which some coefficients are reduced or eliminated.

We first explore the properties of a Fourier series and a Hermite series estimator and then come back to wavelet estimators. We shall explore wavelet estimators based both on the reproducing kernels as in (13.2) and on positive kernels arising from the positive scaling functions $\rho^r(t)$ of Chapter 8.

13.1   Fourier series density estimators

The Fourier series estimator of a continuous density function $f(x)$ is obtained by approximating the delta functions in (13.1) by the delta sequence consisting of the Dirichlet kernel

$$D_m(t) = \frac{\sin\left(m + \frac{1}{2}\right)t}{2\pi \sin \frac{1}{2}t} \tag{13.3}$$

or by the Fejer kernel

$$F_m(t) = \frac{\sin^2(m+1)t/2}{2\pi(m+1)\sin^2 t/2}. \tag{13.4}$$

We have seen in Chapter 4 that both, when multiplied by $\chi_{[-\pi,\pi]}(t)$, are delta sequences and that $\{F_m\}$ is a positive delta sequence. The sorts of densities that can be estimated by using them are ones with support in $[-\pi, \pi]$. The associated estimators are

$$\tilde{f}_m^{(1)}(x) \ : \ = \frac{1}{N}\sum_{i=1}^{N} D_m(x - X_i)$$

$$= \sum_{n=-m}^{m} \tilde{c}_n\, e^{inx}, \qquad |x| \le \pi,\ m = 0, 1, \cdots \tag{13.5}$$

where

$$\tilde{c}_n = \frac{1}{2\pi N}\sum_{i=1}^{N} e^{-inX_i}, \quad n = 0, \pm 1, \pm 2, \cdots,$$

and

$$\tilde{f}_m^{(2)}(x) \ : \ = \frac{1}{N}\sum_{i=1}^{N} F_m(x - X_i), \tag{13.6}$$

$$= \sum_{n=-m}^{m} \tilde{c}_n\left(1 - \frac{|n|}{2m+1}\right)e^{inx}, \quad |x| \le \pi,\ m = 0, 1, \cdots.$$

We would like the estimator of the density $f(x)$ to converge to it in some sense as the sample size N gets larger and as $m \to \infty$. One such sense is mean square convergence given by the requirement that

$$E[\tilde{f}_m(x) - f(x)]^2 \to 0.$$

This expression may be split up into a variance and a bias part

$$E[\tilde{f}_m(x) - f(x)]^2 = E[\tilde{f}_m(x) - f_m(x)]^2 + [f_m(x) - f(x)]^2$$
$$= \nu_m(x) \ + \beta_m^2(x) \tag{13.7}$$

where $f_m(x) = E(\tilde{f}_m(x))$. We have already considered the bias part in Chapter 4. This is just the difference between $f(x)$ and the partial sums (or the Cesàro means) of its Fourier series.

This is clear from (13.5) since

$$E(\tilde{c}_n) = \frac{1}{2\pi N} \sum_{i=1}^{N} \int_{-\pi}^{\pi} e^{-inx} f(x)dx = \frac{1}{N} \sum_{i=1}^{N} c_n = c_n.$$

$$\nu_m(x) = E\left[\tilde{f}_m^2(x) - f_m^2(x)\right]$$

$$= \frac{1}{N} \left[\int_{-\pi}^{\pi} \delta_m^2(x,t)f(t)dt - \left(\int_{-\pi}^{\pi} \delta_m(x,t)f(t)dt\right)^2\right]$$

$$\leq \frac{1}{N}\|f\|_\infty \|\delta_m(x,\cdot)\|_2^2, \tag{13.8}$$

where δ_m is either (13.3) or (13.4). In both cases $\|\delta_m(x,\cdot)\|_2^2 = 0(m)$, which gives a bound

$$\nu_m(x) \leq \frac{c_1 m}{N}, \qquad m, \ N = 1,2,\cdots. \tag{13.9}$$

Since $f(x)$, being a density, cannot be periodic whereas its Fourier series is, we cannot expect the bias $\beta_m(x)$ to converge to zero very rapidly unless the periodic extension of $f(x)$ belongs to the Sobolev space $H_{2\pi}^\alpha$. This will happen, e.g., if the original $f \in C^\alpha$ and $f^{(k)}(\pm\pi) = 0$, $k = 0, 1, \cdots, \alpha - 1$.

Proposition 13.1
Let f be a continuous density function with support on $[-\pi, \pi]$, then

(i) $E\left|\tilde{f}_m^{(2)}(x) - f(x)\right|^2 \to 0$ *uniformly as* $N, m \to \infty$, *and* $\frac{m}{N} \to 0$.

If, moreover, $f \in C^\alpha$, $\alpha > 1$ and $f^{(k)}(\pm\pi) = 0$, $k = 0, 1, \cdots, \alpha - 1$, then

(ii) $E\left|\tilde{f}_m^{(1)}(x) - f(x)\right|^2 = O(N^{-1+1/(2\alpha-1)})$, $m = \left[N^{\frac{1}{2\alpha-1}}\right]$

(iii) $E\left|\tilde{f}_m^{(2)}(x) - f(x)\right|^2 = O(N^{-2/3})$, $m = [N^{1/3}]$.

PROOF   For (i) we need merely invoke the uniform convergence of the Cesàro means (Chapter 4) to deduce that $\beta_m^2(x) \to 0$ uniformly. The same is true for $\nu_m(x)$ by (13.9).

For (ii) we use the crude estimate $|c_n n^\alpha| \leq c$ and hence

$$|\beta_m(x)| = \left| \sum_{|n|>m} c_n e^{inx} \right| \leq \sum_{|n|>m} \frac{c}{n^\alpha} = O(m^{1-\alpha}).$$

Thus by taking $m = [N^{1/(2\alpha-1)}]$ the bias and variance terms are balanced and the mean square error is

$$O(m^{2-2\alpha}) = O(m/N) = O(N^{(2-2\alpha)/(2\alpha-1)}).$$

In the case of (iii) the rate of convergence of the bias is $O(m^{-1})$ no matter how smooth f is [Z, p.122]. The error is given by the same calculation as in (ii). □

REMARK 13.1   If we only know that f is continuous, then

$$E|\tilde{f}_m^{(1)}(x) - f(x)|^2$$

may not converge to zero. This is because the bias does not converge to zero for some continuous function as remarked in Chapter 4. In this case, $\tilde{f}_m^{(2)}$ is the better estimator. On the other hand, if $f \in C^\infty$ the convergence rate for $\tilde{f}_m^{(1)}$ approaches $O(N^{-1})$ while for $\tilde{f}_m^{(2)}$ is never better than $O(N^{-2/3})$. Another sense in which $\tilde{f}_m^{(2)}$ is better is in the fact that it itself is a density, while $\tilde{f}_m^{(1)}$, since it may take negative values, is not. ■

13.2   Hermite series density estimators

Fourier series estimators because of their periodicity are appropriate only when the density has compact support. In general it does not, and a method that gives an estimator on the entire real line is preferred. One such method is based on the Hermite functions considered in Chapter 6, $\{h_n\}$ [Sch], [W14]. They constitute a complete orthogonal system in $L^2(\mathbb{R})$.

The estimator is similar to the Fourier series estimator; it is

$$\tilde{f}_m(x) = \sum_{n=0}^{m} \tilde{a}_n h_n(x) \tag{13.10}$$

where

$$\tilde{a}_n = \frac{1}{N} \sum_{i=1}^{N} h_n(X_i).$$

In order to find the rate of convergence we use the fact (Chapter 6) that

$$|h_n(x)| \le C(n+1)^{-1/12}, \qquad n = 0, 1, \cdots .$$

The coefficients are seen to satisfy

$$E(\tilde{a}_n - a_n)^2 = \frac{1}{N^2} \sum_{i=1}^{N} E(h_n^2(X_i))$$

$$+ \frac{1}{N^2} \sum_{i=1}^{N} \sum_{j \ne i} E(h_n(X_i)h_n(X_j))$$

$$- \frac{2}{N} \sum_{i=1}^{N} E(h_n(X_i))a_n + a_n^2$$

$$= \frac{1}{N^2} \sum_{i=1}^{N} E(h_n^2(X_i)) - a_n^2$$

$$\le \frac{C^2}{N}(n+1)^{-1/6} \qquad\qquad (13.11)$$

since $a_n = \int_{-\infty}^{\infty} f(x)h_n(x)dx = E(h_n(X_i))$.
The mean square error of this estimator may now be calculated.

Proposition 13.2
Let f be a density such that $(x - D)^r f \in L^2(R)$, for $r \ge 1$; then the mean square error of the estimator given by (13.10) satisfies

$$E[\tilde{f}_m(x) - f(x)]^2 = O\left(N^{-1+\frac{6}{6r+1}}\right), \quad m = [N^{\frac{6}{6r+1}}]$$

uniformly for $x \in R$.

PROOF   The variance of the estimator is given by

$$\nu_m(x) = \sum_{n=0}^{m} E(\tilde{a}_n - a_n)^2 h_n^2(x) \le \frac{C^4}{N} \sum_{n=0}^{m} (n+1)^{-\frac{1}{3}}$$

$$\le C_1 \frac{m^{2/3}}{N}$$

by (13.11). The bias estimate uses the fact that $(x + D)h_n = \sqrt{2n}\,h_{n-1}$ (see Chapter 6), which implies that for f satisfying the hypothesis,

$$a_n \quad \sqrt{2n}\sqrt{2n+2}\cdots\sqrt{2n+2r-2}$$

$$= \int_{-\infty}^{\infty} f(x)(x+D)^r h_{n+r}(x)dx$$

$$= \int_{-\infty}^{\infty} (x-D)^r f(x)h_{n+r}(x)dx, \quad n = 0, 1, \cdots.$$

Hence by Bessel's inequality

$$\sum_n a_n^2 n^r < \infty$$

and

$$\beta_m^2(x) = \left| \sum_{n=m+1}^{\infty} a_n h_n(x) \right|^2$$

$$\leq \sum_{n=m+1}^{\infty} a_n^2 n^r \sum_{n=m+1}^{\infty} n^{-r} h_n^2(x) = O\left(m^{\frac{5}{6}-r}\right).$$

We then combine $\nu_m(x)$ and $\beta_m^2(x)$ as in the last section to obtain the conclusion. $\square$

Thus the rate of mean square convergence approaches $O(N^{-1})$ as $r \to \infty$. However, just as with the Fourier series estimator, it is possible to find continuous functions for which the bias does not tend to 0. The hypothesis is satisfied for gamma densities for certain values of r.

13.3   The histogram as a wavelet estimator

In this section we consider an estimator based on the Haar wavelets (Prototype I of Chapter 1). We shall see that this estimator is just the traditional histogram.

The estimator in (13.2) has an alternate expression

$$\tilde{f}_m(x) = \sum_{n=-\infty}^{\infty} \tilde{a}_{mn} 2^{m/2}\phi(2^m x - n),$$
$$\tilde{a}_{mn} = \frac{1}{N}\sum_{i=1}^{N} 2^{m/2}\phi(2^m X_i - n). \tag{13.12}$$

In this form it can be truncated to a finite sum

$$\tilde{f}_{mk}(x) = \sum_{n=-k}^{k} \tilde{a}_{mn} 2^{m/2} \phi(2^m x - n), \qquad (13.13)$$

which has many of the same convergence properties.

The reproducing kernels $\{q_m(x, y)\}$ constitute a quasi-positive delta sequence that satisfies

$$q_m^2(x, y) = 2^{2m} q^2(2^m x, 2^m y) \le 2^{2m}.$$

This is easy to show in this case since $\phi(x)$ and $q_m(x, y)$ are just characteristic functions of intervals (indicator functions in statistics terminology).

The estimator $\tilde{f}_m$, as opposed to f^*, is not an unbiased estimator of f. However since the bias is

$$\beta_m(x) = E\tilde{f}_m(x) - f(x)$$
$$= \int q_m(y, x) f(y) dy - f(x), \qquad (13.14)$$

and this converges to 0 for continuous $f(x)$ by Proposition 1.5, it is *asymptotically unbiased.*

The variance of $\tilde{f}_m$ is defined by

$$\nu_m(x) := E\left[\tilde{f}_m(x) - E\tilde{f}_m(x)\right]^2,$$

which again is easy to show, and satisfies

$$\nu_m(x) \le \frac{1}{N} \int q_m^2(y, x) f(y) dy \le \frac{2^{2m}}{N}.$$

The two expressions may be combined into the mean square error, as in the last two sections,

$$MSE = E[\tilde{f}_m(x) - f(x)]^2 = \nu_m(x) + \beta_m^2(x),$$

which if $m = O(\log N)$ converges to 0 uniformly on bounded sets for continuous $f(x)$.

Another useful measure of error is the *integrated mean square error,*

$$IMSE = \int_{-\infty}^{\infty} E[\tilde{f}_m(x) - f(x)]^2 dx,$$

which also converges to 0 if $m = O(\log N)$. The integral of ν_m satisfies

$$\int_{-\infty}^{\infty} \nu_m(x)dx \leq \frac{1}{N} \int_{-\infty}^{\infty} \int_{-\infty}^{\infty} q_m^2(y,x)f(y)dydx$$

$$= \frac{1}{N} \int_{-\infty}^{\infty} q_m(y,y)f(y)dy \leq \frac{2^m}{N}$$

while that of β_m^2 is

$$\int_{-\infty}^{\infty} \beta_m^2(x)dx = \|f_m - f\|^2,$$

where $f_m = E\tilde{f}_m = \int q_m(y,\cdot)f(y)dy$. Since the MRA $\{V_m\}$ is dense in $L^2(\mathbf{R})$, and f_m is the projection of f on the space V_m, we see that this latter integral converges to 0 for all $f \in L^2(\mathbf{R})$.

Both of these properties are expected of any reasonable density estimator. Another advantage of the wavelet approach is in the calculation of the coefficients in (13.12), which may be expressed as

$$\tilde{a}_{mn} = \frac{2^{m/2}}{N} \Big/ \# \{X_i \in [n2^{-m}, (n+1)2^{-m})\}. \qquad (13.15)$$

We choose $m = \overline{m}$ initially at the finest scale of interest, which would certainly be no finer than that required to put all X_i in different intervals.

We then consider successively coarser choices of scale until we get the coarsest scale of interest, say $m = \underline{m}$ when all X_i belong to the same interval. However we do not use (13.15) to calculate the new $\tilde{a}_{mn}$ but rather use the decomposition algorithm of Mallat. This involves the projection of V_m onto V_{m-1} and W_{m-1} and is given by

$$a_{m-1,k} = 2^{-1/2}(a_{m,2k} + a_{m,2k+1})$$
$$b_{m-1,k} = 2^{-1/2}(a_{m,2k} - a_{m,2k+1}). \qquad (13.16)$$

We may iterate it as often as we want but in practice would stop after we get to $\underline{m}$. In fact we can stop long before it is reached, since the b_{mk}, heretofore unused, give us a measure of the error at each stage. Here's how it works: the finest scale $\overline{m}$ gives us an integrated mean square error

$$e_{\overline{m}} = \int E[\tilde{f}_{\overline{m}}(x) - f(x)]^2 dx,$$

which we cannot improve practically by increasing m. Such an increase would increase the variance without a corresponding reduction in bias.

This may be expressed as

$$
\begin{aligned}
e_m &= \int \mathrm{Var}\, \tilde{f}_m(x)dx + \int \mathrm{Bias}^2 \tilde{f}_m(x)dx \\
&= \int E \frac{1}{N} \sum_{i=1}^{N} q_m^2(x, X_i)dx - \frac{1}{N}\int f_m^2(x)dx + \int (f(x) - f_m(x))^2 dx \\
&= \frac{1}{N}\left[\int q_m(x, x)f(x)dx - \sum_k a_{m,k}^2\right] + \sum_{j=m}^{\infty}\sum_k b_{jk}^2 \\
&= \frac{1}{N}\left[\int 2^m \phi(2^m x - [2^m x])f(x)dx - \sum_k a_{m-1,k}^2 - \sum_k b_{m-1,k}^2\right] \\
&\quad + \sum_{j=m}^{\infty}\sum_k b_{jk}^2 ,
\end{aligned}
$$

where we have used the orthogonality to express

$$
\sum_k a_{m,k}^2 = \sum_k a_{m-1,k}^2 + \sum_k b_{m-1,k}^2 .
$$

Since $\phi(x - [x]) = 1$, the integral is just 2^m, and the change in error in going from one scale to the next coarser scale is

$$
\begin{aligned}
e_m - e_{m-1} &= \frac{1}{N}\left[2^m - \sum_k a_{m-1,k}^2 - \sum_k b_{m-1,k}^2 - 2^{m-1}\right.\\
&\quad \left. + \sum_k a_{m-1,k}^2\right] - \sum_k b_{m-1,k}^2 \\
&= \frac{2^{m-1}}{N} - \frac{N+1}{N}\sum_k b_{m-1,k}^2 .
\end{aligned}
$$

That is, the change in error depends only on the wavelet coefficients $b_{n,k}$. We proceed by starting with $\overline{m}$ and checking how much the error changes in going from $\overline{m}$ to $\overline{m} - 1$. Then we iterate this as long as the error is small until we find an m at which the error increases by a larger amount. This will then be our coarsest scale.

In practice we cannot know these coefficients exactly but must use the estimate given by (13.12) for a_{mn}. This in turn will give us the estimated $\tilde{b}_{m-1,n}$ and $\tilde{e}_m$.

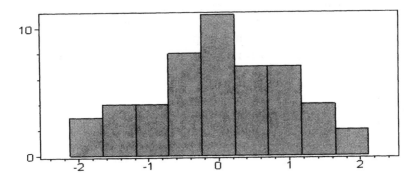

FIGURE 13.1
Histogram for data in Example 13.1.

Example 13.1
A random sample of size 25 based on a $N(0,1)$ distribution was generated. The finest scale was chosen to be $\overline{m} = 3$, and the values of the estimated coefficients $\tilde{a}_{3,k}$ were calculated. These were used to calculate the wavelet coefficients $\tilde{b}_{m,k}$ and the other scaling coefficients $\tilde{a}_{m,k}$ from (13.16). The results are summarized in Table 1. ⬚

Table 1. **Total estimated error for different scales** m

m	3	2	1	0	-1	-2	-3
$\sum_k \tilde{b}_{mk}^2$	—	0.208	0.029	0.059	0.026	0.108	0.650
$\sum_k \tilde{a}_{mk}^2$	0.550	0.352	0.323	0.264	0.238	0.130	0.650
$\tilde{e}_m - \tilde{e}_3$	0	0.056	0.006	0.027	0.034	0.136	0.807

In this case the total error increases most rapidly in going from $m = -1$ to $m = -2$, and hence $m = -1$ is probably the appropriate value. This is just a histogram with nodes at $\cdots, -2, 0, 2, 4, \cdots$.

13.4 Smooth wavelet estimators of density

We now return to the general case and assume that $\phi \in S_r$ and satisfies property Z_λ of Chapter 8. We also assume that $f(x)$, the density in question, has a degree of smoothness by requiring it to belong to the Sobolev space H^α for some $\alpha > \frac{1}{2}$. The estimator will again be given by (13.2) except that $q_m(x, y)$ is the general reproducing kernel of V_m. The other form of the estimator (13.12) and the truncated version (13.13) are equally valid in the general case.

THEOREM 13.1
Let the scaling function $\phi \in S_r$ and satisfy property Z_λ for some $\lambda \geq 1$; let $X_1, X_1, \cdots, X_N$ be an i.i.d. sample from the continuous bounded density $f(x)$; let

$$\tilde{f}_m(x) = \frac{1}{N} \sum_{i=1}^{N} q_m(x, X_i), \qquad x \in \mathbb{R};$$

(i) then $E|\tilde{f}_m(x) - f(x)|^2 \to 0$, uniformly on compact sets as $m \to \infty$ and $m = O(\log$

If, in addition, $f \in H^\alpha$, $\alpha > \lambda + \frac{1}{2}$, $m \approx \frac{\log 2}{2\lambda + 1} \log N$, then

(ii) (a) $E|\tilde{f}_m(x) - f(x)|^2 = O(N^{-2\lambda/2\lambda+1})$,
    (b) $\int E|\tilde{f}_m(x) - f(x)|^2 dx = O(N^{-2\lambda/2\lambda+1})$.

PROOF   The variance of the estimator is given by

$$E|\tilde{f}_m(x) - f_m(x)| \leq \frac{1}{N} \int_{-\infty}^{\infty} q_m^2(x, y) f(y) dy \qquad (13.17)$$

$$\leq \frac{1}{N} \|f\|_\infty q_m(x, x) = \frac{1}{N} \|f\|_\infty 2^m q(2^m x, 2^m x)$$

$$= O(2^m/N)$$

since $q(x, x)$ is continuous and periodic. The bias converges to 0 uniformly on compact sets since $q_m(x, y)$ is a quasi-positive delta sequence. Hence if $m = O(\log N)$ and $m \to \infty$ then both the bias and the variance converge to 0.

The rate of convergence in (ii) is obtained by using Lemma 8.7 and (13.17). If $f \in H^\alpha$, then the bias is given by

$$|E\tilde{f}_m(x) - f(x)| = \left| \int q_m(x, y) f(y) dy - f(x) \right|$$

$$\leq \|q_m(x, \cdot) - \delta(x, \cdot)\|_{-\alpha} \|f\|_\alpha = O(2^{-m\lambda})$$

by Schwarz's inequality for Sobolev spaces. Hence if $2^m/N \approx 2^{-2m\lambda}$ then the mean square error is bounded by $O(N^{-2\lambda/2\lambda+1})$.   □

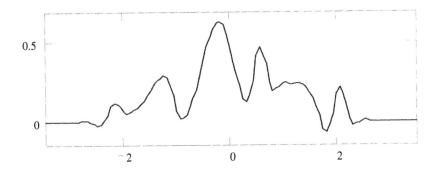

FIGURE 13.2
Smooth wavelet estimator for the same data as in Figure 13.1.

The same inequalities, though not the same proof, hold for the integrated mean square error.

REMARK 13.2   In some wavelet cases, such as the Meyer wavelets, the parameter λ can be taken arbitrarily large and hence the error rate approaches $O(N^{-1})$.   ■

REMARK 13.3   The estimator may also be expressed in truncated form

$$\tilde{f}_{m,k}(x) = \sum_{n=-k}^{k} \left\{ \frac{1}{N} \sum \phi(2^m X_i - n) \right\} 2^m \phi(2^m x - n),$$

for which the same rate is obtained—provided $k = k(m)$ grows sufficiently fast.   ■

REMARK 13.4   The estimator $\tilde{f}_m$ may also be given in terms of the empiric distribution $f^*(x)$ given by (13.1) as

$$\tilde{f}_m(x) = \int_{-\infty}^{\infty} q_m(x, y) f^*(y) dy = - \int_{-\infty}^{\infty} \frac{\partial q_m(x, y)}{\partial y} F^*(y) dy,$$

and hence used with convergence properties of F^*.   ■

Corollary 13.1
Let $f(x)$, $\phi(x)$ be as in Theorem 13.1 with $r \geq 1$. Then $\tilde{f}_m(x)$ is almost surely consistent as $m \to \infty$, $m = O(\log N)$ and

$$\tilde{f}_m(x) - f(x) = O\left(\left(N^{-\lambda/(\lambda+1)} \log \log N \right)^{\frac{1}{2}} \right) \qquad a.s.,$$

for $m = O(\log N / 2(\lambda + 1) \log 2)$.

The proof follows the law of the iterated logarithm [Ki], together with the bound

$$|\tilde{f}_m(x) - f(x)| \leq |\tilde{f}_m(x) - f_m(x)| + |f_m(x) - f(x)|$$

$$\leq \sup_y |F^*(y) - F(y)| \int_{-\infty}^{\infty} \left| \frac{\partial q_m}{\partial y}(x, y) \right| dy + O(2^{-m\lambda})$$

$$= O\left(N^{-\frac{1}{2}} |\log \log N|^{\frac{1}{2}} 2^m \right) + O(2^{-m\lambda}).$$

The results given above are asymptotic and not of much use in choosing an appropriate value of the parameter m. Fortunately the techniques of the prototype Haar system carry over to the general case.

The coefficients of our density estimator

$$\tilde{a}_{mk} = \frac{1}{N} \sum_{i=1}^{N} \phi_m(X_i - k2^{-m})$$

and

$$\tilde{b}_{mk} = \frac{1}{N} \sum_{i=1}^{N} \psi_m(X_i - k2^{-m})$$

are formally the same as in the Haar case. The dilation equations give us our decomposition algorithm,

$$a_{m-1,k} = 2^{-\frac{1}{2}} \sum_j c_j a_{m,2k+j}$$

$$b_{m-1,k} = 2^{-\frac{1}{2}} \sum_j c_{1-j}(-1)^j a_{m,2k+j}. \qquad (13.18)$$

Hence we can again proceed as before. We choose a finest scale $\overline{m}$ of interest; calculate $\tilde{a}_{m,k}$ from the data and then use (13.18) to find $\tilde{a}_{m-1,k}$ and $\tilde{b}_{m-1,k}$. We keep decreasing m until the integrated mean square error takes a jump.

However there is one difficulty. In the case of the prototype, $\tilde{a}_{m,k}$ was obtained by merely counting the number of sampled values. In general $\phi(x)$ is not a constant function, and mere counting will not work. In fact $\phi(x)$ can be a very wild function and is usually not even given by a formula. Rather the properties of ϕ must be obtained from the dilation equation. The values of $\phi(x)$ may be found by beginning with $\phi^{(0)}(x)$, the prototype Haar scaling function, and then approximating $\phi(x)$ by $\phi^{(j)}(x)$ where

$$\phi^{(j+1)}(x) = \sqrt{2}\sum_k c_k \phi^{(j)}(2x - k), \qquad j = 0, 1, 2, \cdots. \qquad (13.19)$$

This scheme works for certain choices of the c_k's (see [D] or [V]), in particular, those that lead to the standard wavelets. The result is that with this scheme,

$$\phi^{(j)}(x) \rightarrow \phi(x) \quad \text{in} \quad L^2(\mathbb{R})$$

and $\{\phi(x - n)\}$ is orthonormal.

We may use (13.19) directly to define the coefficients $\tilde{a}_{mk}$. Indeed we have

$$a_{m,n}^{(j+1)} = 2^{-\frac{1}{2}}\sum_k c_k a_{m-1,2n+k}^{(j)}, \qquad j = 0, 1, \cdots. \qquad (13.20)$$

Thus we begin with $\tilde{a}_{m,k}^{(0)}$, obtained by counting, and use the recursive definitions (13.19) and (13.20) to approximate $\tilde{f}_m(x)$ by

$$\tilde{f}_m^{(j)}(x) = \sum_n \tilde{a}_{m,n}^{(j)}\phi^{(j)}(x - n), \qquad j = 0, 1, \cdots. \qquad (13.21)$$

13.5 Local convergence

If the distribution of the sample does not have a density, and we form the density estimator (13.2), it cannot of course be integrated mean square consistent. However it will be locally mean square consistent at points at which the distribution function $F(x)$ is differentiable. This is a consequence of Lemma 8.4, which we repeat.

Lemma 13.1

Let $q_m(x,t)$ be the RK of V_m, $\phi \in H^r$, $r \geq 1$, $0 \leq \alpha \leq r$; then

$$K_m(x,t) = \frac{(x-t)^\alpha}{\alpha!}\frac{\partial^\alpha}{\partial t^\alpha}q_m(x,t)$$

is a quasi-positive delta sequence.

As an immediate corollary we have

Corollary 13.2

Let the distribution function $F(x)$ have a derivative at x_0. Then

$$E(\tilde{f}_m(x_0)) \to F'(x_0)$$

as $m \to \infty$.

The proof is similar to the previous ones since

$$E\tilde{f}_m(x_0) = \int q_m(x_0,t)dF(t)$$
$$= \int (x_0-t)\frac{\partial q_m(x_0,t)}{\partial t}\frac{F(t)-F(x_0)}{t-x_0}dt.$$

This must converge to

$$\lim_{t\to x_0}\frac{F(t)-F(x_0)}{t-x_0} = F'(x_0).$$

Thus $\tilde{f}_m(x_0)$ is asymptotically unbiased and, since the variance has the same bound as in the last section, is mean square consistent as well for $m = O(\log N)$.

13.6 Positive density estimators based on characteristic functions

Since only the trivial case of the Haar system gives scaling functions that are positive [J1], the estimators of the previous sections have not themselves been density functions. Another estimator based on wavelets is possible however. It uses the scaling function basis $\{\phi_{mn}\}$ of V_m to approximate the characteristic function $\theta(w)$, i.e., the Fourier transform, rather than the density function. If the density function $f(x)$ is bounded, then $\theta(w) \in L^2(\mathbb{R})$ and its projection on V_m may be given by

$$\theta_m(w) = \int_{-\infty}^{\infty} q_m(w,\zeta)\theta(\zeta)d\zeta$$
$$= \sum_n a_{mn} 2^{m/2}\phi(2^m w - n).$$

The Fourier transform of θ_m is

$$\hat{\theta}_m(x) = \sum_n a_{mn} 2^{-m/2}\widehat{\phi}(2^{-m}x)e^{-ixn2^{-m}},$$

which may be used to obtain an approximation to $f(x)$

$$f_m(x) = \frac{1}{2\pi}\hat{\theta}_m(-x)$$
$$= \frac{1}{2\pi}\widehat{\phi}(-2^{-m}x)2^{-m/2}\sum_n a_{mn}e^{inx2^{-m}}. \qquad (13.22)$$

For the scaling function ϕ given by

$$\widehat{\phi^2}(x) = \int_{x-\pi}^{x+\pi} h,$$

for symmetric, positive function h supported on $[-\epsilon, \epsilon]$ (Example 7 of Chapter 3) we have $\phi(x) > 0$ for $|x| < \pi + \epsilon$ and $\phi(x) = 0$ for $|x| > \pi + \epsilon$ and $\phi(x) = \phi(-x)$. Hence in this case $f_m(x)/\widehat{\phi}(-2^{-m}x)$ is a function in $L^2(-2^{m+1}\pi, 2^{m+1}\pi)$. Furthermore, $f_m(x) \geq 0$; to see this we express θ_m as

$$\theta_m(w) = \frac{1}{2\pi}\int \hat{q}_m(w,y)\overline{\hat{\theta}(y)}dy$$

by using the isometry property of the Fourier transform. Then f_m becomes, for $m = 0$,

$$f_0(x) = \frac{1}{2\pi}\int_{-\pi-\epsilon}^{\pi+\epsilon}\sum_n \widehat{\phi}(x)\widehat{\phi}(y)e^{-in(x+y)}f(-y)dy$$
$$= \int_{-\pi-\epsilon}^{\pi+\epsilon}\widehat{\phi}(x)\widehat{\phi}(y)\delta^*(x+y)f(-y)dy$$
$$= \widehat{\phi}(x)\sum_k \widehat{\phi}(x+2\pi k)f(x+2\pi k),$$

each of the terms of which is nonnegative. (Here δ^* is the periodic extension of the delta function.)

We now estimate $f_0(x)$ – and, by a change of scale, $f_m(x)$ – by using the Fejer kernel F_n on $[-\pi, \pi]$ to estimate the periodic part. This leads to the estimator

$$\tilde{f}_{0,n}(x) = \widehat{\phi}(x)\frac{1}{N}\sum_i F_n(x - X_i)\widehat{\phi}(X_i). \qquad (13.23)$$

For $|x| < \pi - \epsilon$, we have

$$E\tilde{f}_{0,n}(x) = \widehat{\phi}(x)\int_{-\infty}^{\infty} F_n(x - t)f(t)\widehat{\phi}(t)dt$$

$$= \widehat{\phi}(x)\sum_x \int_{-\pi}^{\pi} F_n(x - t)f(t + 2\pi k)\widehat{\phi}(t + 2\pi k)dt \to f_0(x)$$

at points of continuity as $n \to \infty$. Since $f_0(x) = f(x)$ for $|x| < \pi - \epsilon$, we have asymptotic unbiasedness. For other values of x we use a change of scale and the resulting estimator

$$\tilde{f}_{m,n}(x) = 2^{-m}\tilde{f}_0(2^{-m}x).$$

Thus we have a nonnegative estimator whose Fourier transform belongs to V_m. The rate of convergence will be relatively slow since even for C^∞ functions $(F_n * f)(x) - f(x) = O(n^{-1})$.

13.7 Positive estimators based on positive wavelets

Another approach which gives us estimators which are themselves densities uses the positive wavelets constructed in Chapter 8 to avoid Gibbs phenomenon. They have properties given by the theorem which we repeat.

THEOREM 13.2
Let $\rho^r(t) = \sum_n r^{|n|}\phi(t - n)$, where ϕ is any scaling function with compact support and r is chosen so large that $\rho^r(t) \geq 0$, for $t \in \mathbb{R}$; then $\{2^{\frac{m}{2}}\rho^r(2^m t - n)\}_{n \in \mathbb{Z}}$ is a Riesz basis of V_m; its dual basis is generated

by $\tilde{\rho}_r$, where

$$\tilde{\rho}_r(t) = \frac{1}{2\pi(1-r^2)}[(1+r^2)\phi(t) - r\{\phi(t+1) + \phi(t-1)\}].$$

However, this biorthogonal system does not give us the positive kernel we need. Rather we use a modification which gives the desired properties. The kernel that gives us the approximation in V_0 to $f \in L^2$ is given by

$$k_r(t,s) = \left(\frac{1-r}{1+r}\right)^2 \sum_{n=-\infty}^{\infty} \rho^r(t-n)\rho^r(s-n), \qquad (13.24)$$

i.e.,

$$f_0^r(t) = \int_{-\infty}^{\infty} k_r(s,t)f(s)ds.$$

This kernel satisfies the conditions needed to generate a positive delta sequence $\{k_{r,m}\}$ where

$$k_{r,m}(s,t) = 2^m k_r(2^m s, 2^m t), \quad m \in \mathbb{R}. \qquad (13.25)$$

(See Chapter 8, Section 8.) We summarize the properties of $k_{r,m}$ in the next proposition.

Proposition 13.3
[W-She4] Let $k_{r,m}$ be as in (4.7), let $f \in L^1 \cap L^2(\mathbb{R})$; let

$$f_m^r(t) = \int_{-\infty}^{\infty} k_{r,m}(t,s)f(s)ds;$$

then $f_m^r \in V_m$ and
   (i) if $M_1 \leq f(t) \leq M_2$ for $t \in \mathbb{R}$, then $M_1 \leq f_m^r(t) \leq M_2$ for $t \in \mathbb{R}, m \in \mathbb{R}$,
   (ii) if $M_3 \leq f(t) \leq M_4$ for $t \in [a,b]$, then for each $\epsilon > 0, \delta > 0$, there is an m_0, such that for $t \in (a+\delta, b-\delta)$, $M_3 - \epsilon \leq f_m^r(t) \leq M_4 + \epsilon$, for $m \geq m_0$.

The linear estimator associated with this kernel is given by the expression

$$\widetilde{f_m^r}(x) = \int_{-\infty}^{\infty} k_{r,m}(x,y)f^*(y)dy = \frac{1}{N}\sum_{i=1}^{N} k_{r,m}(x,X_i) \qquad (13.26)$$

as one might expect.

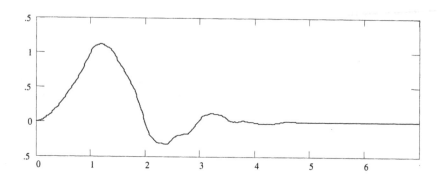

FIGURE 13.3
Daubechies scaling function ($N = 4$).

If our density $f \in H^r$, the Sobolev space, the usual linear wavelet estimator, has bias satisfying $O(2^{(1-r)m})$. If the wavelet is sufficiently regular, this can be used to estimate the bias for our new estimator.

Lemma 13.2
Let the scaling function $\phi \in C^2$, $f_m(x) = E\tilde{f}_m(x)$, where $\tilde{f}_m$ is the usual linear estimator, let the density $f \in H^r$, for $r \geq 3$; then

$$f^r_m(x) - f_m(x) = O(2^{-2m}).$$

See Chapter 8 for an indication of proof.

THEOREM 13.3
[W-She2] Let ϕ be as in the lemma, let $\widetilde{f^r_m}(x)$ be the estimator (13.26); let the density function $f \in H^3$, then the mean square error satisfies

$$MSE = O(N^{-4/5})$$

for a proper choice of m, uniformly on bounded sets.

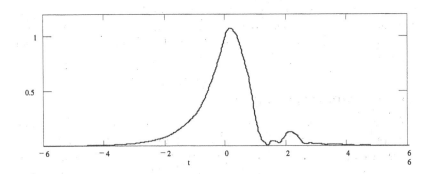

FIGURE 13.4
The associated summability function for the scaling function
in Figure 13.3.

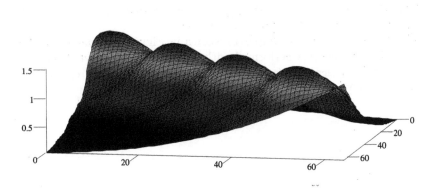

FIGURE 13.5
The positive kernel associate with Daubechies scaling function
$(N = 4)$.

This rate is somewhat better than non-negative estimators based on other orthogonal systems. It is the same as the rate attained for symmetric kernel estimators. However not too much should be made of this. The principal benefit of using wavelet rather than standard kernel estimators lies in the subsequent thresholding [D-J-K-P] as well as in the ease of calculation using the fast wavelet transform (decomposition algorithm).

13.7.1   Numerical experiment

In this subsection, we illustrate how the smooth positive wavelet estimators $\tilde{f}_m^r(x)$ (type b) are used to estimate the density functions of univariate data. The results are compared to the case of the linear projection wavelet estimators $\tilde{f}_m(x)$ (type a). The graphs of Daubechies scaling function ($N = 4$) [D] and its associated non-negative basis function are shown in Figure 13.3 and Figure 13.4.

We base our density estimate of the Old Faithful geyser data. The data for the eruption lengths (in minutes) of 107 eruptions of Old Faithful geyser can be found in Table 2. This data set can be found in [Sil].

We illustrate the result of the experiment with the graphs of the estimators. The results show that the negative values occur for the estimators of type a in some places, while the values for the estimators of type b are always nonnegative. The latter are also smoother than the former.

The results for the example are shown in Figure 5 through Figure 8. The experiments are performed with the same window widths for the two different kernels (see [W-She2]). The results of other experiments may also be found in [W-She2].

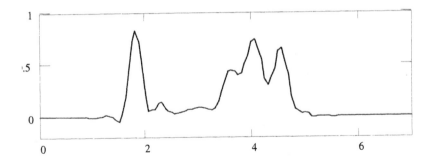

FIGURE 13.6
Density estimate for the Old Faithful geyser data using the reproducing kernel associated with Daubechies scaling function $(N = 4)$.

Table 2. Eruption lengths (in minutes)
of 107 eruptions of Old Faithful geyser.

4.37	4.62	4.42	1.83	3.92	3.92	2.33	3.33	4.18	4.50
4.70	4.33	4.83	3.43	4.33	3.20	4.57	3.73	4.58	1.80
1.68	1.83	3.50	4.93	3.95	4.58	3.58	1.67	3.50	3.70
1.75	4.25	4.65	3.77	2.93	3.50	3.70	4.63	4.62	2.50
4.35	4.40	4.12	4.50	3.68	3.80	4.25	1.83	4.03	2.27
1.77	4.10	1.88	4.00	3.43	3.80	3.58	2.03	1.97	2.93
4.25	4.00	1.82	3.52	4.20	1.80	3.67	2.72	4.60	4.63
4.10	1.90	4.50	1.67	4.08	1.95	1.90	4.03	4.00	4.00
4.05	3.93	4.73	2.25	2.00	1.77	4.13	1.73	3.75	1.97
1.90	4.00	4.07	4.33	1.85	4.28	4.53	3.10	1.73	3.72
1.85	4.60	4.08	3.87	4.13	3.38	3.37			

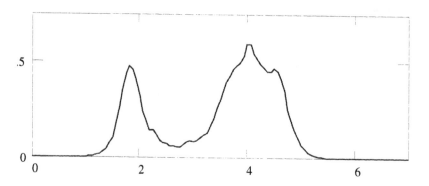

FIGURE 13.7
Density estimate for the Old Faithful geyser data using the
positive kernel of Figure 13.4.

13.8 Density estimation with noisy data

In this section we shall consider the problem of estimating a density
in the presence of additive noise. The observations then are of the form

$$Y_i = X_i + Z_i, \qquad i = 1, 2, \cdots, N,$$

where the $Z_i's$ (the noise) are independent of the $X_i's$ and are i.i.d. random variables with some known density $k(z)$. The actual observations
then have a density of the form

$$g(y) = \int_{\mathbb{R}} k(y - x) f(x) dx. \tag{13.27}$$

We now assume that the unknown density is $f(x)$ and the noisy observations have density $g(y)$ given by (13.27). Our estimate of $g(y)$ can
be obtained by the method of section 13.4. In order to estimate $f(x)$ we
invert the equation (13.27) formally by using the Fourier transform

$$\widehat{f}(w) = \widehat{g}(w)/\widehat{k}(w). \tag{13.28}$$

Unfortunately this $\widehat{f}(w)$ usually will not have an inverse Fourier transform, but it will if $g \in V_m$, and $\widehat{k}(w) \neq 0$ for all w. In that case we
have

Lemma 13.3
[W-She1] Let $k \in L^1(\mathbb{R}), \widehat{k}(w) \neq 0$. If $g \in V_m$, the wavelet subspace associated with a Meyer wavelet, then equation (13.27) has a unique solution $f \in V_{m+1}$.

Since our estimate of g is

$$\widetilde{g}_m(y) = \frac{1}{N} \sum_{i=1}^{N} q_m(y, Y_i)$$

which belongs to V_m, we can use it to obtain an estimate of f from its Fourier transform

$$\widetilde{f}_m(x) = \frac{1}{2\pi} \int [\widehat{\widetilde{g}}_m(w)/\widehat{k}(w)]e^{iwx} dw. \tag{13.29}$$

This function belongs to V_{m+1} by the lemma. We can also find another expression for $\widetilde{f}_m$ similar to the one for $\widetilde{g}_m$ which uses a different kernel. We first define θ_m by

$$\theta_m(x) := \frac{1}{2\pi} \int \widehat{\phi}(\xi)/\widehat{k}(2^m \xi)e^{ix\xi} d\xi. \tag{13.30}$$

Again by the lemma, this function belongs to V_{m+1}. We use it in turn in the kernel $P_m(x, y)$

$$P_m(x, y) := \sum_n 2^m \theta_m(2^m x - n)\phi(2^n y - n). \tag{13.31}$$

This enables us to express $\widetilde{f}_m$ as

$$\widetilde{f}_m(x) = \frac{1}{N} \sum_{j=1}^{N} P_m(y, Y_j). \tag{13.32}$$

This is shown by substituting the expanded version of $\widetilde{g}_m$ in (13.27) and then changing variables. Thus we have

Lemma 13.4
Let $\widetilde{f}_m(x)$ be given by (13.29); then

$$k * \widetilde{f}_m = \widetilde{g}_m.$$

Now we try to find the rate of convergence of (13.27) to f. Of course this depends on the smoothness of f. We initially suppose it to be a bandlimited function, i.e., one whose Fourier transform has compact support. Such functions are entire functions of exponential type which are in $L^1(\mathbb{R})$ when restricted to the real axis.

Proposition 13.4

Let $f(x)$ be a density function which is the restriction to the real axis of an entire function of exponential type; let $\tilde{f}_m(x)$ be the estimator corresponding to a Meyer wavelet given by (13.32) with $\widehat{k}(w) \neq 0$. Then there is m_0 such that for $m \geq m_0$

$$E[\tilde{f}_m(x) - f(x)]^2 \leq C_m \|f\|_\infty / N$$

as $N \longrightarrow \infty$ and $C_m = \frac{1}{\pi} \int_{-\frac{2^{m+2}\pi}{3}}^{\frac{2^{m+2}\pi}{3}} |\widehat{k}(w)|^{-2} dw$.

PROOF Since f is bandlimited, its Fourier transform has support in some interval, say $[-\lambda, \lambda]$. Hence $f \in V_m$ for $\frac{2^{m+1}}{3} \geq \lambda$. This follows from the fact that

$$\widehat{f}(w) = \widehat{f}(w)\widehat{\phi}(w2^{-m})$$
$$= \sum_n f_n e^{iw2^{-mn}} \widehat{\phi}(w2^{-m}) \tag{13.33}$$

since $\widehat{\phi}(w2^{-m}) = 1$ on the support of $\widehat{f}(w)$ and the periodic extension and hence the Fourier series of $\widehat{f}(w)$ equals it on the support of $\widehat{\phi}(w2^{-m})(\subseteq [-\frac{2^{m+2}\pi}{3}, \frac{2^{m+2}\pi}{3}])$.

Since $\tilde{g}_m \in V_m$, it follows that $\tilde{f}_m \in V_{m+1}$, by Lemma 13.3. The same is true for $f_m = E\tilde{f}_m$, and hence for $2^m \geq 3\pi\lambda$, $E\tilde{f}_m - f \in V_m$. Furthermore f_m is the projection of f onto V_m, and hence $f_m = f$. Thus the bias is zero for such m.

The variance calculation is similar to that in section 13.4. We first find that

$$Var[\tilde{f}_m(x)] \leq \frac{\|f\|_\infty}{N} \int P_m^2(x, t) dt \tag{13.34}$$

and by using (13.31) that

$$\int P_m^2(x, t) dt = \sum_n 2^m \theta_m^2(2^m x - n) \tag{13.35}$$

by the orthonormality of $\{2^{m/2}\phi(2^n t - n)]$. Since the right hand side of (13.35) is periodic, we need merely to show it is bounded on a finite interval. By (13.30) we see that

$$2^{\frac{m}{2}}\theta(2^m x - n) = \frac{2^m}{2\pi}\int_{-\frac{4\pi}{3}}^{\frac{4\pi}{3}}\frac{\widehat{\phi}(\xi)}{\widehat{k}(2^m\xi)}e^{i(2^m x - n)\xi}d\xi$$

$$= \frac{2^{m/2}}{2\pi}\left\{\int_0^{2\pi} + \int_{-2\pi}^0\right\}\frac{\widehat{\phi}(\xi)e^{i2^m x\xi}}{\widehat{k}(2^m\xi)}e^{-in\xi}d\xi,$$

and hence is the sum of two Fourier coefficients. We then use Bessel's inequality to calculate that

$$\sum_n 2^m\theta^2(2^m x - n) \leq \frac{2\cdot 2^m}{2\pi}\int_{-\frac{4\pi}{3}}^{\frac{4\pi}{3}}\frac{|\widehat{\phi}(\xi)|^2}{|\widehat{k}(2^m\xi)|^2}d\xi$$

$$\leq \frac{1}{\pi}\int_{-\frac{4\pi}{3}}^{\frac{4\pi}{3}}\frac{2^m}{|\widehat{k}(2^m\xi)|^2}d\xi = \frac{1}{\pi}\int_{-\frac{2^{m+2}\pi}{3}}^{\frac{2^{m+2}\pi}{3}}|\widehat{k}(w)|^{-2}dw$$

$$= C_m. \tag{13.36}$$

This completes the proof. □

This result only works if the density f (or g) is bandlimited. For other densities the bias is nonzero but may converge to 0 as $m \to \infty$. We can still get reasonable rates if the kernel k is not too smooth and f belongs to some Sobolev space.

Lemma 13.5

[W-She1] Let k be continuous and bounded and satisfy

$$|\widehat{k}(w)| \geq C(1 + w^2)^{-\frac{\alpha}{2}};$$

let $g \in H^\beta(\mathbb{R})$, where $\beta \geq \alpha > 0$; let g_m be the projection of g on V_m and $k * f_m = g_m$; then

$$\|f_m - f\|_2 \leq C\|g_m - g\|_{H^\beta}.$$

This lemma enables us to get an estimate for the bias of our estimator; indeed if $g \in H^\gamma, \gamma > \beta$, we get the rate of convergence

$$\|f_m - f\|_2 = O(2^{(\beta-\gamma)m}).$$

This works for any $\beta \geq \alpha$ and hence for $\beta = \alpha$. The same argument holds for pointwise convergence,

$$|f_m(x) - f(x)| = O(2^{(\alpha - \gamma)m}).$$

THEOREM 13.4

[W22] Let $g \in H^\gamma$; k be as in Lemma 13.5, and let $\widetilde{f}_m$ be given by (13.29); then the MSE satisfies

$$E[\widetilde{f}_m(x) - f(x)]^2 = O(N^{-1+\frac{1+2\alpha}{1+2\gamma}}).$$

13.9   Other estimation with wavelets

Other types of nonparametric estimation in statistics involve (*i*) spectral density estimation, and (*ii*) nonparametric regression. Spectral density estimation involves techniques similar to those for probability density, except that periodic estimators must be used. Nonparametric regression can be based on probability density estimation but requires multivariate estimation.

13.9.1   Spectral density estimation

We start with a discrete stationary random process $\{X(t), t \in \mathbb{Z}\}$ with zero mean and finite covariance. We wish to estimate the spectral density

$$f(\lambda) = \frac{1}{2\pi} \sum_{k=-\infty}^{\infty} C_k e^{-ik\lambda}$$

where $C_k = E(X(t)X(t+k))$ is the covariance. We assume it is observed at $t = 1, 2, \cdots, N$ and form the periodogram

$$I_N(\lambda) = \frac{1}{2\pi} \sum_{k=-N+1}^{N-1} \tilde{C}_k e^{-i\lambda k}$$

where

$$\tilde{C}_k = \begin{cases} \frac{1}{N} \sum_{t=1}^{N-|k|} X(t+|k|)X(t), & |k| < N \\ 0, & |k| \geq N. \end{cases}$$

Whereas $I_N(\lambda)$ is an asymptotically unbiased estimator of $f(\lambda)$ since

$$E(I_N(\lambda)) \equiv \frac{1}{2\pi} \sum_{k=-N+1}^{N-1} C_k e^{-i\lambda k},$$

its variance gives problems, and it must be smoothed by a spectral window $W_M(\lambda)$ usually consisting of a trigonometric polynomial. The estimator then is

$$\tilde{f}_{MN}(\lambda) = (W_M * I_N)(\lambda).$$

An estimator of this kind may be based on periodic wavelets since the spectral density is a periodic function. These are constructed from the usual wavelets by periodization as in Chapter 7 or may be constructed directly as shown there as well. This sequence $\{\psi_n\}$ of periodic wavelets is composed of periodic functions of period 1 which form an orthonormal basis of $L^2(0,1)$. The periodized version of V_m becomes V_m^* and has a basis consisting of

$$\psi_0, \psi_1, \cdots, \psi_{2^m-1}.$$

The reproducing kernel for V_m^* is, by the notation in Chapter 7,

$$q_m^*(x,y) = \sum_{k=0}^{2^m-1} \phi_{m,k}^*(x)\phi_{m,k}^*(y) = \sum_{k=0}^{2^m-1} \phi_m^*(x - k2^{-m})\phi_m^*(y - k2^{-m}).$$

It is again fairly easy to show that these kernels form a quasi-positive delta sequence on $(0,1)$ as $m \to \infty$. The associated spectral density estimator is

$$\tilde{f}_{m,N}(\lambda) = \frac{1}{2\pi} \int_0^{2\pi} q_m^* \left(\frac{\lambda}{2\pi}, \frac{y}{2\pi}\right) I_N(y)dy, \quad m = 0, 1, 2, \cdots . \quad (13.37)$$

This does not exactly correspond to the usual spectral window, which involves convolution, but similar arguments to those used for convolution can be used. In particular, $\frac{1}{2\pi}q_m^* \left(\frac{x}{2\pi}, \frac{y}{2\pi}\right)$ is a delta sequence of class denoted as Γ_1 in [W13] (with a slight modification of definition). Hence, as shown therein, the estimator (13.37) will be mean square consistent with rate $O(N^{-2/3})$ for smooth spectral densities of Gaussian processes.

There is no advantage of these estimators over the others as far as convergence is concerned, since most spectral windows also lead to quasi-positive delta sequences. There may be an advantage in estimating spectral density functionals of processes with a periodic component. However, this has not as yet been explored. It may also be possible to use the periodicized positive wavelets which, however, also has not yet been done.

13.9.2   Regression estimators

Any density estimator on $\mathbb{R}^d$ leads to an estimator of the nonparametric regression function

$$r(x) = E(Y \mid X = x).$$

In particular the wavelet estimators can be used, but we need first to consider the multivariate estimators. For $d = 2$ the estimator is

$$\tilde{f}_m(x, y) = \frac{1}{N} \sum_i q_m(x, X_i) q_m(y, Y_i)$$

where $q_m(x, t)$ is the reproducing kernel of V_m. The estimator of $r(x)$ is

$$\tilde{r}_m(x) = \frac{1}{N} \frac{\sum_{i=1}^{N} Y_i q_m(x, X_i)}{\frac{1}{N} \sum_{i=1}^{N} q_m(x, X_i)} = \frac{\tilde{\nu}_m(x)}{\tilde{f}_m^1(x)}, \qquad (13.38)$$

where $\tilde{f}_m^1(x)$ is the linear estimator of section 13.4. The numerator of (13.38) may be expressed as

$$\tilde{\nu}_m(x) = \int y \tilde{f}_m(x, y) dy$$

$$= \frac{1}{N} \sum_i q_m(x, X_i) \int y q_m(y, Y_i) dy$$

$$= \frac{1}{N} \sum_i q_m(x, X_i) Y_i.$$

This follows from the property that

$$\int y^k q_m(x, y) dy = x^k, \qquad k = 0, 1, 2, \cdots, r$$

provided our scaling function $\phi(x)$ is r-regular, $r \geq 1$.

Hence the estimator is similar to kernel estimators of the regression function and by the same arguments can be shown to be mean square consistent [PR].

13.10   Threshold Methods

The estimators we have covered in this chapter are all linear estimators; for the most part they involve low pass filters. These are appropriate when noise is high frequency and the signal, e.g., the regression function, is lower in frequency; the effect of such a filter is to smooth the signal. But some signals are not smooth and some noise is not high frequency. Rather, a common (and more realistic) assumption is that the noise is Gaussian white noise.

A traditional way to remove additive white noise is by using the Fourier transform to convert the noise into a function which is approximately constant. This makes sense since the covariance functional of white noise (a generalized random process [G-S, vol. 4, p. 260] obtained by differentiation of the Wiener process) is just a multiple of the delta function. Its Fourier transform has constant magnitude. Hence if we remove a constant threshold from the Fourier transform of the noisy signal, and then take the inverse Fourier transform, we might expect to remove at least a portion of the noise. This works even when the signal is not itself smooth.

The same idea is very widely used in wavelet based estimation except that the wavelet series coefficients instead of the Fourier transforms are used to get rid of noise. One advantage of using wavelets for denoising is that the wavelets have a very strong localization property as we have observed in Chapter 5. Thus if a signal is very smooth for some portions of its domain, but not for others, then the non-linear wavelet approximation obtained by retaining only those coefficients above a certain threshold will also exhibit the same behavior. See [D-J-K-P] for a justification.

The process we have just mentioned is usually called "hard" thresholding, in which the wavelet coefficients b_{mn} with magnitude below a certain threshold λ are dropped. In "soft" thresholding, they are also dropped, but in addition, the remaining coefficients are shrunk by a factor of $|b_{mn}| - \lambda$. The pointwise convergence of the wavelet series with these new coefficients converges pointwise just as the linear estimators do.

THEOREM 13.5

[T-V] Let $f \in L^2(\mathbb{R})$ have wavelet coefficients $b_{mn} =< f, \psi_{mn} >$, and let $b_{mn}(\lambda)$ denote the coefficients obtained by either hard or soft thresholding. Then

$$\lim_{\lambda \to 0} \sum_{m=-\infty}^{\infty} \sum_{n=-\infty}^{\infty} b_{mn}(\lambda)\psi_{mn}(x) = f(x) \ a.e.$$

These threshold methods seem to be the most useful when trying to estimate a deterministic signal corrupted by noise. However, they are more difficult to apply in density estimation problems since the delta function discontinuities in the empiric distribution are apt to be picked up by the wavelet threshold estimators. These discontinuities are (of necessity) eliminated by a low pass filter such as that obtained by the linear estimators. The threshold estimators probably would be the most effective for density estimation if they are used after the additive noise in the random variable is removed. This can be done by the deconvolution method described in a previous section.

A problem that now arises is the determination of a suitable λ to use for either hard or soft thresholding. Donoho and Johnstone [D-J] have proposed a universal threshold $\lambda = \sqrt{2 \log N} \sigma$, where N is the sample size and σ the variance of the noise. As with any other general prescription, this does not always work, but seems to be as good a starting point as any. Other schemes for choosing the appropriate threshold level may be found in [Vi].

13.11   Problems

1. In the usual kernel estimator of density, a difference kernel is used in place of the reproducing kernel in (13.2). That is, if $k(x)$ is a density function,

$$\tilde{f}_m(x) = \frac{m}{N} \sum_{i=1}^{N} k(m(x - X_i))$$

    is the estimator. If f and k are both continuous with compact support then show that $E|\tilde{f}_m(x) - f(x)|^2 \to 0$ uniformly if $\frac{m}{N} \to 0$ by

    (i)   showing that $mk(m(x - y))$ is a positive delta sequence,
    (ii)   showing that the bias of $\tilde{f}_m(x)$ converges to $f(x)$ uniformly,
    (iii) finding a bound on the variance.

2. Let $f(x) = \frac{1}{4} \cos \frac{x}{2}$, $|x| \le \pi$, $f(x) = 0$, $|x| > \pi$. Generate a random sample $X_1, X_2 \cdots X_N$, drawn from a distribution with this density by

(i) generating random numbers in $[0,1]$: $Y_1, Y_2, \cdots, Y_N$.

(ii) setting $X_i = F^{-1}(y_i)$, where $F(x)$ is the distribution function,

$$F(x) = \int_{-\pi}^{x} f(t)dt = \tfrac{1}{2} + \tfrac{1}{2}\sin x, \quad -\pi \leq x \leq \pi.$$

3. Find the density estimator $\tilde{f}_m^{(1)}(x)$ of (13.5) and $\tilde{f}_m^{(2)}$ of (13.6) for a sample size $N = 50$ from Problem 2 for values of $m = 1, 5, 25$.

4. Use the sample of Problem 2 and the Hermite series density estimator of (13.10) to approximate $f(x)$ (using the same values of m).

5. Generate a sample of size 50 from a normal distribution with mean $\mu = 1$ and variance $\sigma = 1$. Use the sample and find the Hermite series estimator for $m = 1$, 5, and 25.

6. Find the histogram estimator with the sample of Problem 5 for $m = 0, 1, 2$ in (13.12).

7. Use the estimator for $m = 2$ in Problem 5 and the decomposition algorithm (13.17) to obtain the estimators for $m = 0$ and $m = 1$.

8. (If you do not have a computer.) Draw 25 numbers, between 0 and 1, from a random number table. Partition the interval $(0,1)$ into 2^3 equal intervals and use (13.15) to determine $\tilde{a}_{3n}$. Find $\tilde{a}_{2,n}$ and $\tilde{a}_{1,n}$ from (13.16), and sketch the resulting histogram.

Chapter 14

Orthogonal Systems and Stochastic Processes

One of the most elegant uses of orthogonal systems is in conjunction with continuous stochastic processes. Such a process $\{X(t), t \in I\}$ is a family of random variables indexed by the parameter t [Ro, p.91]. For processes that are continuous in mean square, i.e.,

$$E|X(t)|^2 < \infty$$

and

$$E|X(t) - X(s)|^2 \to 0 \quad \text{as} \quad t \to s,$$

there is a theory, the Karhunen-Loève theory, that converts the process into an uncorrelated sequence of random variables. This is done by using a certain orthogonal system, expanding $X(t)$ in terms of it, and showing the coefficients are uncorrelated.

In this chapter we first present this theory. We then consider wavelet expansions of stationary processes and show that for certain wavelets, the coefficients of the expansion have negligible correlation for different scales. We then introduce a modification of the wavelets, which leads to uncorrelated coefficients. Finally we show that for certain nonstationary processes the wavelets may be chosen to give uncorrelated coefficients.

14.1   K-L expansions

In this section we introduce elements of the Karhunen-Loève (K-L) theory. A more complete readable exposition may be found in [Ro, p.203]. We shall assume a background in stochastic processes sufficient for a number of manipulations involved. However they should be understandable, in a formal sense, to persons without such a background.

Let $\{X(t), t \in [0,T]\}$ be a real process that is continuous in mean square and has zero mean $(E(X(t)) = 0)$. Then the covariance function given by

$$r(t,s) = E(X(t)X(s)), \qquad s, t \in [0,T],$$

is a continuous function on $[0,T] \times [0,T]$. If $0 < T < \infty$, then it may be used as the kernel of an integral operator

$$(\mathcal{R}f)(t) := \int_0^T r(t,s)f(s)ds, \tag{14.1}$$

which will be self adjoint and completely continuous (compact). As such it will have a countable set of eigenvalues $\{\lambda_i\}$ each of which has a finite number of associated eigenfunctions $\varphi_i(t)$ that may be assumed to be orthonormal [R-N, p.227]. Then for each g in the range of $\mathcal{R}$ (as an operator on $L^2([0,T])$), we have the convergent expansion

$$g(x) = \sum_i \langle g, \varphi_i \rangle \varphi_i(x), \qquad x \in [0,T]. \tag{14.2}$$

Moreover this series converges uniformly on $[0,T]$ [R-N, p.244].

We now return to our random process $X(t)$ and expand it in terms of the $\{\varphi_i\}$. This involves the random integral

$$X_i = \int_0^T X(t)\varphi_i(t)dt, \tag{14.3}$$

which exists whenever

$$\int_0^T E(|X_t|)dt < \infty, \qquad \text{[Ro, p.205]}.$$

But $E|X_t| \le \{EX_t^2\}^{1/2} = r^{1/2}(t,t)$, which is bounded on $[0,T]$, and hence this integral is finite. The integral (14.3) satisfies the usual properties of integrals.

The X_i in (14.3) are uncorrelated since

$$E(X_iX_j) = E\left\{ \int_0^T X(t)\varphi_i(t)dt \int_0^T X(s)\varphi_j(s)ds \right\}$$

$$= \int_0^T \int_0^T E(X(t)X(s))\varphi_i(t)\varphi_j(s)dt$$

$$= \int_0^T \int_0^T r(t,s)\varphi_i(t)\varphi_j(s)dt = \lambda_j \delta_{ij}, \qquad (14.4)$$

by a version of Fubini's theorem. We also need to know that $X(t)$ can be represented by such a series with these coefficients. It can be, in the mean square sense, since

$$E\left[X(t) - \sum_{i=0}^N X_i \varphi_i(t)\right]^2 = r(t,t) - 2\sum_{i=0}^N E(X(t)X_i)\varphi_i + \sum_{i=0}^N \lambda_i \varphi_i^2(t)$$

$$= r(t,t) - \sum_{i=0}^N \lambda_i \varphi_i^2(t). \qquad (14.5)$$

By (14.4) the $\lambda_j \geq 0$, and hence we can invoke Mercer's theorem [R-N, p.245], which says that

$$r(t,s) = \sum_i \lambda_i \varphi_i(t)\varphi_i(s)$$

where the convergence is uniform. Hence the last expression in (14.5) converges to 0 uniformly and

$$X(t) = \sum_i X_i \varphi_i(t) \qquad (14.6)$$

in the mean square sense.

Example: Brownian Bridge

The Wiener process $W(t)$ or "Brownian motion" is a Gaussian process on $[0, \infty)$ such that $W(0) = 0$ and $r(t,s) = \min(t,s)$, $0 < t, s$. The Brownian bridge is a similar process on [0,1] with both $X(0) = 0$ and $X(1) = 0$ with $r(t,s) = \begin{cases} t(1-s), & t < s \\ s(1-t), & s < t \end{cases}$. The eigenvalue problem $R\varphi = \lambda\varphi$ may, in this case, be converted to a differential equation since

$$\lambda\varphi'(t) = \int_0^t -s\varphi(s)ds + \int_t^1 (1-s)\varphi(s)ds,$$

and

$$\lambda\varphi''(t) = -t\varphi(t) - (1-t)\varphi(t) = -\varphi(t).$$

Since $\varphi(0) = \varphi(1) = 0$, the solution is given by

$$\lambda_n = 1/\pi^2 n^2, \quad \varphi_n(t) = \sqrt{2}\sin\pi nt, \quad n = 1, 2, \cdots.$$

Hence we have

$$X(t) = \sum_{n=1}^{\infty} X_n \sqrt{2}\sin\pi nt$$

where

$$X_n = \int_0^1 X(t)\sqrt{2}\sin\pi nt\, dt.$$

Then the X_n in this case are not only uncorrelated but, since the process is Gaussian, are in fact independent.

While this K-L theory is elegant, its utility is limited by the difficulty of finding the eigenfunctions. Unless they are also, as in the Brownian bridge example, eigenfunctions of a differential operator, only approximations are obtained. Also in some useful cases in which $X(t)$ is stationary, the interval of definition is infinite. Then $r(s, t)$ is no longer a Hilbert-Schmidt kernel (i.e., belongs to $L^2(I \times I)$), and the eigenvalue results do not hold.

14.2   Stationary processes and wavelets

A (wide-sense) stationary process is one for which $E(X(t)) = c$ and

$$E(X(t)X(s)) = E(X(t-s)X(0)),$$

and hence

$$r(t, s) = r(t-s, 0) = R(t-s), \quad t, s \in \mathbb{R},$$

depends only on the difference between t and s. Such processes correspond to many physical phenomena. The function $R(t)$ is bounded since

$$|R(t)| = |E(X(t)X(0))| \leq \{E(X^2(t))E(X^2(0))\}^{\frac{1}{2}} = R^{\frac{1}{2}}(0)R^{\frac{1}{2}}(0).$$

If $X(t)$ is mean square continuous as well, then $R(t)$ is continuous. The power spectrum of $X(t)$ is the Fourier transform $\hat{R}(w)$ of $R(t)$. This may not be an ordinary function but when it is, is called the spectral density.

We will find the wavelet expansion of $X(t)$ with respect to the Meyer type wavelets (Example 7 of Chapter 3). Their scaling function is given by

$$\hat{\phi}(w) = \left\{ \int_{w-\pi}^{w+\pi} dP \right\}^{\frac{1}{2}},$$

where P is a probability measure supported on $[-\epsilon, \epsilon], \epsilon \leq \frac{\pi}{3}$, and the mother wavelet by

$$\hat{\psi}(w) = e^{-iw/2} \left\{ \hat{\phi}(w + 2\pi) + \hat{\phi}(w - 2\pi) \right\} \hat{\phi}\left(\frac{w}{2}\right).$$

First however we observe that the approximation of $X(t)$ by $X_m(t)$ where

$$X_m(t) = \sum_n a_{mn}\phi_{mn}(t), \quad a_{mn} = \int X(t)\phi_{mn}(t)dt$$

is mean square consistent for any $\phi \in S_r$. That is,

$$E[X(t) - X_m(t)]^2 \to 0 \quad \text{as } m \to \infty, \, t \in \mathbb{R}.$$

This follows from the fact that $\{q_m(x, t)\}$ is a quasi-positive delta sequence (Chapter 8). Indeed we have

$$E[X(t) - X_m(t)]^2 = E[X^2(t)] - 2E[X(t) - X_m(t)] + E[X_m^2(t)]$$

$$= R(0) - 2 \int R(t - s)q_m(s, t)ds$$

$$+ \int \int R(u - s)q_m(s, t)q_m(t, u)du\, ds \quad (14.7)$$

where $q_m(s, t)$ is the reproducing kernel of V_m. (Note however that the sample paths are not in $L^2(\mathbb{R})$ [Ms] and hence $X_m \notin V_m$; nonetheless the interchange of integrals in (14.7) is valid since q_m is rapidly decreasing and $X(t)$ is bounded.)

Since $R(u - s)$ is continuous and in $L^2(\mathbb{R})$, we have

$$\int R(u - s)q_m(s, t)ds \to R(u - t)$$

uniformly on bounded sets and boundedly for all $u \in \mathbb{R}$. Therefore (14.7) converges to 0 as $m \to \infty$.

We can also obtain rates of convergence under additional hypotheses on $X(t)$ by methods similar to those in Chapter 8. We express $X_m(t)$ in the wavelet series as

$$X_m(t) = \sum_{k=-\infty}^{m-1} \sum_{n=-\infty}^{\infty} b_{kn} \psi_{kn}(t)$$

where

$$b_{kn} = \int_{-\infty}^{\infty} X(t) \psi_{kn}(t) dt.$$

Then we have

$$E(b_{mn} b_{kj}) \tag{14.8}$$

$$= \int \int R(t - s) \psi(2^m t - n) \psi(2^k s - j) 2^{m/2} 2^{k/2} dt\, ds$$

$$= \frac{1}{2\pi} \int \hat{R}(w) \hat{\psi}(2^{-m} w) e^{-iwk2^{-m}} \overline{\hat{\psi}(2^{-k} w)} e^{iwj2^{-k}} 2^{-(m+k)/2} dw$$

since convolution in the inner integral is transformed into the pointwise product under the Fourier transform. For the Meyer wavelets $\hat{\psi}(w)$ has support on

$$[\pi - \epsilon, 2\pi + 2\epsilon] \bigcup [-2\pi - 2\epsilon, -\pi + \epsilon],$$

and hence $\hat{\psi}(2^{-m} w)$ and $\hat{\psi}(2^{-k} w)$ have disjoint support if $|k - m| > 1$. For $m = k + 1$ we find that

$$E(b_{k+1,n} b_{kj})| \leq \frac{2^{-k+\frac{1}{2}}}{2\pi} \int_{2^{k+1}(\pi-\epsilon)}^{2^{k+1}(\pi+\epsilon)} |\hat{R}(w)| dw$$

$$\leq \|R\|_1 2^{3/2} \frac{\epsilon}{\pi}.$$

For $k = m$ we obtain the result

$$E b_{kn} b_{kj} = \frac{2^{-k}}{2\pi} \int \hat{R}(w) \left| \hat{\psi}(2^{-k} w) \right|^2 e^{-iw(n-j)2^{-k}} dw,$$

which are just Fourier coefficients. If $\hat{R}(w)$ and $\hat{\psi}(w)$ are both C^p, then $E(b_{kn} b_{kj}) = O(|n - j|^{-p})$; we have proved

Proposition 14.1

Let $X(t)$ be a wide-sense stationary process with spectral density $\hat{R}(w) \in C^p$, $p > 1$; let $\psi(t)$ be a mother wavelet of Meyer type such that $\hat{\psi}(w) \in$

C^p; then the discrete process

$$b_{mn} = \int X(t)2^{m/2}\psi(2^m t - n)dt$$

(i)   is uncorrelated at scales differing by more than 1,
(ii) has arbitrarily small correlation at scales differing by 1,
(iii) has correlation $= O(|n - j|^{-p})$ for fixed scale and translation differing by $|n - j|$.

It should also be observed that for fixed m, $\{b_{mn}\}$ is a discrete wide-sense stationary process [Ms]. These results can also be extended to certain nonstationary processes. In particular, since $\hat{\psi}(w) = 0$ for $|w| \leq \pi - \epsilon$, they can be used to analyze $1/f$ processes. These processes have spectral "densities" of the form $\hat{R}(w) = |w|^{-\lambda}$ [Fl].

The results of Proposition 14.1 still hold since the integrals are well defined. However the mean square convergence may not hold in these cases.

14.3   A series with uncorrelated coefficients

The wavelet expansions in the last section represent an approximation to the K-L expansions, but the coefficients are uncorrelated only at different scales. In this section we modify the wavelets to obtain an uncorrelated coefficient sequence. The new system will not be orthogonal but rather have an associated biorthogonal system [Z-W]. We shall also assume in this section that $\hat{R}(w) \neq 0$, $w \in \mathbb{R}$.

We first define a new function corresponding to the scaling function. It is $\theta_0(t)$ given by its Fourier transform

$$\hat{\theta}_0(w) := \hat{\phi}(w)\hat{R}^{-\frac{1}{2}}(w)$$

and its biorthogonal companion $\theta^0(t)$ given by

$$\hat{\theta}^0(w) := \hat{\phi}(w)\hat{R}^{\frac{1}{2}}(w).$$

They satisfy $\int \theta^0(t - n)\theta_0(t - m)dt = \delta_{nm}$. We define U_0 and U^0 to be respectively the closed linear span of $\{\theta_0(t - n)\}$ and of $\{\theta^0(t - n)\}$. If we define X_n by

$$X_n = \int_{-\infty}^{\infty} X(t)\theta_0(t-n)dt \qquad (14.9)$$

then we have

$$\begin{aligned}
E[X_n X_k] &= \int\int R(t-s)\theta_0(t-n)\theta_0(s-k)dt\,ds \\
&= \frac{1}{2\pi}\int \hat{R}(w)|\theta_0(w)|^2 e^{iw(n-k)}dw \\
&= \frac{1}{2\pi}\int |\widehat{\phi}(w)|^2 e^{iw(n-k)}dw \qquad (14.10)\\
&= \int \phi(t-n)\phi(t-k)dt = \delta_{nk}.
\end{aligned}$$

The questions are: does $\{\theta_0(t-n)\}$ form a basis of U_0 and can similar modifications be made to the wavelets?

We first show that these functions induce a multiresolution-like analysis and Riesz bases for the nested subspaces. Furthermore, our wide-sense stationary process can be approximated arbitrarily closely by a K-L-like expansion whose basis is composed of these functions.

Towards that end, define a set of scaling-function-like functions

$$\hat{\theta}_{m,n}(w) = \hat{R}(w)^{-1/2}\widehat{\phi}_{m,n}(w), \qquad (14.11)$$

$$\hat{\theta}^{m,n}(w) = \hat{R}(w)^{1/2}\widehat{\phi}_{m,n}(w), \qquad (14.12)$$

which are shifted versions of the following functions

$$\hat{\theta}_m(w) = \hat{R}(w)^{-1/2}\widehat{\phi}_m(w),$$

$$\hat{\theta}^m(w) = \hat{R}(w)^{1/2}\widehat{\phi}_m(w).$$

Equation (14.11) has an interesting interpretation when $\hat{R}(w)$ decreases as the frequency increases (low-pass): in this case $\hat{R}^{-1/2}(w)$ acts as a high-pass whitening filter. The definitions in (14.11) and (14.12) can also be considered as a pair of linear transformations on $L^2(\mathbb{R})$, denoted by $\mathcal{R}^{-1/2}$ and $\mathcal{R}^{+1/2}$. Then, $\mathcal{R}^{-1/2}$ maps each V_m in the MRA onto a subspace U_m, the closure of the subspace spanned by $\{\theta_{m,n}(t)\}$. In fact, we have

Lemma 14.1

For any m, $\{\theta_{m,n}(t)\} = \{\theta_m(t-2^{-m}n)\}$ is a Riesz basis of U_m, and $\{\theta_{m,n}(t), \theta^{m,n}(t)\}$ is a biorthogonal sequence.

PROOF Without loss of generality, we prove it for the case of $m = 0$. To show that $\{\theta_0(t-n)\}$ forms a Riesz basis of U_0, we only need to show that there exist A and B such that

$$A \sum_n |c_n|^2 \leq \left\| \sum_n c_n \theta_0(t-n) \right\|^2 \leq B \sum_n |c_n|^2$$

where $\{c_n\}$ is any sequence in ℓ^2. By Parseval's equality

$$\left\| \sum_n c_n \theta_0(t-n) \right\|^2$$

$$= \frac{1}{2\pi} \int_{-4\pi/3}^{4\pi/3} |\theta_0(w)|^2 \left| \sum_n c_n e^{-iwn} \right|^2 dw$$

$$= \frac{1}{2\pi} \int_{-4\pi/3}^{4\pi/3} |\widehat{\phi}(w)|^2 \cdot \hat{R}^{-1}(w)] \left| \sum_n c_n e^{-iwn} \right|^2 dw$$

$$\leq \frac{1}{2\pi} \sup_{|w| \leq 4\pi/3} \hat{R}^{-1}(w) \int_{-4\pi/3}^{4\pi/3} |\widehat{\phi}(w)|^2 \left| \sum_n c_n e^{-iwn} \right|^2 dw$$

$$\leq \frac{1}{2\pi} \sup_{|w| \leq 4\pi/3} \hat{R}^{-1}(w) \int_{-2\pi}^{2\pi} \left| \sum_n c_n e^{-iwn} \right|^2 dw$$

$$= 2 \sup_{|w| \leq 4\pi/3} \hat{R}^{-1}(w) \sum_n |c_n|^2.$$

This establishes the second inequality. Similarly, for the first inequality,

$$\left\| \sum_n c_n \theta_0(t-n) \right\|^2$$

$$= \frac{1}{2\pi} \int |\hat{\theta}_0(w)|^2 \left| \sum_n c_n e^{-iwn} \right|^2 dw$$

$$= \frac{1}{2\pi} \int_{-4\pi/3}^{4\pi/3} |\widehat{\phi}(w)|^2 \cdot [\hat{R}^{-1}(w)] \left| \sum_n c_n e^{-iwn} \right|^2 dw$$

$$\geq \frac{1}{2\pi} \inf_{|w| \leq 4\pi/3} \hat{R}^{-1}(w) \int_{-\pi}^{\pi} |\widehat{\phi}(w)|^2 \left| \sum_n c_n e^{-iwn} \right|^2 dw$$

$$\geq \inf_{|w| \leq 4\pi/3} \hat{R}^{-1}(w) \min\{\widehat{\phi}(-\pi), \widehat{\phi}(\pi)\} \sum_n |c_n|^2.$$

Here, we have used the fact that the minimum $\phi(w)$ in $[-\pi, \pi]$ is achieved at the boundary points (notice that when $\hat{\phi}(w)$ is symmetric, $\hat{\phi}(-\pi) = 1/\sqrt{2} = \hat{\phi}(\pi)$).   □

A similar result with exactly the same proof holds for $\{\theta^{mn}\}$ and $U^m = \mathcal{R}^{1/2}V_m$. While these subspaces satisfy many of the properties of the $\{V_m\}$, they are not dilation subspaces. Nonetheless we have

Lemma 14.2

$\{U_m\}$ *(and* $\{U^m\}$*) form a nested sequence of subspaces in* $L^2(\mathbb{R})$ *such that*

$$V_{m-1} \subset U_m(\text{resp. } U^m) \subset V_{m+1}$$

and

$$\overline{\bigcup_m U_m} = \overline{\bigcup_m U^m} = L^2(\mathbb{R}), \qquad (14.13)$$

$$\bigcap_m U_m = \bigcap_m U^m = \{0\}. \qquad (14.14)$$

PROOF For any $x(t) \in U_m$, the inverse of $\mathcal{R}^{-1/2}$ carries $x(t)$ to a unique $y(t) \in V_m$. But $y(t) \in V_{m+1}$, hence $x(t) \in U_{m+1}$, i.e., $\{U_m\}$ is nested. To show (14.14), we can use contradiction. Suppose there is an $x(t) \neq 0$ such that $x(t) \in \bigcap_m U_m$, then its inverse $y(t) \neq 0$ and $y(t) \in \bigcap_m V_m$, which contradicts (3.2.iv of Chapter 3). Finally, without loss of generality, (14.13) can be established if we can show

$$V_0 \subset U_1.$$

Towards that end, consider an arbitrary $x(t) \in V_0$. Then, $x(t) \in U_1$ if and only if

$$\hat{x}(w) = \beta(w)\hat{\theta}_1(w),$$

where $\beta(w)$ is a periodic function with period $2 \cdot 2\pi = 4\pi$. To show this, we notice that $\hat{x}(w)$ can be represented as

$$
\begin{aligned}
\hat{x}(w) &= \alpha(w)\hat{\phi}(w) \\
&= \alpha(w)m_0(w/2)\hat{\phi}(w/2) \\
&= \alpha(w)m_0(w/2)[\hat{R}^{1/2}(w)]^*_{4\pi}[\hat{R}^{-1/2}(w)]^*_{4\pi}\hat{\phi}(w/2) \\
&= \alpha(w)[\hat{R}^{1/2}(w)]^*_{4\pi}m_0(w/2)[\hat{R}^{-1/2}(w)]^*_{4\pi}\hat{\phi}(w/2)
\end{aligned}
$$

where $[\hat{R}^{1/2}(w)]^*_{4\pi}$ is the 4π-periodic extension of the restriction of $\hat{R}^{1/2}(w)$ to $[-4\pi/3, 4\pi/3]$ and similarly for $[\hat{R}^{-1/2}(w)]^*_{4\pi}$. Since $m_0(w/2)$ is zero in $[-8\pi/3, -4\pi/3]$ and $[4\pi/3, 8\pi/3]$, and $\hat{\phi}(w/2)$ is zero outside $[-8\pi/3, 8\pi/3]$, the product of the last three terms can be written as

$$m_0(w/2)[\hat{R}^{-1/2}(w)]^*_{4\pi}\widehat{\phi}(w/2)$$
$$= m_0(w/2)\hat{R}^{-1/2}(w)\widehat{\phi}(w/2)$$
$$= m_0(w/2)\widehat{\phi}_1(w).$$

Hence $\hat{x}(w)$ becomes

$$\hat{x}(w) = \alpha(w)[\hat{R}^{1/2}(w)]^*_{4\pi}m_0(w/2)\hat{\phi}_1(w),$$

where the product of the first three factors is a periodic function with period of 4π. Hence, $V_0 \subset U_1$. The same technique can be used to show $U_0 \subset V_1$ and $V_0 \subset U^1 \subset V_2$. $\square$

Similar to (14.11) and (14.12), we define the following wavelet-like functions $\gamma_{m,n}$ and $\gamma^{m,n}$

$$\hat{\gamma}_{m,n}(w) = \hat{R}^{-1/2}(w)\hat{\psi}_{m,n}(w), \qquad (14.15)$$

$$\hat{\gamma}^{m,n}(w) = \hat{R}^{1/2}(w)\hat{\psi}_{m,n}(w),$$

based on the following functions:

$$\hat{\gamma}_m(w) = \hat{R}^{-1/2}(w)\hat{\psi}_m(w),$$

$$\hat{\gamma}^m(w) = \hat{R}^{1/2}(w)\hat{\psi}_m(w).$$

Then, for any given m it is not difficult to show

Lemma 14.3
The functions given by (14.15),$\{\gamma_{m,n}(t)\} = \{\gamma_m(t-2^{-m}n)\}$, constitute a Riesz basis of M_m, where

$$M_m = \mathcal{R}^{-1/2}W_m,$$

and $\{\gamma_{m,n}(t), \gamma^{m,n}(t)\}$ is a biorthogonal sequence in both m and n.

Together, Lemmas 14.1, 14.2, and 14.3 imply that $L^2(\mathbb{R})$ has a multiresolution-like decomposition

$$L^2(\mathbb{R}) = U_0 \bigoplus_{m \geq 0} M_m,$$

where $\oplus$ represents a direct sum and $\{\theta_{0,n}(t)\}$ and $\{\gamma_{m,n}(t)\}$, $m \geq 0$ form a basis (but not necessarily a Riesz basis) of $L^2(\mathbb{R})$. Hence, we have the following

THEOREM 14.1
If the spectral density of $X(t)$ is strictly positive, then it has a K-L-like expansion

$$X(t) = \sum_n X_{0,n}\theta^0(t - n) + \sum_{m \geq 0}\sum_n X'_{m,n}\gamma^m(t - 2^{-m}n)$$

$$= \sum_m \sum_n X'_{m,n}\gamma^m(t - 2^{-m}n) \quad (14.16)$$

where the expansion coefficients $X_{0,n} = \langle X(t), \theta_0(t - n)\rangle$ and $X'_{m,n} = \langle X(t), \gamma_m(t - 2^{-m}n)\rangle$ are uncorrelated. Furthermore, the convergence in (14.16) is in the mean square sense.

PROOF   The calculations of (14.10) can be used to show the coefficients are uncorrelated. Thus we need only show the mean square convergence. Consider the mean square difference between $X(t)$ and the partial sums of the expansion (14.16)

$$E\left[\left(X(t) - \sum_m \sum_n X'_{m,n}\gamma^m(t - 2^{-m}n)\right)^2\right] \quad (14.17)$$

$$= E\left[\begin{array}{c} X^2(t) - 2\sum_m \sum_n X(t)X'_{m,n}\gamma^m(t - 2^{-m}n) \\ + \left(\sum_m \sum_n X'_{m,n}\gamma^m(t - 2^{-m}n)\right)^2 \end{array}\right]$$

$$= R(0) - 2\sum_m \sum_n [\gamma^m(t - 2^{-m}n)]^2 + \sum_m \sum_n [\gamma^m(t - 2^{-m}n)]^2$$

$$= R(0) - \sum_m \sum_n [\gamma^m(t - 2^{-m}n)]^2 \geq 0.$$

Here we have used the fact that $X'_{m,n}$ are uncorrelated and

$$\int R(t - s)\gamma_m(s - 2^{-m}n)ds = \gamma^m(t - 2^{-m}n).$$

To show that (14.17) is zero in the limit (m, n go from $-\infty$ to $+\infty$), we first notice that the partial sum in the last line is convergent since it is nonnegative and bounded above by $R(0)$. Hence, the inequality in (14.17) also holds in the limit.

Furthermore, notice that $\gamma^m(t - 2^{-m}n)$ can be written as

$$\gamma^m(t - 2^{-m}n) = \frac{1}{2\pi} \int \hat{\gamma}^m(w) e^{iw(t-2^{-m}n)} dw$$

$$= \frac{1}{2\pi} \int \hat{R}^{1/2}(w) \hat{\psi}_m(w) e^{iw(t-2^{-m}n)} dw$$

$$= \langle \hat{R}^{1/2}(w), \hat{\psi}_m(w) e^{iw(t-2^{-m}n)} \rangle$$

and $\{\hat{\psi}_m(w) e^{iw(t-2^{-m}n)}\}$ is an orthonormal basis of all the square integrable functions in the frequency domain ($L^2(\mathbb{R})$). Hence, by Parseval's
equality (for series expansions)

$$\sum_m \sum_n [\gamma^m(t - 2^{-m}n)]^2 = \int |\hat{R}^{1/2}(w)|^2 dw = \int \hat{R}(w) dw = R(0);$$

that is, equality holds in (14.17) in the limit.   □

14.4 Wavelets based on band limited processes

In the last section we repeatedly used the fact that $\hat{R}(w) \neq 0$. However many processes arising in applications are band limited and hence
$\hat{R}(w) = 0$ for $|w| > \sigma$. Thus we need to modify our approach. In fact, we
shall introduce a new family of wavelets based on $R(t)$ for which similar
results hold [W-Z].

Accordingly let $X(t)$ be a wide-sense stationary band limited process
whose spectral density $\hat{R}(w)$ is continuous, $\hat{R}(0) = 1$, and

$$\hat{R}(w) = \begin{cases} 0, & |w| \geq \sigma, \\ > 0, & |w| < \sigma. \end{cases}$$

By changing the time scale if necessary we may take $\sigma = \pi + \epsilon$ where
$0 \leq \epsilon \leq \frac{\pi}{3}$. We denote by $\hat{R}^*(w)$ the 2π periodic extension

$$\hat{R}^*(w) := \sum_k \hat{R}(w + 2\pi k), \quad w \in \mathbb{R}.$$

Clearly $\hat{R}^*$ is bounded, continuous, and positive.

The scaling function $\phi(t)$ is defined as that function whose Fourier
transform $\hat{\phi}$ is given by

$$\hat{\phi}(w) := \hat{R}^{1/2}(w) / (\hat{R}^*(w))^{1/2}. \tag{14.18}$$

In order to qualify as an orthogonal scaling function we need only to show (Chapters 3, 9) that

(i) $\sum |\phi(w + 2\pi k)|^2 = 1,$

(ii) $\widehat{\phi}(w) = m_0 \left(\frac{w}{2}\right) \widehat{\phi}\left(\frac{w}{2}\right),$

(iii) $\widehat{\phi}$ is continuous at 0 and $\widehat{\phi}(0) = 1,$

where $m_0\left(\frac{w}{2}\right)$ is a 4π periodic function in $L^2(\mathbb{R})$. The first comes directly from our definition (14.18) while the second and third may be shown by using the fact that

$$\widehat{\phi}(w) = \begin{cases} 0, & |w| \leq \pi + \epsilon, \\ 1, & |w| \leq \pi - \epsilon. \end{cases}$$

Hence the support of $\widehat{\phi}(w)$ is contained in the interval $[-2\pi + 2\epsilon, 2\pi - 2\epsilon]$ where $\widehat{\phi}\left(\frac{w}{2}\right) = 1$. Thus $m_0\left(\frac{w}{2}\right)$ may be taken as the 4π periodic extension of $\widehat{\phi}(w)$ since, furthermore, the $\widehat{\phi}(w \pm 4\pi)$ are zero in the support of $\widehat{\phi}\left(\frac{w}{2}\right)$.

The mother wavelet may then be constructed from $\phi(t)$ by the standard procedure, i.e.,

$$\widehat{\psi}(w) := e^{-iw/2}\overline{m_0\left(\frac{w}{2} + \pi\right)}\widehat{\phi}\left(\frac{w}{2}\right). \tag{14.19}$$

Lemma 14.4
Let $\phi(t)$ be the function given by (14.18) and $\psi(t)$ the one given by (14.19). Then $\phi(t)$ is a scaling function that generates a multiresolution analysis $\{V_m\}$, and $\psi(t)$ is a mother wavelet such that

$$2^{m/2}\psi(2^m t - n) = \psi_{mn}(t)$$

is an orthonormal basis of $L^2(\mathbb{R})$.

PROOF It remains only to show that $\bigcup_m V_m$ is dense in $L^2(\mathbb{R})$. But this is clear since any $L^2(\mathbb{R})$ function whose Fourier transform has compact support is in V_m for m sufficiently large. Since such functions are dense in $L^2(\mathbb{R})$ so is $\bigcup_m V_m$.   □

The expression for the wavelet may be simplified in our case since only a finite number of terms are needed. It becomes

$$\widehat{\psi}(w) = e^{-iw/2}(\widehat{\phi}(w + 2\pi) + \widehat{\phi}(w - 2\pi))\widehat{\phi}\left(\frac{w}{2}\right).$$

This is the same formula as in the last section, and the wavelets are similar. From this formula, it follows that

$$|\hat{\psi}(w)| = \begin{cases} 1, & \pi + \epsilon \le |w| \le 2\pi - 2\epsilon, \\ 0, & |w| \le \pi - \epsilon \text{ or } |w| > 2\pi - 2\epsilon. \end{cases}$$

The results of the last section thus hold for these wavelet expansions, but now because of the compact support of $\hat{R}(w)$, many coefficients are zero. In fact, the wavelet coefficients of $X(t)$ satisfy

$$|E(b_{mn}b_{mk})| = \left| \frac{2^{-m}}{2\pi} \int \hat{R}(w) \left| \hat{\psi}(w2^{-m}) \right|^2 e^{iw2^{-m}(n-k)} dw \right|$$

$$\le \frac{2^{-m}}{\pi} \int_{2^m(\pi-\epsilon)}^{\pi+\epsilon} \hat{R}(w) \left| \hat{\psi}(w2^{-m}) \right|^2 dw.$$

This is zero for $m > 0$. Hence the wavelet series is truncated, and we have the following

THEOREM 14.2

Let $X(t)$ be a wide-sense stationary band limited process such that $\hat{R}(w)$ has support on $[-\epsilon - \pi, \pi + \epsilon]$, let $\phi(t)$, and $\psi(t)$ be given by (14.18) and (14.19), respectively. Then

$$X(t) = \sum_n a_{1n} 2^{1/2} \phi(2t - n) \quad (14.20)$$

$$= \sum_n a_{-mn} 2^{-m/2} \phi(2^{-m}t - n) + \sum_{k=0}^{m} \sum_n b_{-kn} 2^{-k/2} \psi(2^{-k}t - n)$$

in the mean square sense, where the $\{a_{mn}\}$ and $\{b_{mn}\}$ are the usual coefficients of $X(t)$; the b_{kj} and b_{mn} are uncorrelated for $|k - m| > 1$ and are equal to zero for $m > 0$.

PROOF The uncorrelatedness follows from the last section. Therefore we need only prove the mean square convergence. We first observe that $R(t - s) \in V_1$ as a function of s for each t since

$$(\widehat{R(t - \cdot)})(w) = \hat{R}(w)e^{-iwt} = \hat{R}(w)e^{-iwt}\hat{\phi}(2^{-1}w)$$

$$= \sum_k \hat{R}(w + 4\pi k)e^{-i(w+4\pi k)t}\hat{\phi}(2^{-1}w) \quad (14.21)$$

and since $\hat{R}(w) = 0$ for $|w| > \pi + \epsilon$ and $\hat{\phi}(2^{-1}w) = 1$ for $|w| \le 2\pi - 2\epsilon$. Furthermore, since $\hat{\phi}(2^{-1}w) = 0$ whenever $|w| > 2\pi + 2\epsilon$, the periodic extension in (14.21) is valid. This tells us that $\widehat{R}(t - \cdot)$ is a 4π periodic function times $\widehat{\phi}(2^{-1}w)$, and hence $R(t - s) \in V_1$ as a function of s.

The mean square error in (14.20) is

$$E\left[X(t) - \sum_n a_{1n}\sqrt{2}\,\phi(2t - n)\right]^2$$

$$= R(0) - 2\sum_n \alpha_{1n}(t)\sqrt{2}\,\phi(2t - n)$$

$$+ \sum_k \int \left\{\sum_n \alpha_{1n}(u)\sqrt{2}\phi(2t - n)\right\}\sqrt{2}\phi(2u - k)du\sqrt{2}\phi(2t - k)$$

$$= R(0) - 2R(0) + R(0) = 0,$$

where $\alpha_{1n}(u) = \int R(u - s)\sqrt{2}\phi(2s - n)ds$, and hence the series expansion is

$$R(u - t) = \sum_n \alpha_{1n}(u)\sqrt{2}\phi(2t - n).$$

□

We can also imitate the procedure of the last section and introduce biorthogonal sequences $\{\theta_{mn}\}$ and $\{\theta^{mn}\}$ as well as $\{\gamma_{mn}\}$ and $\{\gamma^{mn}\}$. They are given by the same formulae (14.11) and (14.15) except that they are valid only for $m < 0$ for θ_{mn} and γ_{mn}. For $m = 0$ we may take

$$\hat{\theta}_0(w) = \chi_{[-\pi-\epsilon,\pi+\epsilon]}(w) / [\widehat{R}^*(w)]^{1/2}.$$

Then the expansion coefficients

$$X_n = \int X(t)\theta_0(t - n)dt$$

are uncorrelated and

$$E\left[X(t) - \sum_{n=-N}^{N} X_n\theta^0(t - n)\right]^2 = O(\epsilon) \quad \text{as } N \to \infty.$$

The ϵ in this last statement is the one in the definiton of $\phi(t)$ (and hence of $\theta^0(t)$) and can be made small if the support of $\widehat{R}(w)$ is close to $[-\pi, \pi]$. However, it cannot be made zero, and therefore we have only an approximation in these cases.

14.5   Nonstationary processes

Certain nonstationary processes, e.g., those with stationary increments, can be studied by methods similar to the last sections [Ms].
However there is a class of processes, the cyclostationary processes, that
in general cannot but may sometimes be attacked by other wavelet methods.

A second order process $\{X(t), t \in \mathbb{R}\}$ is cyclostationary(or periodically
correlated) with period $T > 0$) [Pa1],[Hued] if

$$E(X(t+T)) = E(X(t)),$$
$$r(t+T, s+T) = E[X(t+T)X(s+T)]$$
$$= E[X(t)X(s)] = r(t,s). \tag{14.22}$$

We shall assume that T is the smallest number for which (14.22) holds,
which, by changing the scale if necessary, we take it to be 1. We shall
assume that $E(X(t)) \equiv 0$. These processes occur naturally in many
applications involving seasonal or diurnal data [G]. An example of such a
process is the modulation of a deterministic periodic signal by a random
stationary process.

Cyclostationary processes can be constructed from scaling function
series of the form

$$X(t) = \sum C_n \phi(t - n) \tag{14.23}$$

where ϕ is a scaling function, and $\{C_n\}$ is a stationary discrete second
order process with zero mean since

$$r(t,s) = \sum_n \sum_m E(C_n C_m) \phi(t-n) \phi(s-m)$$
$$= \sum_n \sum_m \rho_{n-m} \phi(t-n) \phi(s-m) \tag{14.24}$$
$$= \sum_n \sum_m \rho_{n+1-m-1} \phi(t-1-n) \phi(s-1-m) = r(t-1, s-1).$$

If, in addition, $\{C_n\}$ is assumed to be white noise, i.e., i.i.d. with variance σ^2, then $r(t,s) = \sigma^2 q(t,s)$, where q is the reproducing kernel of
V_0. The general case can also be expressed in terms of this reproducing
kernel

$$r(t,s) = \sum_n \sum_m \rho_{n-m} \phi(t-n) \phi(s-m)$$

$$= \sum_j \sum_m \rho_j \phi(t - m - j)\phi(s - m) = \sum_j \rho_j q(t - j, s).$$

However, the converse is not so easy; given $X(t)$ a cyclostationary process, is it possible to find a scaling function $\phi(t)$ such that (14.23) is satisfied? In fact, one would not expect to find a true scaling function since only one scale is involved, but it might be possible to find a shift-invariant space with a generating function $\phi(t)$, i.e., a space with a Riesz basis $\{\phi(t - n)\}$. Such a generating function does not necessarily have an associated dilation equation.

Example 14.1
We present one example for which a scaling function does arise, the binary transmission process [Pa1]. Here $X(t)$ takes the value ± 1 in each interval $n - 1 \le t < n$ with the choice of signs random. In this case $E(X(t)) = 0$ and

$$r(t, s) = \begin{cases} 1, \text{ if } n - 1 \le t, s < n, \text{ for some } n \in \mathbb{Z}, \\ 0, \text{ otherwise.} \end{cases}$$

Clearly $X(t)$ is nonstationary and $r(t + 1, s + 1) = r(t, s)$. Furthermore, $r(t, s)$ may be expressed as

$$r(t, s) = \phi(t - [s])$$

where $\phi(t)$ is the Haar scaling function (Chapter 1). But in this case

$$q(t, s) = \phi(t - [s]) = \sum_n \phi(t - n)\phi(s - n).$$

The expansion of $X(t)$ with respect to the $\{\phi(t - n)\}$ is

$$C_n = \int_{-\infty}^{\infty} X(t)\phi(t - n)dt = \int_n^{n+1} X(t)dt,$$

and hence

$$E(C_n C_m) = \int_n^{n+1} \int_m^{m+1} E(X(t)X(s))dt\,ds$$

$$= \int_n^{n+1} \int_m^{m+1} r(t, s)dt\,ds = \delta_{nm},$$

i.e., $\{C_n\}$ is uncorrelated.

The mean square convergence is easy to show as well in this case. This results in the representation of $X(t)$,

$$X(t) = \sum_n C_n \phi(t - n)$$

by an uncorrelated sequence $\{C_n\}$. □

For certain types of cyclostationary processes, those which have the form of a (wide sense) stationary process modulated by a periodic deterministic signal, the methods of section 14.3 can be used. If this modulated process is given by

$$X(t) = f(t)Y(t)$$

where f is a non-vanishing periodic function of period 1 in $L^2(0,1)$ and Y is the stationary process whose spectral density $\widehat{R}(w)$ does not vanish, then the definitions of θ_0 and θ^0 in section 14.3 are modified by incorporating f, i.e., $\theta_0^*(t) = f^{-1}(t)\theta_0(t)$, and $\theta^{0*}(t) = f(t)\theta^0(t)$ where $\widehat{\theta_0^*}(w) = \widehat{R}^{-1/2}(w)\phi(w)$ and similarly for θ^{0*}. The results are the same.

An approximation to $X(t)$ is given by

$$X_0(t) = f(t)Y_0(t) = \sum A_n f(t-n)\theta^0(t-n)$$

where the A_n are uncorrelated and are given in terms of the dual basis. Finer scales are not possible in this case however.

Many cyclostationary processes of interest to engineers can be put into a similar form

$$X(t) = \int_{-\infty}^{\infty} \mathbf{h}^T(t, w)\mathbf{Y}(w)dw$$

where $\mathbf{Y}(w)$ is a stationary vector valued process and $\mathbf{h}(t, w)$ which satisfies

$$\mathbf{h}(t + 1, w + 1) = \mathbf{h}(t, w)$$

is also a vector. These include amplitude, frequency, and phase modulation, as well as pulse modulation, which is already on the form (14.23) [G].

If, in the scalar case, $h(t, w) = q(t, w)$, the reproducing kernel of a shift-invariant subspace, then $X(t)$ will have the form (14.23). In the vector case, it is necessary to use multiwavelets, and the result would be a vector version of (14.23).

While it seems difficult to establish a general sufficient condition for a cyclostationary process to have the form (14.23), a necessary condition is easy to find.

Proposition 14.2
Let $X(t)$ be a cyclostationary process given by (14.23) with uncorrelated $\{C_n\}$ and symmetric scaling function ϕ. Then

$$\phi(s) = \frac{1}{2\pi} \int_0^{2\pi} \frac{\widehat{R}^*(w,s)}{[\widehat{R}^*(w,0)]^{1/2}} dw$$

where

$$\widehat{R}^*(w,s) = \sum_k \widehat{R}^*(w + 2\pi k, s).$$

and $\widehat{R}^*(w,s)$ is the Fourier transform of $R(t,s)$ with respect to t.

PROOF The proof depends only on a few calculations beginning with

$$\widehat{R}^*(w,s) = \sum_n \widehat{\phi}(w)e^{-iwn}\phi(s-n).$$

The 2π periodic extension is

$$\widehat{R}^*(w,s) = \sum_k \widehat{\phi}(w + 2\pi k) \sum_n e^{-iwn}\phi(s-n), \tag{14.25}$$

which for $s = 0$ becomes

$$\widehat{R}^*(w,0) = \widehat{\phi}^*(w) \sum_k \phi(n)e^{iwn} = \left[\widehat{\phi}^*(w)\right]^2$$

by the Poisson summation formula. We substitute this expansion in (14.24) to obtain

$$\widehat{R}^*(w,s) = \left[\widehat{R}^*(w,0)\right]^{1/2} \sum_n e^{-iwn}\phi(s+n). \tag{14.26}$$

Thus this periodic function satisfies

$$\frac{\widehat{R}^*(w,s)}{\left[\widehat{R}^*(w,0)\right]^{1/2}} = \sum_n e^{iwn}\phi(s+n)$$

and hence the Fourier coefficients of the left side are given by $\phi(s+n)$, i.e.,

$$\phi(s+n) = \frac{1}{2\pi} \int_0^{2\pi} \frac{\widehat{R}^*(w,s)}{[\widehat{R}^*(w,0)]^{1/2}} e^{-iwn} dw.$$

This result for $n = 0$ gives our conclusion. □

14.6 Problems

1. Let $0 < \alpha < \pi$ and let $X(t)$ be a (Gaussian) stochastic process on $[0,1]$ such that $X(0) = X(1) = 0$ whose covariance function is

$$r(t,s) = \frac{1}{\alpha \sin \alpha} \begin{cases} \sin \alpha t \sin \alpha(1-s), \ 0 < t < s < 1 \\ \sin \alpha s \sin \alpha(1-t), \ 0 < s < t < 1. \end{cases}$$

(i) Convert the eigenvalue problem $\mathcal{R}\phi = \lambda\phi$ to the differential equation $\phi = \lambda(\phi'' + \alpha^2\phi)$ with boundary conditions $\phi(0) = \phi(1) = 0$.

(ii) Solve (i) for the eigenvalues $(\lambda_k = 1/(\alpha^2 - \pi^2 k^2))$ and eigenvectors.

(iii) Expand $X(t)$ in a series with uncorrelated coefficients.

2. Show the Brownian bridge is not a stationary process.

3. Let $\psi(t)$ be the Shannon wavelet; let $X(t)$ satisfy the hypothesis of Proposition 14.1. Show that the wavelet coefficients are uncorrelated at different scales.

4. Let $\phi(t)$ be the scaling function of the Shannon wavelet, let $X(t)$ be as in Problem 3, and let

$$X_n = \int_{-\infty}^{\infty} X(t)\phi(t-n)dt.$$

Show that $\{X_n\}$ is a stationary sequence.

5. Let $X(t)$ be a wide-sense stationary process whose correlation function is $R(t) = [2\pi(t^2+1)]^{-1}$. Find the spectral density and use it to construct $\theta_0(t)$ and $\theta^0(t)$ of Section 14.3 for the scaling function of the Shannon wavelet.

6. Find the wavelet-like functions $\gamma_{mn}(t)$ for the example of Problem 5.

7. Let $X(t)$ be a wide-sense stationary process with $\hat{R}(w) = 1 - \frac{|w|}{\pi+\epsilon}$, $|w| < \pi + \epsilon$; 0 otherwise. Find the Fourier transform of the associated scaling function (14.8).

8. Show that the scaling function in Problem 7 satisfies the conditions needed for a multiresolution analysis.

Bibliography

[A-K] N. Atreas and C. Karanikas, Gibbs phenomenon on sampling series based on Shannon's and Meyer's wavelet analysis. *J. Fourier Anal. Appl.* 5 (1999), 575–588.

[Al] G. Alexits, *Convergence Problems of Orthogonal Series*, Oxford, New York, 1961.

[Ar] N. Aronszajn, Theory of reproducing kernels, *Trans. Amer. Math. Soc.* 68 (1950), 337-404.

[As] R. Askey, *Orthogonal polynomials and special functions*, SIAM, Philadelphia, 1975.

[A-W-W] P. Auscher, G. Weiss, and M. V. Wickerhauser, Local sine and cosine bases of Coifman and Meyer and the construction of smooth wavelets, in *Wavelets—A Tutorial in Theory and Applications*, C. K. Chui (ed.), 237-256, Academic Press, New York, 1992.

[Ba] G. Battle, A block spin construction of ondelettes Part I: Lemarié functions, *Comm. Math. Phys.* 110 (1987), 601-615.

[Be] J. Benedetto, Frame decompositions, sampling, and uncertainty principle inequalities, in *Wavelets: mathematics and applications*, 247–304, Stud. Adv. Math., CRC, Boca Raton, FL, 1994.

[B-C-R] G. Beylkin, R. Coifman, and V. Rokhlin, The fast wavelet transform, in *Wavelets and Their Applications*, M. Ruskai, *et al.* (ed.), 181-210, Jones and Bartlett, Boston, 1991.

[B-C] S. Bochner and K. Chandrashekharan, *Fourier Transforms*, Princeton University, Princeton, 1949.

[Br] H. Bremermann, *Distributions, Complex Variables, and Fourier Transforms*, Addison-Wesley, Reading, MA, 1965.

[B] P. L. Butzer, A survey of the Whittaker-Shannon sampling theorem and some of its extensions, *J. Math. Res. Exposition* 3 (1983), 185-212.

[B-S-S] P. L. Butzer, W. Splettstößer, and R. L. Stens, The sampling theorem and linear predictions in signal analysis, *Jahresber. Deutsch. Math.-Verein.* 90 (1988), 1-60.

[C-Q] C. Canuto and A. Quarteroni, Approximation results for orthogonal polynomials in Sobolev spaces, *Math. Comp.* 38 (1982), 67-86.

[C-I-S] W. Chen, S. Itoh, and J. Shiki, Irregular sampling theorems for wavelet subspaces, *IEEE Trans. Inform. Theory* 44 (1998), 1131–1142.

[Ca] J. S. Chihara, *An Introduction to Orthogonal Polynomials*, Gordon and Breach, New York, 1978.

[Cu] S. T. Chiu, Bandwidth selection for kernel density estimation, *Annals Stat.* 19 (1991), 1883-1905.

[C] C. K. Chui, *An Introduction to Wavelets*, Academic Press, New York, 1992.

[C-L] E. A. Coddington and N. Levinson, *Theory of Ordinary Differential Equations*, McGraw Hill, New York, 1955.

[C-D-F] A. Cohen, I. Daubechies, and J. C. Feaveau, Biorthogonal bases of compactly supported wavelets, *Comm. Pure Appl. Math.* 45 (1992), 485-560.

[C-M] R. R. Coifman and Y. Meyer, Remarques sur l'analyse de Fourier à fenêtre, série I, *C. R. Acad. Sci.*, Paris 312 (1991), 259-261.

[C-M-W] R. Coifman, Y. Meyer, and M. W. Wickerhauser, Size properties of wavelet packets, in *Wavelets and Their Applications*, Ruskai, et al. (eds.), Jones and Bartlett, Boston, 1992.

[C-H] D. Colella and C. Heil, Characterizations of scaling functions I. Continuous solution, *SIAM J. Matrix Anal. Appl.* (1993).

[D1] I. Daubechies, The wavelet transform, time-frequency localization and signal analysis, *IEEE Trans. Info. Th.* 36 (1990), 961-1005.

[D2] I. Daubechies, Orthonormal bases of compactly supported wavelets, *Comm. Pure Appl. Math.* 41 (1988), 909-996.

[D] I. Daubechies, *Ten Lectures on Wavelets, SIAM,* Philadelphia, 1992.

[D-J] D. L. Donoho and I. M. Johnstone, Idea spatial adaptation by wavelet shrinkage, *Biometrika* 81 (1994), 425-455.

[D-J-K-P] D. L. Donoho, I. M. Johnstone, G. Kerkyacharian and D. Picard, Density estimation by wavelet thresholding, *Ann Statist.* 24(1996), 508-539.

[D-L-R] A. Dempster, N. M. Laird, and D. B. Rubin, Maximum likelihood from incomplete data via the EM algorithm (with discussion), *J. Royal Stat. Soc.* (B) 39 (1977), 1-38.

[E] R. E. Edwards, *Fourier Series, A Modern Introduction,* 2 Vol., Holt, Rinehart and Winston, New York, 1967.

[Fl] P. Flandrin, On the spectrum of fractional Brownian motion, *IEEE Trans. Info. Th.* 35 (1989), 197-199.

[F-R] A. Földes and P. Rèvèsz, A general method for density estimation, *Studia Sci. Math. Hungar.* 9 (1974), 443-452.

[G] W. A. Gardner, *Introduction to random processes with applications to signals and systems,* 2nd ed., New York: McGraw-Hill, 1990.

[G-S] I. M. Gelfand and G. E. Shilov, *Generalized Functions,* Vol. 1, 3, Academic Press, New York, 1964, 1967.

[G-H-M] J. Geronimo, D. Hardin, and P. R. Massopust, Fractal functions and wavelet expansions based on several functions, *J. Approx. Th.* 78 (1994), 373-401.

[Gi] J. W. Gibbs, Letter to the editor, *Nature* 59 (1898-1899), 200

[Gi1] J. W. Gibbs, Letter to the editor, *Nature* 59 (1898-1899), 606.

[Gr] H. Griffin, *Elementary Theory of Numbers,* McGraw-Hill, New York, 1954.

[G-M] A. Grossman and J. Morlet, Decomposition of Hardy functions into square integrable wavelets of constant shape, *SIAM J. Math. Anal.* 15 (1984), 723-736.

[Ha]   H. F. Harmuth, *Transmission of Information by Orthogonal Functions*, Springer, New York, 1972.

[He]   C. Heil, Methods of solving dilation equations, *Proc. 1991 NATO ASI on Prob. and Stoch. Methods in Anal. with Appl.* (1992), 15-45.

[H-W] C. E. Heil and D. F. Walnut, Continuous and discrete wavelet transforms, *SIAM Review* 31 (1989), 628-666.

[Hel] G. Helmberg, The Gibbs phenomenon for Fourier interpolation, *J. Approx. Theory* 78 (1994), 41–63.

[H-H] E. Hewitt and R. E. Hewitt, The Gibbs-Wilbraham phenomenon: An episode in Fourier analysis, *Archives for Hist. of Exact Sciences* 21 (1980), 129.

[H]   J. R. Higgins, Five short stories about the cardinal series, *Bull. Amer. Math. Soc.* 12 (1985), 45-89.

[Hi]   E. Hille, Introduction to the general theory of reproducing kernels, *The Rocky Mountain J. Math.* 2 (1972), 321-368.

[Hu]   H. L. Hurd, Representation of strongly harmonizable periodically correlated processes and their covariances, *J. Multivariate Anal.* 29 (1989), 53-67.

[Jc]   D. Jackson, *Fourier Series and Orthogonal Polynomials*, MAA, Oberlin, OH, 1941.

[J]   A. J. E. M. Janssen, The Zak transform and sampling theorem for wavelet subspaces, *IEEE Trans. Signal Proc.* 41 (1993), 3360-3364.

[J1]   A. J. E. M. Janssen, The Smith-Barnwell condition and nonnegative scaling functions, *IEEE Trans. Info. Th.* 38 (1992), 884-886.

[Je]   A. J. Jerri, *The Gibbs phenomenon in Fourier analysis, splines and wavelet approximations.* Mathematics and its Applications, 446. Kluwer Academic Publishers, Dordrecht, 1998.

[Ka]   C. Karanikas, Gibbs phenomenon in wavelet analysis. *Results Math.* 34 (1998), 330–341.

[Ke]   S. Kelly, Pointwise convergence for wavelet expansions, Ph.D. dissertation, Washington University, St. Louis, 1992.

[Ke1]   S. Kelly, Gibbs phenomenon for wavelets, *Appl. Comp. Harmonic Anal.* 3 (1996), 72-81.

[Ke-K-R]   S. E. Kelly, M. A. Kon and L. A. Raphael, Local convergence for wavelet expansions, *J. Funct. Anal* 126 (1994), 102-138.

[Ki]   J. Kiefer, On large deviations of the empiric D.F. of vector chance variables and a law of the iterated logarithm, *Pacific J. of Math.* 11 (1961), 649-660.

[K-W]   T. Koornwinder and G. G. Walter, The finite continuous Jacobi transform and its inverse, *J. Approx. Theory* 60(1990), 83-100.

[K]   J. Korevaar, Distributions defined by fundamental sequences, *Nederli Akad. Wetensch. Proc. Ser.* A 58 (1955), 368-389, 483-503, 663-674.

[K1]   J. Korevaar, *Mathematical Methods*, Academic Press, New York, 1968.

[K-T]   R. Krommal and M. Tarter, The estimation of probability densities and cumulatives by Fourier Series methods, *JASA* 63 (1968), 925-952.

[L-M]   G. Lemarié and Y. Meyer, Ondeletters et bases Hibertiennes, *Revista Matematica Iberoamericana* 2 (1986), 1-18.

[L]   Y-M. Liu, Wavelets and sampling theorems, Ph.D. thesis, University of Wisconsin-Milwaukee, Milwaukee, 1993.

[L1]   Y-M. Liu, Irregular sampling for spline wavelet subspaces, *IEEE Trans. Inform. Theory* 42 (1996), no. 2, 623–627.

[L-W]   Y-M. Liu and G. G. Walter, Irregular sampling in wavelet subspaces, *J. Fourier Anal. Appl.* 2 (1995), no. 2, 181–189.

[Lo]   S. Łojasiewicz, Sur la valeur et la limite d'une distribution dans un point, *Studia Math.* 16 (1957), 1-36.

[Md]   W. R. Madych, Some elementary properties of multiresolution analysis of $L^2(\mathbb{R}^n)$ in w*avlets, A Tutorial in Theory and Applications*, C. K. Chui (ed.), 259-294, Academic Press, New York, 1992.

[Ma]   S. Mallat, Multiresolution approximation and wavelet orthonormal bases of $L^2(\mathbb{R})$, *Trans. AMS* 315 (1989), 69-88.

[Ms] E. Masry, The wavelet transform of stochastic processes with stationary increments and its application to fractional Brownian motion, *IEEE Trans. Info. Th.* 39 (1993), 260-264.

[M1] Y. Meyer, The Franklin wavelets, preprint, 1988.

[M] Y. Meyer, *Ondelettes et opérateurs I*, Herman, Paris, 1990.

[Mo] J. Morlet, G. Arens, I. Fourgeau, and D. Giard, Wave propagation and sampling theory, Part II, *Geophysics* 47 (1982), 203-236.

[N-W] M. Z. Nashed and G. G. Walter, General sampling theorems for functions in reproducing kernel Hilbert spaces, *Math. Control Signals Systems* 4 (1991), 373-412.

[O] A. M. Olevskii, *Fourier Series with Respect to General Orthogonal Systems*, Springer, New York, 1975.

[P-W] R.E.A.C. Paley and N. Wiener, *Fourier Transforms in the Complex Domain*, Colloq. Publ. Vol. 29, Amer. Math. Soc., Providence, RI, 1934.

[Pa] A. Papoulis, *Signal Analysis*, McGraw-Hill, New York, 1977.

[Pa1] A. Papoulis, *Probability, Random Variables and Stochastic Processes*, McGraw-Hill, New York, 1991.

[P] E. Parzen, On estimation of a probability density and mode, *Ann. Math. Statist.* 35 (1962), 1065-1076.

[PR] B.L.S. Prakasa Rao, *Nonparametric Functional Estimation*, Academic Press, Orlando, FL, 1983.

[R-N] F. Riesz and B. Sz. Nagy, *Functional Analysis*, Ungar, New York, 1955.

[Ro] M. Rosenblatt, *Random Processes*, Springer, New York, 1974.

[R] W. Rudin, *Functional Analysis*, McGraw-Hill, New York, 1973.

[Sa] G. Sansone, *Orthogonal Functions*, Interscience, New York, 1955.

[Sc] B. Schomburg, On the approximation of the delta distribution in Sobolev spaces of negative order, *Applicable Analysis* 36 (1990), 89-93.

[S] L. Schwartz, *Théorie des Distributions*, Hermann, Paris, (2 vols.), 1951.

[Sch] S. C. Schwartz, Estimation of a probability density by an orthogonal series, *Ann. Math. Statis.* 38 (1964), 1261-1265.

[Sh] C. E. Shannon, Communication in the presence of noise, *Proc. IRE* 37 (1949), 10-21.

[She] X. Shen, Wavelet based numerical methods, Ph. D. thesis, University of Wisconsin-Milwaukee, Milwaukee, 1997.

[She1] X. Shen, A Galerkin-wavelet method for a singular convolution equation on the real line, *J. Int. Equa. Appl.* 12 (2000), 157-176.

[Sp] H. S. Shapiro, *Topics in Approximation Theory*, Springer-Verlag, Berlin-Heidelberg-New York, 1981.

[Si] H-T. Shim, Gibbs' phenomenon and summability in wavelet subspaces, preprint, 1993.

[Si1] H-T. Shim, On the Gibbs phenomenon for the Shannon sampling series in wavelet subspaces and a way to go around. *Commun. Korean Math. Soc.* 13 (1998), 181–193.

[Si2] H-T. Shim, On Gibbs constant for the Shannon wavelet expansion. *Korean J. Comput. Appl. Math.* 4 (1997), 469–473.

[S-V] H-T. Shim and H. Volkmer, On Gibbs' phenomenon for wavelet expansions, *J. Approx. Th.* 84 (1996), 74-95.

[Sil] B. W. Silverman, *Density Estimation for Statistics and Data Analysis*, Chapman and Hall, 1986.

[Sl] D. Slepian, Some comments on Fourier analysis, uncertainty and modeling, *SIAM Review* 28 (1983), 389-393.

[S-W] E. M. Stein and G. Weiss, *Fourier Analysis on Euclidean Spaces*, Princeton University Press, Princeton, NJ, 1971.

[St] G. Strang, Wavelets and dilation equations, *SIAM Review* 31 (1989), 614-627.

[St1] G. Strang, *Introduction to Applied Mathematics*, Wellesley-Cambridge Press, Wellesley, 1986.

[S-S] V. Strela and G. Strang, Finite element multiwavelets, *Approxima-tion theory, wavelets and applications* (Maratea, 1994), 485–496, NATO Adv. Sci. Inst. Ser. C Math. Phys. Sci., 454, Kluwer Acad. Publ., Dordrecht, 1995.

[Str] R. Strichartz, *A guide to Distribution theory and Fourier trans-forms*, CRC Press, Boca Raton, 1994

[So] C. Stone, Consistent nonparametric regression, *Ann. Statis.* 5 (1977), 595-645.

[S-Z] W. Sun and X. Zhou, Sampling theorem for wavelet subspaces: er-ror estimate and irregular sampling. *IEEE Trans. Signal Process.* 48 (2000), 223–226.

[Sz] G. Szegö, *Orthogonal polynomials*, Colloq. Pub., Vol. 23, *Amer. Math. Soc.*, Providence, RI, 1939.

[T] E. C. Titchmarsh, *Eigenfunction Expansions, Parts I and II*, Ox-ford, London, 1962, 1958.

[Vi] B. Vidakovic, *Statistical Modeling with wavelets,* John Wiley & Sons, Inc., New York, 1999.

[V] H. Volkmer, Distributional and square summable solutions of di-lation equations, preprint, 1990.

[V-W] H. Volkmer and G. G. Walter, Wavelets based on orthogonal basic splines, *J. Appl. Anal.* 47 (1992), 71-86.

[Wh] G. Wahba, Interpolating spline methods for density estimation in equispaced knots, *Ann. Statist.* 3 (1975), 30-48.

[W1] G. G. Walter, Discrete discrete wavelets, *SIAM J. Math. Anal.* 23 (1992), 1004-1014.

[W2] G. G. Walter, Wavelets and generalized functions, in *Wavelets–a Tutorial,* C. Chui (ed.), 51-70, Academic Press, Boston, 1992.

[W3] G. G. Walter, Pointwise convergence of wavelet expansions, *J. Ap-prox. Theory* 80 (1995), 108-118.

[W4] G. G. Walter, Approximation of the delta function by wavelets, *J. Approx. Theory* 71 (1992), 329-343.

[W5] G. G. Walter, Sampling theorems as part of wavelet theory, *Proc. Conf. Info. Sci. Sys.*, Johns Hopkins (1991), 907-912.

[W6] G. G. Walter, Operators which commute with characteristic functions with applications to a lucky accident, *Complex Variables* 18 (1992), 7-12.

[W7] G. G. Walter, Wavelet subspaces with an oversampling property, *Indag. Math.* 4 (1993), 499-507.

[W8] G. G. Walter, Translation and dilation invariance in orthogonal wavelets, *Appl. Comp. Harmonic Anal.* 1 (1994), 344-349.

[W9] G. G. Walter, Negative basic spline wavelets, *J. Math. Anal. and Appl.* 177 (1993), 239-253.

[W10] G. G. Walter, Sampling bandlimited functions of polynomial growth, *SIAM J. Math. Anal.* 19 (1988), 1198-1203.

[W11] G. G. Walter, A sampling theorem for wavelet subspaces, *IEEE Trans. Info. Th.* 38 (1992), 881-884.

[W12] G. G. Walter, An alternative approach to ill-posed problems, *J. Int. Equa. and Appl.* 1 (1988), 287-301.

[W13] G. G. Walter, A class of spectral density estimators, *Ann. Inst. Stat. Math.* 32 (1980), 65-80.

[W14] G. G. Walter, Properties of Hermite series estimation of probability density, *Ann. Stat.* 5 (1977), 1258-1264.

[W15] G. G. Walter, Fourier series and analytic representation of distributions, *SIAM Review* 12 (1970), 272- 276.

[W16] G. G. Walter, Hermite series as boundary values, *Trans. AMS* 218 (1976), 155-171.

[W17] G. G. Walter, On real singularities of Legendre expansions, *Proc. AMS* 19 (1968), 1407-1412.

[W18] G. G. Walter, Analytic representations with wavelet expansions, *Complex Variables Th. Appl.* 26 (1994), 235-243.

[W19] G. G. Walter, Orthogonal finite element multiwavelets, preprint, 1995.

[W20] G. G. Walter, Wavelets and sampling, preprint, 1999.

[W21] G. G. Walter, Hybrid Sampling Series, *Proc of Sampta'99*, Leon, Norway (1999).

[W22] G. G. Walter, Density estimation in the presence of noise, *Stat. Prob. Letters* 41 (1999), 237-246.

[W23] G. G. Walter, Approximation with impulse trains, *Results Math.* 34 (1998), 185-196.

[W24] G. G. Walter, Positive Hybrid Sampling Series, *Proc. of ICASSP' 2000*, Istanbul, Turkey (2000).

[W-B] G. G. Walter and J. Blum, Probability density estimation using delta sequences, *Ann. Stat.* 7 (1979), 328-340.

[W-C] G. G. Walter and L. Cai, Sampling and multiwavelets, *Proc. of Sampta '97*, Aveiro, Portugal (1997), 313-316.

[W-C1] G. G. Walter and L. Cai, Periodic wavelets from scratch, *Comp. Anal. & Appl.* 1(1999), 29-41.

[W-H] G. G. Walter and G. G. Hamedani, Bayes empirical Bayes estimation for natural exponential families with quadratic variance functions, *Ann. Stat.* 19 (1991), 1191-1224.

[W-L] G. G. Walter and Y-M. Liu, A class of bandlimited cardinal wavelets, preprint 1993.

[W-She] G. G. Walter and X. Shen, Positive sampling in wavelet subspaces, preprint, 2000.

[W-She1] G. G. Walter and X. Shen, Deconvolution using Meyer wavelets. *J. Integral Equations Appl.* 11 (1999), 515–534.

[W-She2] G. G. Walter and X. Shen, Continuous non-negative wavelets and their use in density estimation. *Comm. Statist. Theory Methods* 28 (1999), 1–17.

[W-She3] G. G. Walter and X. Shen, A substitute for summability in wavelet expansions, *Appl. Numer. Harmon. Anal.*, Birkhäuser Boston, Boston, MA (1999), 51–63,

[W-She4] G. G. Walter and X. Shen, Positive estimation with wavelets. Wavelets, multiwavelets, and their applications (San Diego, CA, 1997), *Contemp. Math.*, 216, Amer. Math. Soc., Providence, RI (1998), 63–79

[W-Si] G. G. Walter and H-T. Shim, Gibbs phenomenon for sampling series and what to do about it, *J. Fourier Anal. & Appl.* 4 (1998), 357-375.

[W-Z] G. G. Walter and J. Zhang, Wavelets based on band-limited processes with a K-L type property, *Proc SPIE Conf. Math. Imagining* (1993).

[W-Z1] G. G. Walter and J. Zhang, Orthonormal wavelets with simple closed form expressions, *IEEE Trans on Sig. Proc.* 46 (1998) 2248-2251.

[W-Za] G. G. Walter and A. Zayed, Multiresolution Analysis with Wavelet Subspaces, *Frac. Calc & Appl. Anal.* 1 (1998), 109-124.

[We] W. Wertz, *Statistical Density Estimation, A Survey*, Van den Hoeck and Ruprecht, 1978.

[Wi] H. Wilbraham, On a certain periodic function, *Cambridge & Dublin Math. J.* 3 (1848), 198.

[X-Z] X-G. Xia and Z. Zhang, On sampling theorems, wavelets, and wavelet transforms, *IEEE Trans. on Sig. Proc.* 42 (1994).

[Y] R. M. Young, *An Introduction to Non-Harmonic Fourier Series*, Academic Press, New York, 1980.

[Yo] K. Yosida, *Functional Analysis*, Academic Press, New York, 1965.

[Z-H-B] A. Zayed, G. Hinsen, and P. L. Butzer, On Lagrange interpolation and Kramer-type sampling theorems associated with Sturm-Liouville problems, *SIAM J. Appl. Math.* 50 (1990), 893-909.

[Za] A. Zayed, *Advances in Sampling Theory and Applications*, CRC Press, Boca Raton, 1993.

[Za-W] A. Zayed and G. G. Walter, Characterization of analytic functions in terms of their wavelet coefficients. *Complex Variables Theory Appl.* 29 (1996), 265–276.

[Ze] A. Zemanian, *Distribution theory and transform analysis*, McGraw-Hill, New York, 1965.

[Z-W] J. Zhang and G. G. Walter, A wavelet based K-L-like expansion for wide-sense stationary processes, *IEEE Trans. Sig. Proc.* 42 (1994), 1737-1745.

[Z] A. Zygmund, *Trigonometric Series*, Cambridge University Press, New York, 1957.

Index

Milton Keynes UK
Ingram Content Group UK Ltd.
UKHW021822071024
449327UK00021B/1395